dust

dust

the modern world in a trillion particles

jay owens

abrams press, new york

First published in Great Britain in 2023 by Hodder & Stoughton,
a Hachette UK company

Library of Congress Control Number: 2022950053

ISBN: 978-1-4197-6416-5
eISBN: 978-1-64700-809-3

Printed and bound in the United States
10 9 8 7 6 5 4 3 2 1

Abrams books are available at special discounts when purchased in quantity
for premiums and promotions as well as fundraising or educational
use. Special editions can also be created to specification. For details,
contact specialsales@abramsbooks.com or the address below.

Abrams Press® is a registered trademark of Harry N. Abrams, Inc.

ABRAMS The Art of Books
195 Broadway, New York, NY 10007
abramsbooks.com

To my friends

Contents

Introduction

We were driving into a mushroom cloud. It was a little after 3 p.m. on 25 July 2015, and a plume of smoke was rising over Sierra National Forest. The smoke was rising fast, the plume growing. As it reached above the top of the mountainside, it was being caught by the wind and blown sideways, spreading out. Yet the centre billowed on upwards, catching the sunlight and dazzling a radiant white. We drove on towards it.

Brad flicked through local radio stations to find out what this was – an explosion? A forest fire? There was no mention, and this silence was strangely eerie. Well, we'd ask at the town ahead, assuming the town wasn't on fire.

The cashier at the Shell gas station just outside North Forks wasn't worried. Forest fire, she said. Fifty acres. One valley over from North Forks, towards Bass Lake. Yes, we'd be able to get through to Mammoth Pool, our campground destination for the night.

We stood on the forecourt and watched planes fly into the smoke to dump fire retardant. They were tiny, their cargo a momentary flare of red powder, enveloped by the cloud faster than it could fall.

The road into Mammoth Pool was fractal, a non-linear straight line. It's twelve miles as the crow flies from North Forks and 38 miles by road. Twisting and turning along the contours of the San Jacinto rivershed through heavy pine forest, my sense of direction gone, the only way was forward. Smoke hung heavy in the sky, the fire behind the hills one ridge away. We kept on.

Then all the trees were dead. We drove through fifteen miles of burnt-out pine forest, remnants of a conflagration two years old: the Aspen Fire of 2013. Tree trunks still stood tall, blackened charcoal totems stretching away as far as the eye could see – 22,992 acres burnt, the Fire Department reported. A disaster bigger than I could picture in my mind's eye – the whole valley aflame. Two years on, little had regrown.

Meanwhile the North Fork blaze stained the sky yellow, the sun red. Fear licked gently across my collarbones: one road in meant one road out. But oh, the smell: sun-warmed pine sap; the hot dry earth, tar, woodsmoke and burning. Each breath drew it deeper into my lungs.

At the Mammoth Pool campground, ash rained down gently on our heads. Brad said it was going to be alright: the forest rangers would come and tell us if we had to evacuate. We'd hear the helicopters up and down the valley. I let my anxieties recede, and we stayed the night – I climbed river-sculpted boulders barefoot in a bikini; we drank Jameson Gold Reserve as the sun set scarlet as if through blood. We slept out on a smooth stone platform next to the river, under the stars.

A day later in Yosemite, the smoke filled the valleys. Two days later, out past Reno, it soured the sunset. The next week, in Death Valley, it still saturated the landscape with haze. The fire was still burning. Took two months to put out.

***

I'd come to California looking for something, but I couldn't have said quite what. Seven years before, I'd graduated into the nadir of recession, and so had played my twenties safe – and by most accounts succeeded. I had a good job in media research and a just-about structurally sound flat in north London. But I was bored.

Fortunately, I had some interesting friends.

'Hey everyone,' Brad had greeted the mailing list of the Institute of Atemporal Studies six weeks earlier. 'After three years of being bound up in the British legal system over place hacking London, they have finally returned my passport – I'm coming back to LA for the summer! Anyone want to go adventuring around LA/ Mojave/Las Vegas/Palm Springs?'

I'd last seen Brad – an American geographer, urban explorer, and author of a book titled *Explore Everything: Place-Hacking the City* – at a party on top of an abandoned grain silo in Oxford the summer before. I didn't know him well, but one thing I understood: if Brad invited you on an adventure, you said yes. So I bought a plane ticket to Los Angeles.

We were joined by another friend from the mailing list, Wayne Chambliss – a corporate strategist and speculative geophysicist who lived and worked down in the Inland Empire, the industrial and logistical sprawl east of LA. The three of us shared a fascination with places as layered, their history continuously overwritten but never quite rubbed out. If you looked at somewhere slantwise, tired and liminal, in the pale light of a new morning, there might be a way of seeing otherwise, of catching a glimpse of a place's past and its future, just there, imminent, lying in waiting.

'If we take the new experience, we stop time – and that's all I really care about,' Brad said.

And so the boys sketched out a line on the map which became a narrative: tracing the footsteps of Jack Parsons, notorious occultist and father of the NASA Jet Propulsion Lab via the strange geological uncanny of California and Nevada. The Devils Postpile to Death Valley and the Devil's Golf Course, to the ghost grid of California City and then the Devil's Gate. That was the story I pitched to the travel journal *Roads & Kingdoms*, anyhow: a road trip in search of Space Age dreams and failed utopias.

Dust had other opinions.

I had first started thinking about dust in spring 2008, lying on the sofa in my flat while procrastinating on putting together a dissertation proposal. I was studying for a Masters degree in geography, on a course named 'Modernity, Space and Place' at University College London. As my eyes cast around for anything to do other than study, I noticed that an enormous amount of dust had gathered under the table. Where trapped by chair legs, it was forming dust bunnies, tangles of an oddly purplish fuzz and hair which were a prodigious size, seeing as I had swept only a couple of days previously. It didn't seem fair. I had to be the one to blame: my previous flat had been excessively dusty too, but then I was sharing it with two engineering students. Now, living alone, this dust had to be my responsibility – yet I was neither balding nor scrofulous, and my flat's soft furnishings weren't becoming threadbare. Where was this material coming from? I was disturbed.

The problem of dust was clearly a more interesting question than any possible dissertation. How frequently did I need to clean in order to prevent dust? Daily? But that would be silly. The dust was quite a pleasing colour and safely out of the way, so perhaps I should just leave it? After all, dust wasn't really yucky – it just sat there quietly – and cleaning to avoid my dissertation was clearly avoidant behaviour. I had thought I lived in every inch of my tiny flat, yet on reflection dust marked all the spaces in which my presence was absent, the out of the way places my feet didn't tread and my body didn't occupy. Dust sat in corners, under the bed and on top of the cooker hood, while the rest of the flat acquired discarded clothes and coffee stains, traces instead of use. And paradoxically this dust marking all the places I wasn't was nonetheless partly made of me: made of my skin and hair no more dead than that attached to my head. It suggested a sort of hidden embodiment apart from the body, dispersed over space and time. The accumulation of dust was like the accumulation of the past, which must at some point become suffocating to the present – housework the only thing preventing the transformation of the home into a nightmarish haunted house. What if dust could be sentient, like the animated soot particles in the film I'd watched recently, *Spirited Away*? This stuff was weird, intriguingly so.

I decided I wanted to do some serious thinking about dust.

The topic was simultaneously unorthodox and oddly fashionable, fitting into my department's interests in waste, dirt and marginal 'subnatures'. Yet I decided against a PhD. Dust, for fuck's sake! It was too small, too trivial, too utterly pointless to devote three or four years of my life to writing some pretty evocative nonsense about haunting and ruination. The intellectual zeitgeist would move on, and then where would I be? I couldn't do it: I tried, half-heartedly, to come up with another proposal topic, but never finished the applications. Academia was a pyramid scheme anyway. So I walked out of the UCL quad and (eventually) got a job.

We drove 2,300 miles on that road trip and dust was always and everywhere. The smoke in the air, the ash drifting through the sunroof and settling over the interior of the car. The patina of history at Bodie ghost town in the Eastern Sierra, carefully preserved on tabletops and glass as a sign of its historic 'authenticity'. The lack of dust at some goldmine cabins farther out in the Great Basin, letting slip a tale of presences long after the occupants had moved out. It even rose from the earth: dust devils dancing alongside the road, animate and uncanny, as we drove through the bunker landscape of Hawthorne Army Depot, Nevada.

Out surveying the geology of Rainbow Basin near Barstow in the Mojave, our phones simultaneously vibrated with a government weather alert – 'Warning: dust storms'.

I got the message.

Dust was the story of this landscape and the throughline of this journey, the thin grey thread linking together people and place, past and present, histories and futures. The failed utopia I was tracing wasn't just one man's dream of space flight but modernity itself, and the disaster of its dream of domination over nature.

I saw what I hadn't seen before: dust was fundamentally political. The Sierra forest was on fire because the climate was changing: California was getting hotter and drier, and the biosphere couldn't adapt. Fire season used to be a few months in the early autumn but now in parts of the state it's damn near all year round. The dust in the mining cabins, meanwhile, revealed the politics of memory: the way these places were safely rendered 'heritage' by their official keepers, their histories of settlement and colonialism, extractivism and collapse obscured by a comforting haze of national nostalgia. I saw the ghost of stories I might come to tell – of nuclear fallout, volcanic ash and air pollution, of climate modelling, space science and geoengineering. Over the years of writing that followed, dust offered a route into facing up to the catastrophe of our time: the climate crisis and all its kindred ruptures in the biogeochemical systems of our planet. It offered a path to connect people to the very geology under our feet and the deep-time histories of how we came to be living atop this fragile crust.

The world in its wounding can seem too much to take in. But perhaps, by following something very tiny, a fragment of insight might be gained.

***

Before we go any further, I should define my terms. What do I mean by dust?

I want to say 'everything': almost everything can become dust, given time. I write of the orange haze in the sky over Europe in the spring, and the pale fur that grows on my writing desk and the black grime I wipe from my face in the evening after a day traversing the city. Dust gains its identity not from a singular material origin but instead through its form (tiny solid particles), its mode of transport (airborne) and, perhaps, a certain loss of context, an inherent formlessness: if we knew precisely what it was made of, we might not call it dust but instead dander or cement or pollen.

Dust is a boundary-crosser, a transgressor: the philosopher Michael Marder calls it 'a breath of matter on the brink of spirit', both solid and yet insubstantial, an element as much of air as it is earth.[1] Dust is matter at the very limit-point of form-lessness, the closest 'stuff' gets to nothing. This paradoxical nature of dust is, of course, a main reason I am interested in it: how rich to think with a material that isn't one thing or another and doesn't stay where it's put – a substance that blurs bounda-ries and refuses black and white certainties for the subtleties of the grey areas.

The word stems from the proto-Indo-European root '*dheu–*': dust, vapour, smoke, which gives us fumes and typhoons, and also typhus. Dust is a matter of smoke and storms and poisons, as we shall see.

'Tiny flying particles', though: this might suffice as a practical starting definition. What kinds of particles might we be talking about?

Dust devils and dust storms are creatures of soil and sand lifted from the earth and raised into the air by strong winds. Sand grains are mostly quartz, silica dioxide – the dust tinted various colours by the other minerals it contains. The dust-laden winds that move north over Europe in the spring are rust-red from iron-rich sands in the Sahara, while China suffers from sandstorms of 'yellow dust sweeping down from the Gobi Desert. Between 8 and 36 million tonnes of mineral dust are aloft in the atmosphere at any one time, making it the most prevalent type of dust on Earth by mass – and it is consequently the type of dust we shall encounter most in this book.[2] Most mineral dust originates from the 'dust belt' extending from the Sahara in North Africa, through the Middle East and Central Asia, to India and the Gobi Desert in northern China. As we'll see in the chapters that follow, prime dust sources are often former lakebeds or flat, low-lying areas of land that have repeatedly flooded over tens of thousands of years. Sediments carried by the water form deep deposits of fine clays and silt soils which provide the tiny, weakly bonded particles that, once dried out and exposed to the wind, are most ready and eager to become dust.

7

The planet also generates dust in the form of volcanic ash, the fragments of rock, volcanic glass and crystallised minerals ejected by volcanoes. The eruption of Krakatau (Krakatoa) in Indonesia in 1883 ejected so much ash and sulphates into the atmosphere that it produced a 'volcanic winter': temperatures in the northern hemisphere dropped 0.4C that summer. For many months the sunsets around the world were strange and spectacular as these particles diffracted the dusk light to dazzling and uncanny effect. In Oslo, a young man wrote in his diary:

'I was walking along the road with two friends – then the Sun set – all at once the sky became blood red – and I felt overcome with melancholy. I stood still and leaned against the railing, dead tired – clouds like blood and tongues of fire hung above the blue-black fjord and the city.'

The sight disturbed him deeply: 'I felt a great, unending scream piercing through nature,' he wrote, and this image stayed with him forever more. His name was Edvard Munch, and a decade later he painted this scene as *The Scream*.[3]

The particles in the smoke cloud that moved me to write, however, were instead the soot and wood ash from a burning pine forest – the air scented like incense by aldehydes and aromatic hydrocarbons. Today, 8.5 million tonnes of this burnt 'black carbon' are emitted around the world each year, most not of natural origin but instead humanmade combustion: diesel engines, wood-fuelled cooking stoves, and burning to clear land for agriculture.[4] In everyday life in London, I breathe in soot emitted by my neighbours' woodfired stoves, which have become one of the top sources of small particle air pollution in the UK, substantially worse than road traffic.[5]

Despite there being far less of it in the atmosphere than mineral dust, black carbon is important because it's an intensely powerful 'climate forcer', absorbing warmth from the sun and contributing substantially to global heating. It's also a major component of fine particle air pollution, known as PM2.5s (particles under 2.5 micrometres in size) – alongside organic carbon compounds, and sulphates, nitrates and ammonium condensing on to it for the ride.

These tiny particles are easily inhaled deep into the lungs. Their even-smaller cousins, ultrafine PM0.1s, can pass through the air sacs in the lungs into the bloodstream, where they can be transported to every organ and can harm potentially every cell in the human body. Tissue damage can occur both directly from pollutant toxicity, as elements such as lead and arsenic hitchhike on soot particles formed by combustion – or indirectly, as the body tries to fight off this foreign material, triggering systematic inflammation.[6] Particulate air pollution causes not just respiratory illnesses but heart disease, cancers, infertility – even neurodegenerative diseases such as Alzheimer's. Altogether, it's the fifth biggest cause of death in the world, the *Global Burden of Disease Report* finds, accounting for 4.2 million lives lost each year, alongside a profound and limiting burden of sickness and disability among the living. If London's air was compliant with World Health Organization standards for PM2.5s, we'd all gain on average an extra 2.5 months of life.[7]

Some would gain far more.

In Lewisham, south-east London, a young girl named Ella Adoo Kissi-Debrah lived with her mother, Rosamund, just 25 metres from the city's busy, often gridlocked South Circular road. She was a bright child, a singer, dancer and performer with model aircraft hanging from her bedroom ceiling: when she grew up she wanted to be a pilot. But in 2010, aged seven, Ella started to develop a strange and persistent cough. All too rapidly, she got sicker and sicker. Sometimes she would just stop breathing and have to be rushed to intensive care. She must have been terrified: each time it would have felt as if she were drowning as her lungs struggled for air – yet her mother describes her as stubborn in the face of her trials. But it wasn't enough. In February 2013, nine-year-old Ella died of respiratory failure. Rosamund recalled later: 'Someone from the neighbourhood who was looking at pollution measurements in the area told me that in the evening when she had her last asthma attack, Lewisham had one of the worst air pollution episodes ever.'[8] Yet no one at the time ever told them this was a risk.

For years, Rosamund Kissi-Debrah fought to expose the real cause of her daughter's death through judicial inquiries and legal appeals, back and forth between lawyers and the courts. Finally, in December 2020, she won a second inquest – where expert doctors testified that Ella's final, fatal asthma attack was the cumulative effect of the toxic air she'd breathed all her too short life. And so Ella made legal history as the first person in the UK to have air pollution listed as a cause of death. In his remarks, coroner Phillip Barlow said there is 'no safe level of particulate matter' in the air and called for national pollution limits to be reduced.[9] As I write this introduction, 'Ella's Law' has passed the House of Lords and now heads back to the Commons to be scrutinised by MPs. Seventy-one years after the Great Smog that killed perhaps twelve thousand Londoners, Britain might finally declare that clean air is a human right.[10]

Urban dust is much more than simply carbon soot from combustion, though: there's friction between people and the environment at every turn. On cars, buses and trains, brakes rub against tyres and tyres press against roads and rails many millions of times a day, each time generating shear forces, stressing materials, and abrading tiny little pieces of metal and rubber and asphalt as they go. This is dust I'm all too intimately familiar with: as a cyclist, I know it as 'road grime'. After a day in central London I cleanse my face and the cotton wool's visibly grey with dirt.

In 2019, a *Financial Times* investigation dubbed the London Underground 'the dirtiest place in the city', with parts of the Central Line between Bond Street and Notting Hill Gate having over eight times the WHO limit for PM2.5 particles. Tube dust is particularly high in iron oxide from the metal brakes and rails, but it's not only mechanical. 'A lot of the dust in this environment is coming from the passengers themselves,' Alno Lesch, operational manager for track cleaning, told the *Financial Times*, pulling out a black tangle from under the train platform. Human hair.

Over a thousand people work night shifts in the tunnels underground while the trains rest, brushing and hoovering the surfaces to remove dust, and spraying a fixative to keep what's left in place. But it doesn't always entirely work: dusting is, after all, a process of stirring up particles that have previously been minding their own business. When Transport for London cleaned the Bakerloo line, they removed 6.4 tonnes of filth and fluff – yet, once they were done, PM2.5 levels at nine of the fifteen stations tested higher rather than lower.[11]

Cleaning up the mess we make is rarely a matter of neat little technofixes, as we shall see throughout this book. When it comes to road dust, electric cars prove to be no cleaner than the petrol polluters they replace. By using regenerative braking, electric vehicles do produce about 75 per cent less brake dust than petrol cars – but they generate more tyre dust, road wear and resuspension of other road debris, because their batteries make them on average heavier.[12] Road dust is a major global source of microplastics, the tiny plastic particles under 5 millimetres in size that have become an increasingly recognised environmental pollution problem in the last decade. About 6.1 million tonnes of tyre wear particles are generated each year, representing a not insignificant 1.8 per cent of global plastic production – plus there's a further 0.5 million tonnes of brake wear particles, too.[13] This makes road dust the source of fully a third of the microplastics in our oceans: it's as big a problem as synthetic textile fibres, the moulting fleeces and fake furs that tend to get more press.[14]

We might even think of microplastics as essentially another type of dust – they are, after all, messily humanmade, plural, toxic and tricky to deal with, just like all the other dusts in this book. Like soot, microplastic particles offer a vehicle to transport other environmental contaminants such as 6PPD-quinone, an antioxidant added to tyre rubber that researchers have found is producing mass die-offs of coho salmon in the Pacific Northwest.[15] Microplastics are also often relatively dark in colour, making them absorb warmth from the sun and therefore accelerate ice melt in the polar and glaciated regions of this world.

Most black carbon emissions are also human-generated through wood and fossil fuel combustion, as are the sulphate and nitrate air pollutants that occasionally take particulate form. (Sulphates and nitrates are mercurial, however: they can also be gases, depending on temperature, humidity and other chemicals present in the atmosphere. This makes them unreliably 'dust', as far as I'm concerned – and as such, they're not a focus of this book.) Beyond combustion, a quarter of mineral dust is anthropogenic too.[16] Dust blowing in the wind in the desert is a natural phenomenon, but human actions are in many places exacerbating this process. Deforestation wipes out the plant life that holds the earth together via its root systems. Arid areas and deserts may seem sparsely vegetated but their soil is in fact alive with delicate cryptobiotic crusts of algae, fungi and microscopic cyanobacteria – all too easily destroyed by expanding agriculture, ploughing and construction. Water pumping is another major mineral dust generator, as it drains down the water table, destroying the meadows and grasslands that relied on this water for life. In dryland landscapes, one single plant might hold down the soil for many metres around it as it stretches its roots far and wide to find moisture. Once the water table falls, these plants cannot survive (as we shall see in the Owens Valley in California, in Chapter 9) – and then the dust rises.

I haven't yet even got to the dust in my flat that distracted me into this project so many years ago. To all the above we can add skin flakes, pet dander, hair, textile fibres, disintegrating bits of particleboard furniture and sofa foam and all the chemicals *that* can contain – such as flame retardants with impenetrable acronyms: PBDEs and TPHPs and chlorinated OPFRs; chemicals designed to keep you safe but which also cause cancer, decrease fertility, impact cognitive ability and cause thyroid disease.[17] Road dust and construction dust blow into your home through the windows and walk in on the soles of your shoes – alongside fragments of mineral dust from distant deserts, and perhaps even the odd radioactive particle, the lingering legacy of nuclear testing and nuclear accidents. A doormat is only so much use: the dust under your sofa contains the world.

Meanwhile – absurdly – we keep setting fire to things inside our homes: we cook on gas, burn candles and incense for 'atmosphere', or light a cosy log fire on a winter's evening. Unlike outdoors, where (after centuries of effort) there are established air quality standards, indoor air goes unmeasured, under-researched, and unregulated. But combustion is combustion: though gas may be cleaner than wood or coal, when hydrocarbons burn, soot and PM2.5s are the inevitable consequence. Scientists have found that during some cooking activities, PM2.5 concentrations in the home can exceed 250 micrograms per cubic metre (measured in $\mu g/m^3$: μ or 'mu' is the Greek symbol for 'micro', millionth), which is over seven times the US Environmental Protection Agency's safe 24-hour level exposure. Cooking is admittedly a short-term activity, but still: 250 $\mu g/m^3$ is not good air. Roast dinners are a particular hazard; so is burnt toast. During a simulated Thanksgiving Day, a holidaymaker may inhale 149 μg of tiny particles.[18] We spend 90 per cent of our time indoors, but the home is sadly not always a place of safety. I write these words during an energy crisis in which millions of people in the UK and across Europe can't afford to heat their homes despite snow on the ground and freezing nights. In the cold and the damp, mould blooms, infiltrating through damp windows and spreading its way up walls. Its spores are airborne: it travels as dust – causing and exacerbating asthma, particularly in children.[19]

'Our most basic common link is that we all inhabit this small planet,' John F. Kennedy once said, and yet we do not all breathe the same air. Indoors as well as out, dust is a medium of deep environmental injustice.

***

So this is a book about all that. The particles may be tiny, but the problems they usher in their wake are planetary.

My focus is on humanmade dust: an anthropocentric approach, certainly, but how else to understand the Anthropocene? We are world-makers, now, and this is a new geological age. The world's biogeophysical systems have been altered by human action: not just the carbon cycle, but nitrogen and phosphorous too; the freshwater cycle, the vast erosive flows of sand and soil; the air and water and rock of the entire globe. Gradually and then suddenly – and not at all equally – our species has transformed from passive passengers on this rock in space to modern Prometheans, setting fire to the planet and finding that it does indeed burn.

The scale of this shift – and its consequences, the present and future harm – is incredibly hard to think about. We find it easiest to imagine things at human scale, at the size of our body parts – centimetres, feet, metres – and ideally from around one to a dozen of these things, numbers we can count on our fingers. Most people have an intuitive sense of scale down to a thousand times smaller (millimetres), and a thousand times larger (kilometres). But beyond that, instinct tends to fail us.

So this is one reason to think with dust: to challenge ourselves to try to see the world at scales beyond our easy imaginings.

The dust we're talking about in this book is almost always sub-millimetre in scale, and very often a hundred or thousand times smaller than that – i.e. 1 to 10 micrometres in size. Invariably dust particles get compared to a human hair in scale – which averages 70–90 micrometres in diameter depending on the origins of the owner.[20] But dust is often a hundred times smaller. Dust dances right at the limit of our vision, or more precisely our visual acuity: the ability of an unaided human eye to perceive an object as distinct. Brightly lit, we can discern individual particles – as when the sun shines on a smooth surface and reveals much previously unseen fuzz. Most of the time, though, as with a cloud of smoke, we perceive not the individual particles but the collective dimming effect. Some of the smallest dust in this book is uranium nanodust that blows off the unprotected mining waste piles in Laguna Pueblo, New Mexico – a substance measured not in millionths of a metre, but nanometres, billionths. But we'll also meet Dr Claire Ryder, an aerosol scientist at the University of Reading, who argues for the importance of 'giant dust' as a neglected but crucial factor in climate modelling – though 'giant' can be defined as merely over 20 micrometres in size, so it's all relative.

Yet through the very small, we also gain a lens on the very large, the very long ago – and thus ultimately the future of our planet.* Each of the place histories that make up the first half of the book – from Los Angeles, in the early 1900s, to the 1930s Dust Bowl and the drying up of the Aral Sea – reveal how dust is made not just locally, in the places where it actually develops and causes problems, but very often far away, in the capitals and colonial centres that make the decisions about where water should flow, and who and where should be denied it. In Chapter 6, on nuclear testing, following the trail of radioactive fallout forces us to think ahead, through the thousand-year half-lives of radioactive nuclei, if we are to understand the true consequences of irradiating the desert and Pacific Island peoples of this world in the name of deterring war. Then Chapter 7, examining dust as a medium of dating ice cores, lets us look both backwards into deep time – into the dust and air trapped in million-year-old ice cores – and forwards, to what this data on the Earth's past climates reveals about the future trajectories of our heating planet.

Eco-philosopher Timothy Morton has argued global warming is a 'hyperobject', something so massive and so widely distributed across time and space that it becomes too large to understand.[21]

---

* Space dust is the definitive example of such extreme scale, of course, and I regret that I can only write about it in passing. The beginning of everything in intergalactic clouds of colliding atoms is beautiful and it is extraordinary, yet it is not *political* in the way that humanmade particulates are – and so it is a matter for other writers and another time.

The coronavirus pandemic was another hyperobject, so is capitalism: impossible to really perceive as wholes, in full totality, yet, nonetheless, we speak of them as things. We might try to get to grips with their effects and their impact through data and statistics or use a lot of big abstract nouns in order to try to gesture at the scale of the concepts we think we're dealing with. But this doesn't always succeed and so we are left anxious, struggling to situate ourselves – and, crucially, our capacity for action, for change-making – amid such monstrously vast challenges.

Their scale is a challenge not only for imagination – but also for politics and justice. Environmental humanities scholar Rob Nixon writes of how 'Climate change, the thawing cryosphere, toxic drift, biomagnification, deforestation, the radioactive aftermaths of wars, acidifying oceans, and a host of other slowly unfolding environmental catastrophes present formidable representational obstacles that can hinder our efforts to mobilize and act decisively'.[22] While their consequences are 'cataclysmic', they are 'scientifically convoluted cataclysms in which casualties are postponed, often for generations'. When public policy is shaped around a four- or five-year electoral cycle, and the news media operates within a time-horizon of just a day or two, how can we make anybody interested in what Nixon describes as these 'disasters that are anonymous and that star nobody'? How can we make visible the harm that is being done – even if it is, as he says, a strange, 'slow violence', that is, 'a violence that occurs gradually and out of sight, a violence of delayed destruction that is dispersed across time and space, an attritional violence that is typically not viewed as violence at all'?

This is why I write about dust. Through this collision of scales – the microscopic and the planetary – I hope to average back out to an approachably human-sized story again. A human-sized story, and an inhabited, embodied one.

I write from experience, both my own travels and interviews, and the diaries, oral histories and stories of many others who have lived lives intimately and sometimes devastatingly entangled with dust. I don't just want to write about how a dust storm happens: I want to tell you how a dust storm *tastes*, how the electricity in the air sings on the fence-wires and your shoulder blades stiffen as the dirt rolls in and the sky goes black. I write of how the dust enters into people's bodies, day after day, year after year – and how the tiniest particles cross from people's lungs into their bloodstreams and cells, and the slow havoc it wreaks there. I write of communities who see they're being sickened, and who start to follow the dust back to where it comes from and who caused it, and who start to fight back. These processes – these sicknesses and fights for environmental justice – are slow. They take place over not just years but generations. By starting from specific places – not 'America' but the Owens Valley in the eastern Sierra, California, not 'the former Soviet Union' but Moynaq, a fishing village on the shores of what was once the Aral Sea – and by following the grey thread of dust through time and place, we might manage to approach matters as vast as 'modernity' and 'environmental crisis' and see spaces within them for community, agency, resistance and change.

Dust is in this way a method. An emergent academic field called 'Discard Studies' seeks to study trash of all kinds – waste, pollution, dirt – in order to understand how social and economic systems really work. As Max Liboiron and Josh Lepawsky put it: 'What must be discarded for this or that system to be created and to carry on?'[23] Rich nations passing the buck with plastic recycling is an obvious example: for all our consumer efforts and regulation and loudly espoused good intentions, there still exists the Great Pacific Garbage Patch – though the approach also applies to recognising the people discarded by capitalism, as much as the things. By studying what's usually excluded or disavowed, Discard Studies seeks to remind us of who and what mustn't be forgotten if we seek true environmental justice.

I think of dust in this way as the shadow of modernity, disregarded but always nonetheless present, as though haunting our too-neat dreams of progress and perfection. I believe that if we want to understand what it means to be modern – and for some reason, I really do – we can't just admire the fruits of modernity – the iPhones, the Teslas, the staggering abundance of consumer entertainment – but should follow that tree down to its roots: the environmental resources and hard human labour that make it so. Mining, construction, manufacturing; industrialised agriculture and global transportation: this is dirty and dusty work.

During the early 2010s, dozens of workers died in dust explosions at iPhone factories in China, as aluminium particles produced by polishing the cases built up in badly ventilated workshops, then detonated.[24] Meanwhile, in the 'Lithium Triangle' of Bolivia, Chile and Argentina, booming global demand for batteries is leading to enormous volumes of groundwater pumping for mineral lithium extraction from underground brine stores. Every ton of lithium carbonate produced requires over two million litres of water to simply disappear into plain air. The water table is dropping, drying up lakes, rivers and wetlands – so much so that conservationists warn that the lithium boom risks turning the region's delicate ecosystems into desert. And where there's desert – and especially where there are dried-up lakebeds – there will be dust.

As I wrote, I came to realise this book is also an absent history of water. Some of the environmental disasters I shall be writing about – most obviously the construction of the Los Angeles Aqueduct in the early 1900s, and how it turned the Owens Valley to dust – are generally told as matters of water politics (or neo-noir intrigue, as in Polanski's *Chinatown*). But few hang around afterwards to ask: what happens when the water is gone? What happens to the people who still live there? What happens next?

It's all too easy to talk of these places as though they are 'ruined' or 'dead', but these phrases miss the fact that land *lives*. Nature isn't passive: it doesn't just receive all that we throw at it, but it also reacts, responds and adapts. Ecosystems are strange, stretchy things. Adaptation may not be infinite – but to talk of

people 'destroying' a place is to award ourselves far too m
lone agency. We fuck 'em up, that's for sure. But in the course
reporting this book, I have been to some of what you mignt
think to be the bleakest places on Earth – and found them
strangely vital. In the words of architectural historian Samia
Henni, 'deserts are not empty'.[25] Framing arid lands as uninhabited was a colonial plot to render them ownable, exploitable and
later *nuclear bombable*. Those of us raised in the global North
might learn to try to see them otherwise, and we need to practise the discipline of hope, to imagine and then fight for what
these places could become.

'I wouldn't call it ruined,' says activist and community organiser Teri Red Owl of another desert landscape, namely
Payahuunadü (also known as the Owens Valley in California),
later in this book. 'I would call it damaged. But, you know,
damages can be repaired.' Hold that idea in mind.

The anthropologist Anna Tsing writes of disturbed landscapes. In
her 2015 ethnography *The Mushroom at the End of the World*,
Tsing traces the world-spanning travels of the matsutake mushroom from the forests of Oregon to the tables of luxury restaurants
in Japan, where it is thought to epitomise the essence of autumn.[26]
It's an influential book in many ways, but it's personally important
to me for the way Tsing's writing revealed how you might write a
big book about a very little thing. Through exploring both the
hyperlocal ecological connections of a fungus in a forest and the
globe-spanning commodity trade built up around it, Tsing seeks to
'illuminate the cracks in the global political economy', demonstrating how capitalism is not some monolithic totality, as so commonly
claimed, but rather something 'patchy', partial and contingent.
Something in which there might be space for other worlds, for other
ways of being and relating between people and planet. Matsutake
turn out to flourish on human-disturbed ground, symbolising
possibilities for renewal and symbiosis in places we might have

written off as lost. Capitalist carelessness may have wrecked so many places, but the possibility of life remains in all of them.

Dust, in its own tiny, world-encompassing way, might come to reveal a similar lesson. It reminds us of the endless geological cycles of destruction and remaking, as wind gradually erodes the earth only for the carried dust to be laid down and over eons be compressed into new sedimentary rock known as loess, which then erodes again in turn. Dust is also a critical component of atmospheric processes, ocean systems, and biological and human processes and systems too. As we'll see later in this book, world-traversing dust flows melt ice on mountain glaciers, fertilise forests and feed plankton blooms in the ocean – making dust a part of the water, iron and nitrogen cycles. Dust in the atmosphere also both absorbs and reflects the sun's energy, making it a major driver of temperature, climate and ultimately global warming.

None of this happens at a remove. Dust ties us intimately into that metabolism, as human activities have planet-scale geochemical impacts – and, in turn, tiny particles get inside our bodies and impact us too. And so dust both nourishes life and takes it. A staggering one in every five deaths is as a result of fine-particle air pollution from burning fossil fuels, scientists reported in 2021.[27] Eight million deaths a year – that's pretty much the entire population of London or New York City – could be avoided if we switched to renewable energy sources.

As we grapple with understanding the Anthropocene and unlearning some of the logics that got us here, I hope dust might be a guide to seeing otherwise. To recognising our place within deep geological expanses of time, and Earth systems much vaster than ourselves. To finding new (old) ways of being in relation to land and planet, not as owners or masters but as something more like custodians or caretakers. And also for recognising the wonder of it all. In the course of writing this book I went looking for environmental disaster and instead found a strange and staggering beauty. A fissured, serrated ice sheet collapsing into a fjord at the end of the world. A sky-filling sorbet of a sunrise over the last remaining silver sliver of the Aral Sea. The green chirruping vibrant lushness of a spring-fed Sierra valley in a place once called 'the land of little rain'.

And I found an array of people – environmentalists and activists, lab scientists, NASA physicists and polar glaciologists, government and tribal employees, radical lawyers – young and old, rural and urban, settler and Indigenous – who knew and believed in this wonder. And were fighting like hell for it to survive.

Following the traces of dust – seemingly the formless, the forgotten, the out of sight – is not, as it might seem, an exercise in eco-grief and mourning.

It is in the end a story about connection.

Chapter 1
The Suburbs of Hell

To say that the modern world began at a specific time and in a specific place is absurd. Historians could hardly disagree more: there's almost a five-hundred-year difference between the earliest and latest suggested dates for the earliest beginnings of the modern era. Was the *now* birthed in the continent-spanning catastrophe of the Black Death (1347–53), which overturned the medieval world order, or in other calamities around this time? Perhaps the fall of Constantinople in 1453, or the European conquest of the Americas starting in 1492? Was modernity the consequence of technology – and if so, which one? Was it the printing press, invented by Gutenberg in 1450 or James Watt's condensing steam engine, launched on to the market in 1776? Or was the modern an idea? Did Machiavelli's *The Prince* (1513) shepherd in the modern political order, or was it Martin Luther's *95 Theses*, four years later? Or must we wait for the French Revolution in 1789 to give us the modern subject: a person of *liberté* and *égalité*, not hierarchy and descent? Did modernity necessarily begin in the West at all? If we look East, to the Mughal Empire or to China, which other cities, social innovations and technologies might we regard as turning points to today?

Yet, by following the trail of dust, we find ourselves in a particular time and place: 1570, in Tudor London, when the air began to fill with filthy fossil smoke in a way that hasn't ceased since.

Air pollution existed before this date. Seneca grumbled about 'the oppressive atmosphere of the city and that reek of smoking cookers' in first-century Rome, and the 'clouds of ashes' they left behind.[1] The fires that fed the Mayan Empire's lime kilns, in what is modern Central America, consumed an entire civilisation, as the vast deforestation required to build their cities – twenty trees to make the lime plaster for a single square metre of cityscape – changed rainfall patterns, exacerbated drought and created catastrophic crop failure.[2] Yet environmental historian William M. Cavert argues that the smoky air of Elizabethan London is yet more significant, for one reason: it marks the beginning of the fossil fuel age. Before 1570, wood was the main source of fuel everywhere in the world (except in the high Arctic, where Inuit and Yupik people's lamps ran on animal fat). For ten thousand years, settled human society had been built on only the energy a tree could store in its lifetime. The widespread adoption of coal was a rupture, a blast of ancient black, mineral solar power beamed straight from the planet's core. This changed everything. As Cavert writes: 'From the perspective of energy regimes . . . early modern England was, or became, "the first modern society".'[3] If we want to understand the worldview that has produced the crisis we face today, it makes sense to start at its beginning.

The shift that occurred in London in the late sixteenth century was not merely a step up in particulate pollution metrics, or even a qualitative shift in types of fuel. It was epochal. So let us witness the birth of this new fossil energy regime – and the trail of the dust it left in its wake.

***

Three hundred and fifty million years ago, in a time before the dinosaurs, our planet was warm and lush. The air was rich with oxygen and, in steady warm sunshine near the equator, forests of fern-like trees grew tall, their spindly trunks stretching thirty to fifty metres in height above shallow root systems bathed by vast tropical swamps. These trees had tough stems made almost entirely of lignin-rich bark, in order to resist the vast insects that preyed on them – yet they toppled over easily, falling on top of each other, tree after tree, season after season, and landing in water that carried insufficient oxygen for their bodies to rot. Floods covered this wood with silt and these layers compressed slowly, over countless generations, into layers of peat.

Deep underground, carried on plumes of molten metal, the continents were on the move. At this time the Laurussian plate (which would become Europe and North America) was colliding with Gondwana (now the southern hemisphere continents) to form the supercontinent of Pangea. Forced together, the plates buckled up into mountain ranges, and neighbouring regions of crust were forced down. Crushed, slowly and relentlessly, the bodies of these prehistoric trees received a strange, undead burial where they did not rot, or turn to dust – but transmuted into something new. The lignin that had given them their strength compacted into lignite; brown peat became bituminous, then black. They became coal.

The geological era was named for the one element that survived this transformation: it's known as the Carboniferous era. Three hundred million years later, people noticed nuggets of this black gold washing up on the beaches of Northumberland.

People first started to make use of coal in the Bronze Age, with the earliest mining occurring in China over 5,000 years ago.[4] The Romans knew about coal as a useful alternative to charcoal and were mining extensively in Britain by the end of the second century CE. Coal smelted iron and fed the hypocaust heating systems in public baths and the villas of the wealthy. At Aquae Sulis (now Bath), it fed a 'perpetual fire' at the temple of Minerva, goddess of wisdom.

After the Romans left Britain in 410 CE, people returned to using just readily available wood for fuel – and continued to do so for the next 750 years. Coal only reappears in the historical record in 1180, in a survey of the lands of the Bishopric of Durham.[5] Coal lay close to the surface in this part of north-east England, making it easy to mine – which started to happen in greater quantities. At this time, forests were being cut down across Europe as populations grew and societies became wealthier, making wood scarcer and more expensive. Dyers, brewers and other trades that required large fires – such as the lime kilns that made plaster and mortar for construction – felt the pinch. Coal was a more affordable alternative and, in the stable years after Magna Carta, a trade emerged, with the fuel being shipped by sea from Newcastle to London in sufficient quantities such that by 1228 the capital had acquired a Seacoal Lane.[6]

London quickly came to know the scourge of its pollution. This northern coal was a soft, bituminous fuel, high in sulphur and inefficient: seacoal fires produce smoke as enthusiastically as heat. The stench was intolerable, and nobles and gentry complained to the king about the public nuisance – yet trades requiring large amounts of energy kept on using coal, however noisome it was. In 1285, King Edward I ordered an investigation into this smoky menace – the world's first air pollution commission – but no solutions nor laws followed, and so London's sooty air saw little improvement. Only the cataclysm of the Black Death, arriving in England in 1348, would create change, as mass death meant manufacturing production plunged for a time. But, then as now, the capital's appetite for beer and brightly coloured cloth came roaring back. Soon the city's seacoal furnaces were back in business – and the city would get sootier still in the years ahead.

***

Let us jump ahead to the 1570s, a time when London was transforming. The medieval post-plague city of 30,000 people had grown rapidly throughout the 1400s and the Tudor period into a bustling metropolis of around 150,000, making it by far the largest city in England – though not of the scale of Beijing or Constantinople, then the biggest cities in the world. A distinctly urban life developed.

Early modern London was a mercantile city and the centre of a web of trade and colonial plunder that spanned half the globe.

Ships arrived at the city docks from the Middle East, Russia and the Americas carrying gunpowder and tobacco; furs, dried fish and whale oil; spices, sugar and luxuries. Education and literacy were rapidly expanding, theatres and poetry flourished, and London printers were publishing a continual flow of books and pamphlets testing the limits of free speech and religious dissent. The streets would have hummed with dozens of languages, as traders, sailors, travelling scholars and Huguenot refugees from France all went about their daily affairs. The city was expanding, new buildings nibbling away at the last green fields that lay between the cities of London and Westminster. London was starting to sprawl, a sea of houses as far as the eye could see.

Historians describe this period as the 'early modern' era, because it was a time when the characteristics of our present day start to be visible in the texture of people's lives. Environmental historians particularly look for ways in which this modernity was 'embodied in its physical environment' – that is, the dark, thick, sooty smoke from all the coal that fuelled this bustle, which produced a distinctive urban atmosphere.[7]

Coal use grew fast at this time, doubling every decade as shortages of wood pushed people to find new sources of fuel. (England faced a very real threat of seaborne invasion, and Queen Elizabeth I was concerned that heavy usage of wood for fuel might leave too few trees to mature into sufficiently mighty warship timbers.) By the winter of 1605–6, London was importing 144,000 tons of coal, or roughly one ton per person, per year.[8] A fleet of four hundred ships ferried fuel around Britain's coastline and down the River Thames to keep the city's hungry fires fed.[9] Coal use kept rising as London grew: each new city dweller meant another ton of ancient carbon turned to ash and dust each year, the city's hearths glowing with energy from the sun's rays, 300 million years ago.

It was a drastic change in just a single generation: historian Ruth Goodman calls it a 'domestic revolution'.[10] Some two hundred years later, during the Industrial Revolution, air pollution would come to England's Midlands and the North, as the soot and smoke of countless factory chimneys turned these towns into the 'dark Satanic mills' of familiar imagery. But in the London of the 1600s, coal warmed the homes of the city's residents, the poorest in particular. Rich households continued to buy more expensive charcoal and wood for use in their living rooms, but most Londoners couldn't avoid seacoal's sulphurous stink and staining, ashy residues. It warmed their homes and their workplaces, the pubs and coffee houses where they gathered, the churches they attended each Sunday. Fossil fires fuelled not only economic production but also provided the basis for health, comfort, family and cultural life. Coal enabled 'social stability, commercial progress, and state power,' as Cavert states, and it was at this time that 'coal burning became an essential aspect of London life'.[11] It produced the very atmosphere of the modern city.

Everyone hated it, of course. Historical records are rich with bickering and complaints about the city air, whether people lived next to a fuel-intensive trade with furnaces burning all day and all night, or their neighbour was just a cheapskate who bought bad fuel that gave off a lot of ash.

In 1664 the poet Alexander Brome complained that it stifled even art:

'Alas! Sir, London is no place for verse;
Ingenious harmless thoughts, polite and terse: Our age
   admits not, we are wrapped in smoke;
And sin, and business, which the muses choke.'[12]

Elizabeth I had tried to tackle the problem in 1579 when, 'greatly grieved and annoyed with the taste and smoke', she banned coal-burning in London when Parliament was sitting, and then attempted to jail a dozen of the city's leading polluters for breaching this rule. (Oh for such assertiveness from our political rulers today.) But the increasingly domestic causes of

London's sooty air would soon render such vigorous enforcement impossible. Most Londoners weren't polluting their environment for profit or from carelessness but because they couldn't afford to do otherwise (those who could afford to use less smoky fuels did). They had to heat their homes in winter and, at that time, winters were especially harsh as the 'Little Ice Age' – between 1300 and 1850 – produced a period of exceptional cold in Europe and North America. (One potential cause is dust blasted far into the stratosphere by volcanic eruptions, which weakened the sun's warmth.)[13] The Thames froze over in 1608 and the city held a Frost Fair on the river. You, too, would have put another lump of coal on the hearth to fend off the chill.

By the early 1600s, a generation of coal fires had left their mark on the city. They changed its form: every home required a chimney in order to make coal fires somewhat more bearable. And they changed its colour, as airborne soot stained and darkened buildings. In 1620 the English king, James I (VI of Scotland) 'was moved with compassion for the decayed fabric of St. Paul's ... near approaching ruin by the corroding quality of coal smoke ... where unto it had been long subject'.[14]

A generation later, in 1661, a man named John Evelyn wrote an entire pamphlet of complaints about London's filthy air: *Fumifugium: or, The inconveniencie of the aer and smoak of London dissipated: Together with some remedies humbly proposed.*

Evelyn is the other great diarist of the seventeenth century. His friend and contemporary Samuel Pepys may be better-known today, having been fortunate that the decade on which he focused his efforts included such blockbuster events as the Great Plague in 1665, and the Great Fire of London in the year following. Evelyn, by contrast, wrote his entire life, composing a half-million-word diary spanning the years 1640 to 1706, alongside no fewer than thirty books and pamphlets on topics ranging from gardening, theology, art, the character of England, famous imposters, how best to construct a library – and the first recorded recipe book of salads. Evelyn was a founding member of the Royal Society, close to the intellectual ferment taking place as two thousand years of ancient Greek thought was being swept away by a scientific revolution to birth a new, rational, quantitative world.

In 1661, this modern man would turn his attention to the problem of London's atrocious atmosphere. *Fumifugium* is a remarkable treatise, florid in style and yet scientifically and architecturally progressive. Addressed to King Charles II, the book describes the ruin smoke was making of the capital, and all that lay within it, and begins in inimitable style:

> 'IT was one day, as I was Walking in Your MAJESTIES Palace at *WHITE-HALL* (where I have sometimes the honour to refresh my self with the Sight of Your Illustrious Presence, which is the Joy of Your Peoples hearts) that a presumptuous Smoake issuing from one or two Tunnels neer *Northumberland-House,* and not far from *Scotland-yard,* did so invade the Court; that all the Rooms, Galleries, and Places about it were fill'd and infested with it; and that to such a degree, as Men could hardly discern one another for the Clowd.'[15]

Evelyn thought this fogginess outrageous, and he knew precisely what was at fault.

> 'And what is the culprit? It is a hellish and Dismal cloud of SEA-COAL that not only hovers over her permanently but as the poet Virgil said: "*Conditur in tenebris altum caligine Coelum [The high heaven was hidden in darkest clouds]*".'

He was writing at a time of restoration and renewal: King Charles II had only just ascended the throne after an eleven-year interregnum of civil war and republicanism. Evelyn found it shameful that the capital and the court was thus overshadowed. He had a plan to fix this. But first he describes the calamitous impact this seacoal smoke is having on the capital. While he may not much use the word 'dust' directly, *Fumifugium* is thick with it nonetheless, as he describes the damage wrought by dirt settling upon the fabric of the city, both out-of-doors and in.

Evelyn may be writing about environmental disaster, yet his language is glorious. 'This is that pernicious Smoake,' he rails, 'which fullyes all her Glory, superinducing a sooty Crust or furr upon all that it lights, spoyling the moveables, tarnishing the Plate, Gildings and Furniture, and corroding the very Iron-bars and hardest stones with those piercing and acrimonious Spirits which accompany its Sulphure.' The dirt and decay it generates is an insult to a houseproud and socially aspirant man – and everything about it disgusts him. Even the life-giving air becomes instead 'this impure vapour which, with its black and tenacious quality, spots and contaminates whatsoever is expos'd to it'.

Evelyn believed that London was a disgrace to the new scientific man starting to dwell in her, and to the King himself. He writes, 'The City of *London* resembles the face rather of *Mount Aetna,* the *Court of Vulcan, Stromboli,* or the Suburbs of *Hell,* then an Assembly of Rational Creatures, and the Imperial seat of our incomparable *Monarch*.' The capital, which ought to be the greatest city in the kingdom, seemed more like the lowest, something dark, shaded, almost subterranean.

Fumifugium may have inspired Milton's descriptions of Hell in his epic poem *Paradise Lost*, published six years later, in 1667. The imagery is the same: both are places of stench and smoke, oppressive darkness, fires and brimstone (the old word for sulphur), as upstart trades change the social order of England away from a primarily agricultural, medieval economy towards a new world of industry and manufacturing.[16] Before industrialisation has even got underway, both Evelyn and Milton are visionaries, able to see in these earliest of days the enormous environmental cost.

Fumifugium is an astonishing work, and not only for its marvellous prose style: it's one of the most detailed treatises on environmental pollution written before the twentieth century. People had known for thousands of years that coal smoke smelled bad, but that was thought to be the limit of its detriment – and sometimes those foul vapours had even been thought to be helpful. The environmental chemist Peter Brimblecombe notes that 'In Roman times sulfur was burnt in religious rites and in Anglo-Saxon England the smoke from coal was thought to drive away bad spirits.' John Evelyn – and his contemporary John Graunt (more about whom in a moment) – took a great leap forwards: it was only their 'closer observations' and 'more scientific approach [that] could recognize the damaging effects of air pollution among the disguising influence of poor public health' more generally.[17]

Amid its baroque phrasings, *Fumifugium* is also a work of scientific significance, warning how living among 'thick, dirty, smoggy air' makes people 'vulnerable to thousands of diseases, corrupting their lungs and disordering their bodies, so that catarrh, coughts [sic] and tuberculosis are more prevalent in this city than any other in the world'. Evelyn is not quite there yet on the medical detail (he describes bodies 'either dryed up or enflam'd, the humours being exasperated and made apt to putrifie') but he has established cause and effect. Three hundred and sixty years later, the world's cities are still assembling air pollution commissions to ask what is to be done: the science may have progressed enormously, but the lack of political appetite to address it seems a constant.

Evelyn is also prescient for identifying the sulphur as the elemental cause of these air pollution woes. He was writing during the very earliest years of chemistry as a science: it was only that same year that Robert Boyle had argued that the elements were not merely the alchemical quartet of earth, air, fire and water, but instead the 'perfectly unmingled bodies' of pure matter that we know today.[18] Evelyn could not have known that the soft seacoal plaguing London at the time was indeed chemically high in sulphur, or that the sulphur dioxide and sulphates produced when it burns are one of the major forms of urban air pollution to this day. He called it 'sulphurous' most probably because he was guided by his nose: it smelled hellish.

The pollution produced by this seacoal was staggering. During the early modern period, London's air was as polluted as the dirtiest cities in the world today.[19]

In 2008, the chemists Peter Brimblecombe and Carlota Grossi modelled 900 years of London's air pollution to understand how it has blackened, weathered and eroded the city's buildings.[20] Their work analyses each type of pollution separately, such that we don't just have to talk generically of airborne particles but can specify out sulphate pollution, black carbon soot and PM10 particulates, the tiny specks under 10 micrometres in size, an eighth of the width of a human hair.

For each one of these tiny toxins, 1575 is the year things start to change. Sulphate pollution starts to rise, shooting up from a baseline of 5 micrograms per cubic metre to 20 $\mu g/m^3$, the same as the UK annual target level today. Sulphate pollution would rise more than twenty times above base level during the 1600s, to a filthy annual average of 120 $\mu g/m^3$. Environmental historian William Cavert writes: 'Early modern London's air contained concentrations of SO_2 that were as much as seventy times its current level, and even exceeded extremely polluted contemporary cities like Beijing.'[21] The amount of black carbon soot in the air more than doubles, then doubles again. Particulate pollution had already risen slightly in the thirteenth century when coal first came to London, before the loss of life in the Black Death sent PM levels back down to baseline for another 250 years. But in 1575, PM10s start to rise again, and this time they keep rising for centuries. Particulate air pollution would end up exceeding the UK's current standard, 40 $\mu g/m^3$, for over 350 years, and this begins not with the Industrial Revolution but two hundred years earlier. Our air became modern well before the economy ever did.

These estimates are annualised, and so they conceal what must have been vastly greater spikes in short-term particle pollution on cold winter days when the damp city air was especially still, when the Thames fog would have hung in London's streets like a malignant spirit. In 2020, the Centre for Cities think-tank estimaded that 6.4 per cent of adult deaths in the capital are linked to particle pollution, as tiny PM2.5s sneak through the lungs into the bloodstream, causing cancers and heart attacks.[22] The cost to human life of these much higher particulate levels in the 1600s and onward is barely fathomable.

We perhaps imagine early modern Londoners dying of fevers, plagues and in childbirth – but they also very often died coughing. Around this time, a man called John Graunt – a prosperous haberdasher, city government official and militia commander – invented the science of demography through his study of the death records that London parish churches had been keeping since 1532. (The 1600s was a remarkable century for polymaths and amateur scholars.) He published his findings in 1662 in *Natural and Political Observations Made upon the Bills of Mortality*, which makes for a compendium of all kinds of misfortune.[23]

Graunt discovered that only 7 per cent of Londoners were fortunate enough to die of 'age' – that is, over the age of seventy, their Biblical three score years and ten. In 1632, thirty-eight Londoners died of the 'King's Evil', another thirty-eight of 'Purples and spotted fever', and two of 'Lethargy'. One was 'Afrighted', nine had 'Scurvy and Itch' and sixty-two died simply 'Suddenly'.

Twenty-two per cent of Londoners died of '*acute* and *Epidemical* Diseases' (excepting the Plague) which Graunt blames on 'corruptions and alterations in the Air,' according to the miasmatic theory of disease then prevalent.[*] Other diseases

---

* Miasma theory was an idea inherited from classical Greece and Rome (and based on the Ancient Greek word for 'pollution'), which held that infectious diseases were caused by 'bad air' – poisonous, foul-smelling vapour containing particles from rotting organic matter. It was overturned in the mid-nineteenth century, through John Snow's studies of a cholera outbreak in London's Soho, and by Louis Pasteur's scientific experiments in Paris.

surged in particular years: consumption (or tuberculosis, as it's now known) accounted for 19 per cent of deaths in 1632. While tuberculosis is a bacterial condition, research today finds that exposure to ambient air pollution significantly increases the risk of infection, o we should count some part of these deaths in our dusty tally.[24] The mysterious-sounding 'rising of the lights' (which killed 98 Londoners in 1632) is another culprit – 'lights' being an archaic term for lungs, and the term referring to illnesses characterised by a hoarse cough and breathing difficulties, including asthma, emphysema and pneumonia. That is, more disease we now know as either caused or exacerbated by dirty, dusty, sooty air.

Though men like Graunt were at the heart of the Scientific Revolution of the seventeenth century, which gave us the foundations of modern reason and the Enlightenment, the science of the day might not give us an exact number we can say were killed by air pollution as we would now understand it. Yet the statistics Graunt calculated nonetheless paint a portrait of a city choking on its own effluent. The threat was apparent to people at the time. Of course, 'the *Smoaks, Stinks*, and close *Air*, are less healthful than that of the Country; otherwise why do sickly Persons remove into the Country-*Air*?' Graunt wrote, and he directly attributes the high death rate in the city as an effect of coal smoke. As each new resident the city gained meant another tonne of sooty seacoal burnt, the capital found itself suffocated by its own success. In Graunt's words, 'London, the Metropolis of England, is perhaps a Head too big for the Body, and possibly too strong'.

Something had to be done, and our friend John Evelyn had just the plan. The most polluting industries – the '*Brewers, Diers, Sope* and *Salt-boylers, Lime-burners,* and the like' – should 'be removed five or six miles distant from *London*', south of the Thames and some distance east of Greenwich, so as not to 'infect' the King's palaces there. The brewers, if they insisted on having fresh water and not brackish, might reasonably be located in Bow. 'Thousands of able *Watermen* may be employed in bringing Commodities into the City,' a network of busy boats and sleds connecting businesses with their buyers. Surrounding the city, Evelyn envisaged a ring of large gardens, 'elegantly planted, diligently kept and supply'd, with such *Shrubs,* as yield the most fragrant and odoriferous *Flowers,* and are aptes to tinge the *Aer* upon every gentle emission at a great distance'. Finally, London's impoverished should be evicted from their 'poor and nasty Cottages near the City', for reasons less of health than simply 'eyesore'.

That classism aside, there is one major environmental error in Evelyn's book – which is otherwise a thorough and sophisticated analysis of urban air pollution and its potential remedies. He explicitly lets London's 'Culinary fires' in the home off the hook as 'hardly at all discernible' and puts all the blame on industry. From more recent analyses of London's coal consumption, we know that domestic fires were in fact the main culprits, as may have been evident to Evelyn himself, walking down a Westminster street on a winter's day as each household's chimneys puffed their malodorous vapours into the air. But then, as now, a man seeking reform may have had to focus on the easiest targets.

Yet, in the end, nothing was done. Historians speculate as to why. Was Evelyn's plan too ambitious and thus unworkable? Or was it a political failing? Was he a man ahead of his time, lacking the allies and advocates in court to nurture his plans to fruition? Perhaps the King's energy and attention was simply elsewhere?

Fumifugium may seem like an early environmental manifesto but it was, environmental historian William Cavert argues, really a 'last gasp' attempt to try to reform London's air. Subsequent rulers would continue 'to dislike London's smoky atmosphere but . . . Rather than reduce pollution they chose to remove themselves from it.'[25] Three hundred and fifty years later, the rich still live in the west of the city, away from the historically polluting industry in the east.

***

Why does this matter?

The 1570s were a moment of brilliant and terrible innovation. Until that point, all the energy on Earth that people used had come from the sun that had shone during their lifetimes or thereabout – its light being converted into useful, chemical energy through the photosynthesis of plants. Human beings unlocked this stored energy through consuming these plants, and the animals that had fed on them. The discovery of fire, deep in the Paleolithic Era, allowed early humans to access more of this chemical energy by making our food more easily digestible. No longer would we be fuelled solely by living biomass: we learnt how to tap the energy reserves in dead matter too, in the 'detrital carbon' of wood and leaves and animal dung, and could use the energy accumulated over not just one growing season, but years or decades.[26] The reservoir of energy available to us suddenly grew vastly deeper.

The transition from wood to fossil coal is a shift of equally epic significance. Life would no longer be powered by the stored sun of a single human lifetime, as embedded in a tree's cellulose, but by the blackened, compressed light of ancient suns brought to the surface from deep, planetary time. It's a power of quite another scale, and it would transform the world – and the atmosphere – into the modern world we see today. Londoners were the first people on the planet to enter, en masse, into this new mode of relationship to the planet's energy resources. 'There was indeed dirty air in ancient Rome and deforestation in the Mayan Empire, but it was only in early modern England that environmental challenges were solved in a way that ultimately led to a new global energy regime,' writes historian William Cavert.[27] It was the beginning of the end of a clean, temperate planet. And dust was there – as soot, smoke and air pollution particulates; as a choking sign of this environmental overreach – right from the start.

'With the availability of vast quantities of cheap coal, the amount of heat energy accessible to humans became virtually limitless,' historian Edmund Burke III writes.[28] It is this excess which will, in a hundred more years, enable the exponential growth of capitalism, beginning with the Industrial Revolution in the late 1700s and perhaps ending, one way or another, within the next fifty or one hundred years – be its successors sustainable 'degrowth', 'doughnut capitalism', a 'circular economy' or some rather less orderly form of collapse. It is tempting, in hindsight, to see disaster baked in from the start: fossil fuels offered great power, and yet they have not been managed with great responsibility. But more than that: fossil fuel is deeply uncanny.

Historians of the Industrial Revolution talk about the 'photosynthetic constraint'. Solar power captured in wood might be renewable, but it's also slow to grow, constrained by climate, and takes up land that could have been used for other purposes.[29] As a result, a wood-powered society has limits to its growth, both economic and in terms of population – limits that the discovery of coal overturns. Coal's energy may ultimately derive from the same process as wood fuel, the photosynthesis of sunlight – but it is vastly compressed. A tonne of coal, 0.75 cubic metres in size, contains as much energy as the wood harvested from around 70 square metres of clear-cut forest.[30] The growth in the coal trade from Newcastle and Northumberland was as though vast new reserves of land had been found, far back in time, and requisitioned as an economic resource – a St George's flag staked in the warm, muddy swamps of the Carboniferous period. Historians of the Industrial Revolution speak of 'ghost acres', the overseas land in the colonies that enabled Britain to accumulate sufficient wealth to invest in machine production and take the great leap forward of the Industrial Revolution.[31] But these ancient tropical forests that produce these dense, black bombs of potential energy that launch the fossil fuel age: these, to me, are ghost lands too.

We might take this metaphor further. 'Modernity is haunted,' claims Jesse Oak Taylor, an environmental humanities' scholar at the University of Washington. 'Since the Industrial Revolution, all economic growth has been shadowed by the specter of atmospheric carbon. If the market has invisible hands, its feet reinscribe a disinterred fossil record upon the sky. Incorporating this ghostly residue demands a profound rethinking of what it means to be modern.'[32]

It is precisely this reimagining of modernity through its dusty residues and disavowed consequences that this book aims to achieve. There is rarely progress without its sooty shadow – that, in a sentence, is the argument we will develop through the next nine chapters. Not unreasonably, we enjoy what these centuries have brought us, but it is now time to reckon also with the cost.

***

But why start this dusty story at this time – in the late 1500s and 1600s, in the city of London? Why was this the time and place that fossil fuels – which had been known about for thousands of years – suddenly transformed into the dominant fuel? And why is this transformation so significant in shaping the world we live in today?

The transition from wood and charcoal to fossil fuels is of such importance in human history for the two processes it would go on to enable: the Industrial Revolution and the rise of capitalism.

For some big-picture thinkers, such as Edmund Burke III, this is sufficient to divide history into before and after. As he writes, 'If we rethink modernity in terms of its bioenergetics, it is clear that there have been only two major energy regimes in human history: the Age of Solar Energy (a renewable resource) from 10,000 BP to 1800 CE, and the Age of Fossil Fuels (a non-renewable resource) from 1800 CE to the present, which includes coal, petroleum, natural gas, and nuclear power.'[33]

Yet fossil fuels on their own might merely have been

environmentally very unpleasant (as we have seen), not a global ecological catastrophe – had they not been yoked to another technology, the steam engine.

For Andreas Malm, associate professor in human ecology at Lund University, Sweden, what was happening in London in the 1600s was not, as yet, revolutionary: 'as long as coal was mostly used in the domestic production of heat, fossil fuels remained unattached to an engine of self-sustaining economic growth'.[34] He argues that the transformative moment came two centuries later with James Watt's steam engine, released for commercial sale in 1776, which coupled the enormous pent-up energy in a lump of coal with a device for the rotation of a wheel, thus enabling fossil fuels to be put to productive, industrial work. For Malm, it was this moment, where fossil fuels were yoked to the new, industrial capitalism and would start to produce a historically unprecedented era of exponential economic growth, that was 'the fatal breakthrough into a warmer world' – that is, the start of the Anthropocene.

What's most valuable about Malm's work, though, is that he argues convincingly against the idea that this is the Anthropocene at all. After all, it was not *anthropos* – i.e. humankind – who kickstarted the adoption of the steam engine, but specifically the owners of the factories, mines, mills and land who stood to profit from its power. Much popular Anthropocene discourse speaks of this turning point in history as though it is 'our' error, something 'we' all got wrong – and perhaps the inevitable consequence of human nature. Malm reminds us this is mystification: 'The historical origins of anthropogenic climate change were predicated on highly inequitable global processes from the start.' Capital owners invested in steam technology in order to exploit a specific economic moment: American lands depopulated by colonial conquest and disease; a cheap workforce from African American slavery and an urbanising British working class; and global demand for cotton textiles. 'Steam engines were not adopted by some natural-born deputies of the human species,' he writes, 'by the nature of the social order of things, they could only be installed by the owners of the means of production.'

Environmental historians such as Jason W. Moore call this present moment not the Anthropocene but the Capitalocene, a specifically capitalist environmental crisis.[35] I would too were we not about to spend a couple of chapters exploring dust-ridden environmental overreach in the socialist Soviet Union as well. Capitalism may be the name of the logic that drives ecological despoliation today, it may be the economic system that has lasted – but sadly it has not been unique in those ambitions. And so I must write about the common factor between capitalism and socialism – that is, the logic of modernity.

We need to bear in mind, however, that this is at all times a human relation. As Malm writes, 'fossil fuels should, by their very definition, be understood as a social relation: no piece of coal or drop of oil has yet turned itself into fuel'.[36] We should take equal care to see dust not simply as a natural 'given', but the consequence of specific social relations (often inequitable) and people's choices (often bad).

There is also a significant element of geography and historical chance. The Capitalocene nearly started not in London but elsewhere – China, too, had many of the necessary prerequisites during the medieval and early modern period. During the Song Dynasty, the country had a large and sophisticated coal and iron industry in the north-west of the country, which was by 1080 producing more iron than the whole of non-Russian Europe centuries later, in 1700.[37] Beijing-based Jesuit missionary Ferdinand Verbiest presented a steam-propelled model trolley to the Kangxi Emperor in 1672, quite possibly the first working machine-powered 'auto-mobile' anywhere in the world. Yet during the twelfth to fourteenth centuries, northern China would be harried by Mongol invaders, plague and Yellow River floods, and the economic and social centre of gravity would shift south, towards the more hospitable Yangtze Delta region (present-day Shanghai). Far from the coal fields, the majority of iron production (and domestic heating) would remain wood fuelled. A cloud of coal dust would indeed come to settle on China, but not until the late twentieth and early twenty-first centuries, when China becomes the manufacturing workshop of the world – and the site of some of its most choking air pollution. But all this is later: by quirks of geography and happenstance, China escapes this fate in the early modern period, and it is London, instead, where dust proves to be a harbinger of the global environmental catastrophe to come.

The period explored so far in this chapter is, in a way, the moment before everything exploded. In the nineteenth century, with industrialisation fully underway, London's air would get filthier and dirtier yet.

***

'When the Great Fire of London came out of its hiding-places and took life in the air of day, it made ashes of more evident and conspicuous things, but it can hardly have made more ashes and cinders than it makes daily under cover. London is not destroyed again, but it has become the place of immeasurable destruction.'
– Alice Meynell, *The Smouldering City*, 1898

Evelyn failed in his task of reforming London's smoggy skies, and the capital continued to be cursed by an atmosphere of gloom for the next three centuries. It would be joined by the industrial cities of Britain's Midlands and the North, where the air was, if anything, worse.

Elizabeth Gaskell's novel *North and South*, published in 1854, is set in the fictional cotton manufacturing town of Milton (a stand-in for Manchester) where the air is choked by dust and smoke as a symbol of all that is hard and constraining about the new industrial modernity of the North – and the new industrialist class who built it. The southern incomers feel out-of-sorts. 'I don't like this Milton,' says Mrs Hale. 'Edith is right enough in saying it's the smoke that has made me so ill.'[38] The factory workers are sick. Young Bessy Higgins speaks of being 'poisoned' by the tiny fibres of cotton dust that hang in the air in the textile factory where she works.

Dust is a vicious thing to breathe – and each industry that produces a lot of it produces its own lung disease as well. For cotton workers this disease is called byssinosis, colloquially known as brown lung. Fibres from the textile irritate and inflame the alveoli, the tiny air sacs in the lungs, stimulating the release of histamine which constricts the air passages. Your chest feels tight. It feels hard to breathe. Endotoxins produced by bacteria living in the cotton fibres make it worse. You develop a horrible, persistent, bronchitic cough.

Like all industrial illnesses, it can of course be mitigated – should the factory bosses choose. They often did not, as Gaskell laid out in her novel. 'Some folk have a great wheel at one end o' their carding-rooms to make a draught, and carry off th' dust; but that wheel costs a deal of money – five or six hundred pounds, maybe, and brings in no profit; so it's but a few of th' masters as will put 'em up.'

Digging for coal was also, of course, a dusty business, and the miners whose labour ultimately powered this epochal change had their own lung disease too: pneumoconiosis. Immune cells in the lungs would form lesions around the invading particles, which, in the worst cases, would necrotise and die, leaving the lungs riddled with voids. It was commonly known as black lung, after the colour of men's bodies when they were autopsied. The damage coal dust did to miners' health had been apparent since at least the 1800s, but the National Coal Board's regulatory standards didn't come into operation until 1970, by which time the British mining industry was nearly dead anyway.

The double damage of coal: once in its production, and once again in its consumption. In 1859, the Scottish chemist Robert Angus Smith noticed that the air in some cities was so acidic that it would turn blue litmus paper red in just ten minutes. Cities grew murkier as the century progressed. Tiny PM10 particles peaked at

$200\ \mu g/m^3$ in 1890, five times dustier than UK air pollution limits today, and dustier than Beijing in its peak sandstorm years a decade ago. Only Delhi today is worse.[39]

In his book, *The Storm-Cloud of the Nineteenth Century*, Victorian art critic John Ruskin complained: 'The Empire of England, on which formerly the sun never set, has become one on which he never rises.'[40] He grumbled about how in 'those old days, when weather was fine, it was luxuriously fine; when it was bad – it was often abominably bad, but it had its fit of temper and was done with it – it didn't sulk for three months without letting you see the sun'. Condemning 'the sulphurous chimney-pot vomit of blackguardly cloud' in Manchester's 'devil darkness', Ruskin questioned whether the great upset to the natural world that he could see all around him was really progress that should be celebrated. Victorian Britain was the 'workshop of the world', rich on the twin exploitations of coal and empire – and yet the triumphant architecture which this prosperity had wrought wore a blanket of soot like a shroud.

***

For two centuries, London's architecture was all the same colour: black. Sulphurous soot from coal fires and the famous London 'pea souper' fogs generated a thin layer of carbon coating every surface in the city. London was so dirty that there was no memory that it might ever have been any other way. During the restoration of 10 Downing Street in 1954, it was discovered that the familiar dark façade was not actually black at all, but originally yellow brick. The shock was too much for the country to take, and the newly clean building was painted black to maintain its previous, familiar appearance.[41]

But then, in the late 1980s and early 1990s, a change: a great clean-up. Scaffolding surrounded landmarks like St Paul's Cathedral for fifteen years as power washers hosed the grime down into the sewers and out of sight. These days the city is russet and pale grey, silver-mirrored and blue green – brick and limestone and glass. The pollution is now polychrome: these days, the primary residue adhering to buildings is not the black of carbon soot but a warmer browny-yellow colour from the organic hydrocarbons in petrol and diesel fuel. As sulphate emissions from traffic fall, buildings may yet turn green as mosses and lichens grow back.[42]

Yet, whether black, brown, yellow or green, you can't just blast dust and grime off London's buildings. In 2007, architectural conservators assessed the Palace of Westminster, the seat of the British Parliament, and found the building to be suffering from centuries of pollution. Westminster Hall had withstood much, and seen more, as it was built nine hundred years ago by William Rufus, son of the Norman conqueror. If any building in the capital could be said to have historical patina, it was this one, the conservators reckoning it hadn't been cleaned in two hundred years. Yet that patina was eating the place: its walls were being corroded by air pollution and penetrated by moisture. It needed a clean.

But how to do this while maintaining respect for the building's fabric? Limestone is porous, soluble stuff. Turn a high-pressure washer on it and you'll melt holes into it. Fortunately, modern-day conservation methods are ingenious and very gentle. Instead of using water, which could dissolve the limestone, the masonry can be brushed. Delicate carving is cleaned using 'poulticing', akin to a clay facemask for the stone, which draws out deep-seated salts and staining. Latex films can be brushed or sprayed on, then left to absorb the grime from the stone before being peeled off, taking the dirt with them.

News of this epic spring-cleaning project reached New York – and the ears of Jorge Otero-Pailos, director of historical conservation at Columbia University. Otero-Pailos is not only a scholar of preservation but an artist too. After several years of discussion with the Parliamentary estate and its conservators, he gained permission to work alongside them. His method was simple: to preserve and display the latex sheets the conservationists had used to clean the stonework. It is these cleaning materials – akin to the dirty rags used to wash down a floor – that make up the artwork exhibited.

In June 2016, I walked into Westminster Hall and inhaled the unmistakable smell of latex rubber. A translucent, glowing curtain fifty metres long and five metres tall hung from the hammerbeam roof of ancient Westminster Hall, a patchwork skin encrusted in the grime of the entire city. It was lit a rich, sandy yellow – John Ruskin's 'golden stain of time' brought to embodiment in front of me. The dust hung like a ghost, suspended four metres away from the wall that had housed it for so many centuries. The doubled image of the wall on the artwork was precise – each block of stone, each divot and mark in the surface marked accordingly on the rubber sheets. It was a mirror, a grimy reflecting pool. A psychoanalyst would note that the double is uncanny: in literature and mythology, the doppelgänger is a harbinger of injury or death.

The show opened five days after the Brexit referendum and during the longest and most eventful week in politics Britain had known for decades. An artwork about dirt, there, then, was almost too much – metaphorically overdetermined, excessive – and this shaped many reviewers' responses. Adrian Searle, writing in the *Guardian* newspaper, described the work in terms of surgical dressings: 'an acre of flayed skin', 'a latex sheet drawn over a corpse before it is trundled out of sight'.[43] The body politic laid bare.

I got a few minutes at the press preview talking to Otero-Pailos, and began with a question close to my heart: 'Why dust?' He referred back to Victorian art critic John Ruskin's *The Ethics of the Dust*, the text from which he took the exhibition's name. In this book – a rather strange, didactic dialogue between an old lecturer and a group of young schoolgirls – Ruskin constructs elaborate mineralogical metaphors to talk about morality and social order. This is often tedious, but what's more interesting is the way that Ruskin argues that everything is dust, given sufficient time. Stone is simply the crystallisation of the dust into some-thing else, for a moment – but unto dust shall it ultimately return. Otero-Pailos talked about how working with dust and stone brings into focus the long reach of geologic time: both the deep future and the deep past. It puts our moment of 'now' into a much broader context.

Otero-Pailos was interested too in pollution's modernity – and from that, the question of when modernity begins. He mentioned one of his previous projects, *The Ethics of Dust: Carthago Nova*, which used this same latex transfer method on the walls of ancient silver mines in Cartagena in south-eastern Spain. These mines once provided the bulk of the Roman Empire's silvery currency and, though long ago, the pollution from this metalworking is still identifiable today, in ice cores deep in the Greenland ice sheet. When he told me this, I shiv-ered: more parallels and doubling with my own work than I had expected.

So why do we clean buildings, I asked? Why take away the traces of dust and soot and pollution, the traces of touch and bodies and use? Why do we perceive this as necessary? Why must these traces be erased?

Good question, Otero-Pailos replied. 'Cleaning is a means of taking care.' It's something we need to do for ourselves, for our bodies, and we transfer this practice onto the objects around us. 'But it doesn't do anything for the object, to clean it!' he exclaimed. The object doesn't *mind* being dusty, and the dust doesn't (gener-ally) harm it – the two things are quite able to sit together and coexist. But we are, somehow, unable to live with this. Habit and norms dictate that the dirt is by definition 'out of place' and must be removed.

Though some critics disagree, I see Otero-Pailos' work as an act of care. Not so much towards the building – the stone doesn't have feelings! Instead, towards people and processes that are normally invisible in art and culture. Otero-Pailos makes art out of the waste from cleaning, with a double residue. That which would be thrown away, he preserves and uses to talk about what we discard and what we conserve.

He forces us to confront a central failing of the Anthropocene. Since modernity began, people have complained consistently and vigorously about airborne dust – but the measures required to control it have come decades or centuries after, if at all. Dust is the definitive externality – a consequence of industrial or commercial activity that affects other parties, without this impact being reflected in market pricing. Bees pollinating other people's crops for free is a positive externality. Dust, as you may guess, is a negative one. The coal mines and factories that powered Britain's Industrial Revolution made a capitalist class very rich, while the cost was borne by their workers in their bodies and lungs and blood.

'The Ethics of Dust' is, for me, about human presence made present – about the building rewritten as not only limestone and glass and a wood-beamed roof, or as big abstract nouns like 'history' and 'tradition' and 'power', but the material traces of millions of bodies, their labours and their livelihoods. It brings the *polis*, the people, right into the heart of Parliament – and it brings a reckoning with the source of Britain's historical prosperity, too. Nobody normally thinks about dust, what it might be doing or where it should go: it is so tiny, so totally, absolutely, mundane that it slips beneath the limits of vision. Otero-Pailos says it is important, that it contains multitudes, that we should pay attention. As do I.

While editing this book manuscript, I finally got round to insulating the loft of my 1890s terraced flat. As the tradesman sawed a hole in the ceiling to gain access a cascade of filth, the dust of ages, rained down on us through collapsing lath and plaster. After some astonished expletives, I was of course fascinated: this wasn't ordinary house dust, but much finer and darker and indisputably sooty, dating back to the days of coal fires many decades before. An authentic Victorian urban atmosphere settled on every surface in the house, staining the walls and annihilating my hallway carpet. I vacuumed and scrubbed for hours but still haven't fully removed its traces. The environmental consequences of the past aren't easily erased. But we must try to clean up after ourselves, nonetheless. The counter to modernity's massive carelessness, as we shall see throughout the course of this book, starts with such tiny acts of taking care.

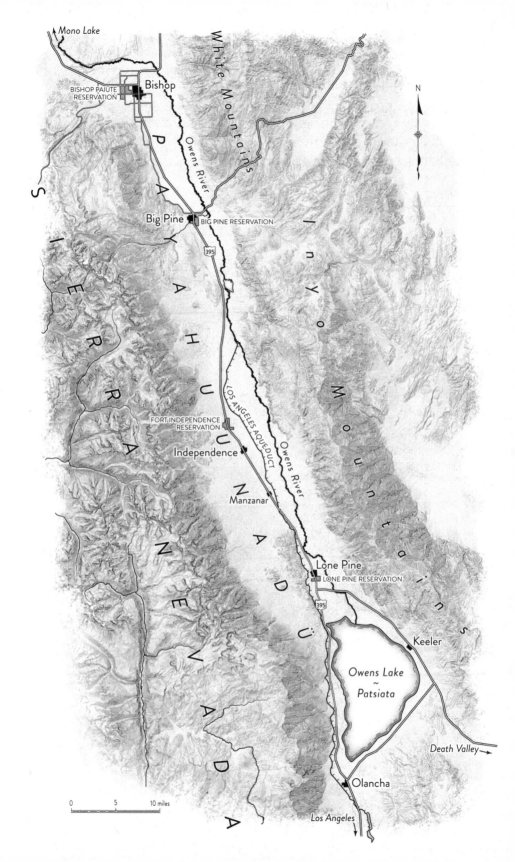

Chapter 2

Turn That Country Dry

London may have birthed modernity, but Los Angeles gave us the world we know today. The city was the twentieth-century writ large: a metropolis of the motorcar; a capital of film and TV. The host of Boeing, Lockheed Martin and the Jet Propulsion Lab, of an economy built on flight and war and going to the Moon, a city of immigration and segregation, of hope, riots and dreams. Los Angeles – not New York – is the definitive American city because it was the first truly *American* city, the first not to look back towards Europe for its street plans and structure but to wholly give itself over to a radically new topology of freeways and exurban sprawl: the spatial logic of a new mass technology. A city sliced up by its planners into what mid-century California writer Carey McWilliams called an archipelago of disparate islands – a luxuriant low-rise efflorescence without a centre; a new, post-modern form of place.[1] A self-mythologising city that turned the photochemical smog its car culture created into a haze of noirish atmosphere. Hike the hills of Runyon Canyon at sunset, and LA is for a moment as rose-tinted as any newcomer could hope.

'Facing the ocean, Southern California is inclined to forget the desert, but the desert is always there and it haunts the imagination of the region,' McWilliams wrote.[2] It haunts the imagination like something undead: because Los Angeles is a vampire that grew by sucking the lifeblood from the land for a thousand miles around. Extract all the water from the land and you're left with one thing: dust. The landscape of desiccation this would produce – a vast dried-up lake-bed and hundreds of square miles of bare dirt and half-dead vegetation – would become, for much of the twentieth century, the largest single dust source in the United States of America.

This chapter is the story of how this environmental catastrophe came to be. What is the logic that makes creating such a sacrifice zone apparently *rational*?

***

The City of Los Angeles as a legal entity is only a small part of the economic and social force that is Los Angeles in total. In its practical operations the city stretches over almost 34,000 square miles, the same land area as Austria or Portugal – yet its environmental footprint is vastly greater still. The true Los Angeles extends deep into the desert, cannibalising water supplies from lesser municipalities and lurching from one water crisis to the next. Los Angeles as a hydrological entity stretches from the Rocky Mountains to Mexico, as it sustains a population of 18 million people not just on water from the (relatively puny) Los Angeles River basin, but from snowfall on the Sierra Nevada and the flow of the Colorado: a river so exhausted by competing water claims that it dries up before it even reaches the sea.

But before the Colorado Aqueduct was built in the 1930s, there was the Los Angeles Aqueduct, running 233 miles from the desert east of California, beyond the Sierras, to supply an endlessly thirsty city. Each step of its construction was legal, and yet it has stuck in cultural memory as a theft, an act 'enshrined in the imagination of Los Angeles as its Original Sin,' says William Alexander McClung, in his book on the mythology of the city, *Landscapes of Desire*. 'Whatever the facts of the matter,' he writes, 'they have been superseded by a will to understand the bringing of water to the Valley as the rupture with both Nature and Law that irrevocably brands LA as a type of Babylon.'[3]

Out in the Eastern Sierra the landscape is big, open, sun-bleached – and rendered pale by a layer of white, talcum-fine dust. Yet the story of how it got that way is unmistakably noir. In the pages that follow, I try to tell it fair. I want to take you first to the Owens Valley in the early twentieth century – and then revisit it today, in Chapter 9 – so we can see the perhaps familiar city of Los Angeles from another angle. To understand just how big the city really is, and what it had to consume to get that way. Because modernity isn't just a matter of bright lights and busy roads and bustling stock exchanges: it's made equally out in the hinterlands that provide the fuel for such wealth and commerce, too. We're taught to see these arid desert landscapes as remote and empty, a space of possibility there for the taking, an asset to be 'utilised' more 'productively'. But land isn't just a 'resource'. It's places. It's people. It's alive.

***

The Sierra Nevada mountain range is the spine of California. Its tallest mountains touch 14,000 feet and stay snow-capped all year round as their summits trap the clouds blowing in off the Pacific Ocean and claim their moisture. To the east, therefore, is a water shadow, the Great Basin: an arid, desert landscape spanning this far fringe of California, nearly all Nevada, much of Utah, and up into Oregon and Idaho. The California counties that lie inside this bioregion feel like the American West. It's high desert here, a big, empty landscape of dun and beige, sand and bone. The land of little rain.[4] I've roamed this landscape with my friend Joel Childers, seeking its abandoned mines and ghost towns. He lives up near Reno and works on an army base; his colleagues say it reminds them of Afghanistan. A landscape more mineral than vegetal, a place of dry washes and alluvial fans, sagebrush steppe and saltbush. The plants that grow here must be halophytes, salt-lovers, as the Great Basin region is literally that, a basin. It holds what little water it gets, and not a single river flows from it to the sea. All the sediment, all the salt, all the heavy metals that the region's rivers have leached from the rocks and carried down-stream over so many thousands of years, all the nickel and selenium and arsenic: they all stay where the water has carried them, resting in the sediments of these endorheic lakes. And there they might have stayed forever, until the men from Los Angeles decided they wanted the Owens Valley's water.

In those days the Owens River was abundant, running rich in the early summer months from the Sierras' snowmelt, a current fifty feet wide and fifteen feet deep. The long, narrow valley – a vivid line of green between the Sierras to the west and the White Mountains in the east – was home to around a thousand Northern Paiute people and a smaller Shoshone community who hunted, trapped its fish and harvested the seeds, nuts and grasses that grew luxuriantly in the river silt. They called it Payahuunadü, 'land of the flowing water', and Owens Lake, Patsiata; they call themselves Nüümü: 'people'. Back in those days the Paiute would seasonally dam the creeks and direct the water into irrigation channels, encouraging the growth of useful plants such as hyacinth tubers for food and willows for basket-making. 'There was a lot of water in this valley and a lot of food,' says Kathy Jefferson Bancroft, historic preservation officer for Lone Pine Paiute Tribe, recalling the stories her grandmother used to tell her about her childhood, back when the lake was full. 'It took time to get it, took time to gather and prepare it, but it was a good life.'[5]

It's important to imagine the valley as it was, so we can understand the magnitude of the destruction wrought upon it. Today only a few fragments of this lush, abundant world exist in the valley, around the springs that seep from the valley sides and the creeks that run with snowmelt, which nurture narrow, one hundred-yard-wide bands of green, generous, vivid *life* in luscious flowering, singing and buzzing profusion. I stood in the gathering dusk one April evening and breathed it all in with surprise and delight. But the valley floor where the Owens River used to run is now almost a monoculture of sparse dryland plant cover: saltbush, rabbitbrush and Russian thistles – that is, tumbleweed.

The first white Americans to see the valley were 'mountain men', the explorers and trappers of the American fur trade. In 1845, the valley was named for one Richard Lemon Owings, whose name transmuted into Owens (no relation of mine). Three years later, gold was discovered in California – then silver, in Nevada, in 1859. The Owens Valley first became a thoroughfare, as stockmen drove great herds of cattle to supply the mining camps, then a settlement. The ranchers fenced the range, cut down prized piñon trees for firewood and prevented the Paiute people from accessing their hunting and fishing grounds. Starving, the Paiute turned to cattle raiding. The settlers responded with war, massacres and ethnic cleansing. In 1863, a thousand defeated Paiute were taken from their homeland and marched through the Mojave to Fort Tejon on the other side of the Sierras, in an attempt to kill off their way of life in the valley entirely. In twenty years, California's Indigenous population fell 80 per cent. The Owens Valley was one piece in a state-wide jigsaw of genocide.[6]

The settlers turned their stolen land to profit. They took over the Paiutes' irrigation systems and expanded them, and soon the valley was filled with orchards and small farms. Mexican and Chinese miners came to the valley to sweat in the Cerro Gordo silver mines, producing the wealth that fuelled the city of Los Angeles' boom. ('First our silver and then our water,' as one local resident would later tell me. Los Angeles would have amounted to nothing without this little valley on the edge of a desert.) The Owens Lake was 50 feet deep and saw daily steamboat crossings. In 1902, President Roosevelt signed the Reclamation Act for irrigation projects across thirteen arid Western states, and a project was duly set up in the valley. The future seemed bright for the settlers: the city of Bishop had incorporated in 1903, banks had opened up, and there was word of railway improvements to come, promising a swift route to market for the products of the valley's labour.

Some thought this narrow valley was the next big thing in California.

There was no future for Los Angeles, thought William E. Smythe, a journalist and chair of the National Irrigation Congress. He thought the city was 'charming', but had grown as far as it could. What was once the home of 5,000 Tongva people had been colonised by the Spanish, grown into an outlaw 'cow town' in the gold rush years around 1850 and then boomed in the 1870s when the railroads arrived. A port developed, banks arrived and land speculators got into business. By the turn of the century, the city was growing rapidly, the population doubling in just four years. Southern California was the fruit basket of America – but land values had rocketed, and the region's water supply was being nearly fully utilised for agriculture. Los Angeles had reached the limits of its watershed, people thought – a constraint exacerbated by drought.

Instead Smythe's gaze turned northwards, to the far side of the Sierra Nevada mountain range and to Owens Valley in particular, where a 110-square-mile lake provided abundant water. 'There can be no question,' he wrote, 'that during the next century they will become the homes of hundreds of thousands of people and the seat of a manifold industrial life.'[7]

Other men had other plans.

***

William Mulholland was born in Belfast in Northern Ireland, the son of a postal guard. He grew up in the south, in Dublin, a bright student but also a quarrelsome one. Dropping out of school, Mulholland left home at fifteen, crisscrossing the Atlantic in the British merchant navy and then roaming America's extractive fringes – Michigan lumber camps, Arizona mining prospects, Southern California oil wells – acquiring bawdy stories and a vigorous character. Yet, by night, he'd been diligently teaching himself geology, fascinated by the fossil layers he encountered while working on the drilling crew: Mulholland was determined to become an engineer.[8] In 1878, aged 22 or 23, he was a *zanjero*, responsible for supervising a ditch for the newly formed Los Angeles City Water Company, working under Frederick Eaton as superintendent. Just eight years later, Mulholland was running the shop and Eaton was working his way up in city government, becoming Mayor of Los Angeles in 1898 on a platform of expanding the city's water supply.

Mulholland started looking for opportunities to expand the city's reach. Nearby rivers – the Kern and Mojave to the north, and the Santa Ana and San Luis Rey to the south – were, he claimed, either too small to be of any use or bogged down by prior water rights claims. Meanwhile, 165 miles to the north was this great torrent of the Owens River, out on what was still practically the frontier and at an altitude of 4,000 feet. If an enterprising water authority superintendent were to build an aqueduct, the water might flow downhill all the way to the city powered by nothing more than gravity.[9]

Fred Eaton was a man with an eye for an advantage. He may have turned the private Los Angeles City Water Company into a municipal authority, the Los Angeles Department of Water and Power (LADWP), but his goal was always personal profit. He envisaged the aqueduct carrying 500 cubic feet per second (14 m³/s) of water, half of which would be for domestic usage in the city, and half he would control himself and distribute privately outside the city limits, at commercial rates. A revised version of this proposal increased Los Angeles' share to three-quarters of the flow, for which Eaton sought to receive an annual fee of $1.5 million – a hefty $43 million at present value.[10] And all while he wasn't paying a cent for the aqueduct's construction. Meanwhile, friends of his had formed a syndicate, the Suburban Homes Company, to buy land in the San Fernando Valley to the north of Los Angeles for just $35 an acre, trading on Eaton's inside information about the water to come. With members including the publisher of the *Los Angeles Times*, Harrison Gray Otis, they were well-placed to exaggerate (some say fabricate) concerns about the drought facing the city and push forward the monumental aqueduct programme as the city's only hope of survival. If all went to plan, Eaton would get rich on both ends of the pipeline, from source to spigot.

Mulholland, on the other hand, was motivated by a loftier vision: progress. Where water flowed, so too would people.

Los Angeles was then a city of 100,000 but this water could make it a metropolis of two million. It might be the greatest city America had ever seen: the best planned, the most relentlessly modern.

'I do think William Mulholland was a purist in this sense that he saw the aqueduct as the greatest good for the greatest number of people,' says his biographer Margaret Leslie Davis.[11] 'He was more interested in the benefit of the city, the benefit of the average man, the benefit of growth and economic commerce.' Unlike Eaton, Mulholland would never really personally profit from rewriting the course of California's rivers. He lived relatively modestly right up until his death. It was merely his vision that was immense.

In 1904, Mulholland and Eaton headed up to the Owens Valley by horse-drawn carriage, leaving – it is said – a merry trail of liquor bottles in their wake.[12] It would be no good having an aqueduct if it didn't have any water. Eaton and Mullholland needed to acquire water rights on the Owens River, and good ones, the 'senior' rights that would give Los Angeles first preference to use up to its full allocation of water a year. In the drought-ridden West, 'junior' rights to use the water others left behind wouldn't be enough to keep an aqueduct flowing. Some years, the people holding them wouldn't get any water at all. The Owens Valley farmers and ranchers who held these senior water rights, however, weren't keen on selling to the city. They were willing to sell land to the Reclamation Service for its reservoir, but that was because its irrigation plans offered the prospect of growth right there in the valley. The irrigation water would let them expand their farms, and then perhaps develop land for housing as the forecasted 'hundreds of thousands' of new residents arrived. Why give away that growth to Los Angeles?

Fred Eaton was a smart man, if not an entirely honest one. Like Robert Moses in New York, a couple of decades later, Eaton and Mulholland used their deep knowledge of seemingly mundane municipal legal process to achieve extraordinarily bold things.

Eaton started by hiring Joseph B. Lippincott of the Owens Valley Reclamation Service, who was running crucial survey work mapping the valley's water flow and who held its land and water rights. These maps made their way into Eaton's hands, who used them as a guide to identify which properties he needed to buy to gain the land and rights needed for the aqueduct. People in the valley thought Eaton was working for the Reclamation Service: he had insider knowledge and the official documents, after all, and had regularly been seen touring the area with Lippincott. Eaton let them keep thinking this and agreed the option to personally purchase the land for $500,000.[13]

It was speculative swizz. In March 1905, Eaton struck a convoluted deal with the city, in which Los Angeles got the rights to some land in Long Valley he'd previously bought, while he kept control of half the land, five thousand cattle, and half a million dollars of farm equipment, all for a personal outlay of just $15,000. The city would pay him another $100,000 in commissions for other lands he'd helped them acquire, often through coercion or fearmongering.[14] Eaton had told some ranchers that they'd better sell up because the Reclamation Service was about to withdraw from the irrigation project and their land would otherwise be worthless; he'd threatened others that their only choice was 'to sell out or dry out'. It was all rather fast dealing, but it got the job done. Owens Valley sold out its water rights to the city without fully realising what had happened.

The City of Los Angeles didn't actually have the money allocated to buy land for the aqueduct at all. Its plan was to borrow the funds against a future bond issue, a plan which could well fail if voters didn't approve it. The water commission was not even permitted to spend any money outside city limits – let alone 235 miles away. But all that could be negotiated. As a member of the commission, J.M. Elliott, put it, 'At the time I was on the Water Board I did not altogether comply with the law strictly; when I saw that an act tended in a good business way to make money for the city, I did it.' Meanwhile Mulholland had been telling friends not to invest in Owens Valley: 'Do not go to Inyo County. We are going to turn that country dry.'

While Eaton was making land deals, Mulholland had been working with Lippincott on a report for the Los Angeles Water Commission assessing the sources of additional water available to the city – an analysis Lippincott delivered just nine days later for a comfortable fee of $2,500 (approximately $85,000 in 2023). The report explored every option other than the Owens Valley and conveniently found that none of these were viable – leading the commission, as intended, to conclude the aqueduct was the best and only option. They approved the project that same day.

A few weeks later, on 27 July 1905, a panel met to decide the fate of the Owens Valley Reclamation Project, part of President Roosevelt's plan to irrigate the West and settle it with family farms. Lippincott – who was still supervising engineer for this project, despite his work with the city – testified that though there was a 'good project' available in the Owens Valley, the water would be far better used in Los Angeles. Besides, there was not in fact enough money in the reclamation fund to do the work. And the land and water rights that Los Angeles had bought up made it impossible, anyway. These were his objective opinions, he swore, and he declared he was in no way selling the valley out: if he collaborated with the City of Los Angeles it was 'because I thought it was for the public good'.

A year later, US President Theodore Roosevelt would echo Lippincott's rationale: 'It is a hundred- or a thousand-fold more important to state that this water is more valuable to the people as a whole if used by the city than if used by the people of Owens Valley.'[15]

It would be more rational, it would be more progressive, it would be more modern. The violence of utilitarianism. Not just to turn a valley to dust, but to insist that it was for the greatest good.

***

Construction of the aqueduct began in 1907, with a labour force of up to six thousand men living and labouring out in the California deserts for six years, followed by a ragtag crew of itinerant labourers, bootleggers and brawlers. 'Whiskey built the aqueduct,' Mulholland later said, and Mojave, the main city along the route, became known as the 'wickedest town in the West'. It was a Panama-sized project on a pint-sized budget – yet from his years of hands-on labour himself, Mulholland had the gift of knowing how to motivate and lead these men, and they broke records for how fast they drilled 53 miles of tunnel through the San Andreas Fault, some of the most geologically volatile terrain in the United States.[16]

On 5 November 1913, a wave of white water rushed down the last sluiceway into the San Fernando Valley in front of a crowd of forty thousand people gathered to celebrate this momentous day for the city's ambitions. From a podium decked in bunting and flags, he spoke in the language of manifest destiny, the belief that white American settlers were God-destined to claim dominion over the entire North American continent. 'This rude platform is an altar, and on it we are here consecrating this water supply and dedicating this aqueduct to you and your children and your children's children – for all time.'[17]

Then – perhaps overwhelmed by the crowd, the noise, and the occasion – Mulholland cut short his speech. 'There it is. Take it,' he simply said.

The syndicate members sold their holdings in the San Fernando Valley for between $500 and $1,000 per acre, a twenty-fold or more return. The aqueduct stopped there, one basin before Los Angeles.

The water didn't even reach the city that had paid for it.[18]

***

Some Owens River water would eventually reach Los Angeles via a roundabout path of irrigation, aquifer recharge and groundwater pumping in the San Fernando Valley to the north. It turned the city green: the dusty, sun-baked Santa Monica Boulevard became shaded with palm trees, new suburban bungalows reclining among bright, verdant lawns.[19]

The water also brought whiteness, argues Karen Piper in her 2006 book, *Left in the Dust*.[20] The Los Angeles River, the city's original water source, was by the early 1900s an abject place, a sewage outlet and overgrown thicket associated with vagrancy and lawlessness – an untrustworthy place, ever-shifting its banks. The neighbourhoods adjoining the river were stigmatised by the stink, and as such, the people who lived there were those who couldn't live anywhere else, mostly Indigenous Americans, Mexicans and Chinese, plus Italians, Jews and Slavs, who may have had European heritage but were, at that time, perceived as non-white.

Since the 1880s, affluent white Americans had been trying to escape downtown Los Angeles by moving west to Bunker Hill. Owens River water would further facilitate this flight by providing the essential fuel that would allow the city to expand. 'The Owens River would promise to bring whiteness, purity and beauty to the Los Angeles suburbs,' Piper writes. It would enable 'new suburban developments [that] attempted to recreate the atmosphere of the Anglo-European 'gentleman's' estate with lawns, gardens, fountains and pools'. Black Americans would be kept out through race-restricted covenants that forbade them to use or own properties and by the 1920s these suburbs were strongholds of the Ku Klux Klan.[21]

As they say, water is power.

In the Owens Valley, the full force of that power would not be seen for the first years the aqueduct was in operation. Rainfall was good, and an unofficial agreement held where Los Angeles took its water primarily from the lower, less-populated end of the valley, letting Inyo County residents continue to use local water pretty much as usual. But in 1920, drought arrived and worsened – in 1924 the valley received just 6.7 inches of rainfall – all while the city kept growing and demanding its full allocation of water,

400 cubic feet per second (11 m³/s). There was now only one way to achieve this, by extracting ground water. Owens Lake started visibly shrinking, leaving lifeless, salt-encrusted ground where water once had been.

The Los Angeles DWP started inching up the valley, buying up more land for water rights. The malfeasance is hard to pinpoint. 'In practically every case its prices were above the valley market,' the California historian Remi Nadeau reports in his 1950 book, *The Water Seekers*. 'Nor did the city take a single piece of land or water by condemnation or unlawful means.' Further, he added, 'In cases where isolated ranchers did not sell, their full share of water was scrupulously delivered by the Los Angeles Water Department.'[22] In total, over a million dollars in water options changed hands, and some ranchers were set for life by the money they made. So why did many in the valley still hate the city?

Some landowners held out from selling: they felt an attachment to their farms that was stronger than profit, for all they'd held the land a scant two generations. But the DWP followed a 'checker-boarding' strategy, buying parcels of land from the willing in such a way that forced holdouts to sell up too, lest they be left having to maintain the entire Big Pine Canal irrigation system single-handedly. The treachery was subtle, but the sense of trickery wasn't. Families felt strong-armed into selling their land to a body they knew was going to destroy it. Journalist Frederick Faulkner reported for the *Sacramento Union* that the city immediately diverted the water 'from the ditches into the aqueduct. It dug wells and installed pumps to exhaust the underground water supply.' Los Angeles had 'bought to destroy, to withdraw from production' – and many the valley hated them for this, a lifetime's work turned to dust before their eyes.[23]

The galling thing was the city wasn't even using the water. For one particularly wet year in the 1910s, the entire flow of the aqueduct was used to irrigate farmland in the San Fernando Valley: Los Angeles itself didn't touch a drop. It was one thing to sacrifice their land for the sake of growing the next great American city

– but quite another to be sucked dry to profit only a syndicate who wished to turn another desert valley dollar-green. Resentment bloomed.

***

Summer temperatures in Lone Pine regularly touch 40°C (104°F). In May 1924, the simmering tension between city and valley was ignited when Los Angeles filed a lawsuit claiming that the valley's canal companies and farmers were illegally diverting water from the aqueduct. The city demanded they stopped using water entirely; a demand, if met, which would destroy them. Instead, they decided to destroy the aqueduct.

A little after one in the morning of 21 May 1924, a tremendous explosion reverberated across the valley, five miles north of the town of Lone Pine. Five hundred pounds of dynamite had blasted a hole in the side of the aqueduct near the Alabama Gates, the point where the Owens River was diverted into an engineered concrete channel and became the city's water. The spillway gates were hurled fifty feet up the hillside, and chunks of concrete tossed up to a quarter of a mile, bringing down power cables and telephone lines. Huge amounts of soil and debris fell back into the river channel, preventing much water from being released.[24]

As DWP workers fought to repair the channel and Los Angeles newspapermen speculated energetically about the explosion's cause, city investigators started to establish what had happened in this distant desert valley. The dynamite was traced to a warehouse owned by two brothers, the affable and affluent Wilfred and Mark Watterson, owners of the Inyo County Bank and directors of the Owens Valley Irrigation District, the organisation trying to save the valley's water for the valley's own use. But although Mulholland's investigators offered a $10,000 reward for information, they heard crickets. Footprints and tyre tracks at the site of the explosion indicated forty or more people had been involved: yet, as detective Jack Dymond later remarked, 'Every resident of

64

the Owens Valley knows who did the dynamiting, but no one will tell.'[25]

It was a turbulent summer. The legal wrangles dragged on, accompanied by protest rallies by day and afterhours planning meetings at Wilfred Watterson's ranch, as the resistance desperately sought a way for the valley to keep its water. Outside forces also saw an opportunity to further their own ends through manipulating the water conflict. Across America, the Ku Klux Klan revival was gaining ground, and they saw in the Owens Valley the potential to escalate local opposition to the Los Angeles water authority into a bigger fight against cities, liberalism, big business – and Jews. A recruiter set to work in the valley to build a local chapter.[26]

Yet the district was by no means united against Los Angeles. Some of the farmers and businessmen had seen the money to be made by selling their water options to the city and took it gladly. This inflamed tensions and divided families: one of the chief 'option-takers' working for the city was the banker George Watterson, uncle to the two brothers leading the resistance. Klansmen started knocking on his door by night, telling him to leave the valley if he valued his life.

On 27 August 1924, another of the leading sell-outs, lawyer Leicester C. Hall, was sitting at the counter of a restaurant in Bishop when a handful of men burst in and grabbed him, strongarming him out the back door into a car. They raced south, a man's arm tight about his neck, Hall choked to near unconsciousness. The car stopped by a cottonwood tree; a crowd of 25 men gathered round. A rope was brought out. Everyone knew what was about to happen next. 'I've done nothing to be ashamed of,' Hall said. 'Give my regards to the Wattersons. They're the ones behind this.' Then, as a last resort – according to one of the more colourful accounts, even as he hung, legs dangling in the air – he made a strange, deliberate gesture: the Freemasons' Grand Hailing Sign of Distress. The bonds of this brotherhood outranked the Klan's, and a fellow Mason stepped forward from the crowd to insist Hall's life was spared. He was released, revived – and run out of Inyo County on orders never to return. He never revealed his attackers' names.[27]

Over the next three years, the valley resistance alternated between words and actions, one Watterson brother leading each approach. Mark's lobbying and legal campaign sought to make a case for restitution, arguing that if Los Angeles was to have the water, then either the city must purchase all the farms in the valley at a fair, independently agreed price, or guarantee the valley enough water to restore its agricultural prosperity.[28] To ensure Los Angeles was paying attention, Wilfred Watterson coordinated a more direct line of attack. On 16 November 1924, seventy men from Bishop drove down to the Alabama Gates spillway, opened up the gates, and disabled the mechanism by which they could be closed again. A torrent of water cascaded out onto the desert floor, eight cubic metres per second – enough to fill a swimming pool in just ten seconds flat. The city's aqueduct ran dry.[29]

When the sheriff came to serve the men with restraining orders, they threw his papers into the rushing water and stayed put. By noon the next day, twenty women had arrived from Bishop to serve their husbands a picnic lunch and scores more people had turned up either to offer their support or simply goggle at the scene. Campfires were lit, and tents and beds set up for a more permanent camp. It was half activist occupation, half county summer fair: Bishop's butchers and groceries supplied food for a collective barbecue; the movie director Lynn Reynolds, who was shooting a Western nearby sent over his orchestra; people gathered from nearby towns to make a day of it.[30]

'Below them the water continued to roar down the spillway for the fourth consecutive day,' Remi Nadeau writes. 'On the edge of the aqueduct near the wheelhouse a woman was silently watching the flow when someone pointed out the tiny blades of grass that had begun to sprout along the edge of the stream. "Yes, that is the lifeblood of this valley," she observed thoughtfully, "and if they'd just let it circulate the valley would come back to life." '[31]

It was a huge PR success: the story of doughty smalltown farmers defending their land from the rapacious big city won sympathetic coverage in newspapers across the country. In May 1925 the courts passed a reparations act and valley residents soon filed for $3 million in damages against the city. Yet the claims process dragged on for months, then years, the DWP refusing to pay a cent without a court order – so by mid 1926, it was back to more direct action. The aqueduct was bombed repeatedly over the next year: better its concrete channel be turned to rubble than the valley turned to dust. Valley resistance dynamited No Name Siphon, shattered sixty feet of pipe on Big Pine Creek and blew up the Cottonwood Power house with blasting gelatine. A bomb was dug into the hillside in an attempt to detonate a landslide and bury the aqueduct with dirt. Following four attacks in June 1927 alone, the city sent one hundred armed guards to defend the aqueduct, then a hundred more – putting the valley under de facto martial law.

The reason I am telling this story in some detail is because, one hundred years later, environmental activists are once again considering the possibility that perhaps you have to blow up a pipeline to achieve action. What has changed?

Andreas Malm, the historian we met in the last chapter, wrote *How To Blow Up A Pipeline* in 2019. In it, he entreats the Western environmental movement to go beyond a dogmatic but ineffectual pacifism and embrace 'intelligent sabotage': that is, strategically damaging the objects and infrastructures that are rapidly, profoundly and irrevocably damaging our planet. Fifty years of protest marches and sit-ins has left us still on the edge of a precipice. 'At what point do we escalate?' Malm asks.

He draws on a long history of confrontational tactics, from the Haitian Revolt in the 1790s to British suffragettes fighting policemen and planting bombs along the route of royal visits in the early twentieth century. Such militancy was an essential part of the struggle for decolonisation. Nelson Mandela famously asserted, 'I called for non-violent protest for as long as it was effective.' Pipeline sabotage was a key tactic of the ANC (African National Congress), who 'considered oil an Achilles' heel of apartheid', Malm writes, and targeted state oil company Sasol in 1980, an action that the ANC's Frene Ginwala said, 'shattered the myth of white invulnerability' the apartheid state depended upon.[32] Other resistance movements in Palestine and Nigeria used similar such tactics.

Yet for some reason the bombing of the Los Angeles aqueduct doesn't tend to make it into Malm's list of precursors and role models. Perhaps it's overlooked because it is a water pipeline, not an oil or gas one – even though the extractivist logic is the same: treating the natural environment as a purely economic 'resource' to be commoditised and turned to profit, regardless of the long-term costs. Or perhaps it's overlooked because as morality tales go, it's by no means a neat one – the frontier violence of the time seemingly no lesson for today. The way the Ku Klux Klan manipulated tensions in the valley to preach antisemitism and race hate is in fact a cautionary warning: environmentalists can sometimes find ourselves with supposed far right 'allies' we should not tolerate for an instant.

Blowing up a pipeline is also harder than Malm fully acknowledges. One reason it's less popular than hashtags and demonstrations is that fact that protestors in America face increasing criminalisation for even peaceful resistance. In 2016, the Dakota Access Pipeline protests saw members of the Standing Rock Indian Reservation (alongside Indigenous and eco-activist supporters from across the Americas) campaign to stop the Keystone XL oil pipeline which they feared would pollute their land and water. They used tactics not of destruction but merely obstruction, 'locking on' to construction equipment to stop its operation and seizing the site up with roadblocks. Police responded with rubber bullets and water canon in icy November conditions: hundreds were injured and treated for hypothermia.[33] Had anyone at Standing Rock tried to blow anything up, it's unlikely they'd have made it out alive.

Some things change, some things stay the same. For historians and journalists, 'water is power' is one of the great lessons of the California Water Wars, yet the slogan protestors rallied around at Standing Rock is one they would have recognised in the Owens Valley a hundred years before – 'Water Is Life' ('*Mní Wičóni*' in Lakota, the language of the region). It really is. And it remains a thing worth fighting for.

⁂

Back to the story of a valley in the shadow of the Sierra Mountains. In April 1927, Mark Watterson was optimistic, writing to one newspaper journalist: 'I feel we have the city on the defensive, and we must strike hard and often now and not give them time to recover themselves.'[34]

His uncle George had other ideas. The water dispute had harmed his business interests and cleaved the Watterson family in two, and so he began digging – with the support and encouragement of the DWP – into the accounts of his nephews' Inyo County Bank. He found a financial black hole. For two years or more, the bank had been using the money it had on deposit – the collective life savings of the valley – to shore up their other business enterprises: a mine, a mineral water company and a resort. Their accounts were an utter fiction of balances they did not in fact hold and loans liable that had already been repaid – all in, the Watterson brothers were in $2.3 million deep (about $38 million in 2023 dollars) and were charged with no less than 36 counts of embezzlement and grand theft. No two ways about it, they were going to go to jail.

Almost all business in the Owens Valley ran through the Inyo County Bank, and its collapse paralysed the place. Families who'd sold their ranches to the city had the proceeds wiped out; friends who'd left securities on deposit discovered their valuables had been sold; shops were left with just the cash in their tills. Despite everything, many local people kept faith. The Wattersons had done it to save the valley, they said. And so a community who'd taken a devastating financial hit dug deep, mortgaged the property they had left, and offered up another million dollars to try to keep the Watterson's enterprises – and thus the valley's economy – going.

But such solidarity couldn't save the Watterson brothers. They were sentenced to ten years in San Quentin State Prison. Even as the train taking them to San Francisco passed Bishop, a man was hammering in a new sign: 'Los Angeles City Limits'.[35] The city owned the place now, quite literally: to this day, Los Angeles owns 89 per cent of private land in Owens Valley.[36] The fight was over. The little guy had lost.

Did anyone really win?

Frederick Eaton tried to sell his landholdings in Long Valley for the city to use as a storage reservoir – but he wanted a million dollars and Mulholland considered this extortion. The two men, partners for fifty years, grew estranged, Eaton bitter. When the Wattersons' bank collapsed, Eaton's heavily mortgaged finances went down with it.[37]

Mulholland built his reservoir elsewhere, on the soft, reddish rock of the San Francisquito Canyon to the north-west of Los Angeles. On the night of 12–13 March 1928, the dam failed catastrophically. A forty-metre wall of water swept down the valley, killing 431 people. Earlier that day, Mulholland himself had inspected the dam and made the call that it was safe, an observed leak not out of the ordinary. No evacuation was needed, he judged. At the inquest he avowed, 'Whether it is good or bad, don't blame anyone else, you just fasten it on me. If there was an error in human judgment, I was the human.'[38]

That error broke him. A man who had not for decades taken so much as a week's holiday finally retired and turned inward in his grief, refusing to speak, scarcely eating, unable to sleep. 'What's the matter with me?' he exclaimed to his daughter, Rose. But he gave his own answer: 'I see things, but they don't interest me. The zest for living is gone.'[39]

But we might say the city of Los Angeles won – or at least it grew propulsively, and isn't that much the same thing under capitalism? Greater Los Angeles now has a population of 18.7 million and is the twentieth-largest megacity in the world.

***

Without water flowing through the irrigation ditches, the orchards and wheatfields that had sustained the valley for fifty years were doomed. Their destruction was rapid. In 1920, the Manzanar area had nearly 5,000 acres of apple, peach and pear orchards in cultivation, alongside vegetable and flower gardens and fields of potato and corn. Its fruit won prizes at the State Fair in Sacramento, and sustained a flourishing little community of 25 houses, a two-room school, and a general store. By the 1930s, it had all been abandoned.[40] The Paiute, forced off their land two generations before, had got by as agricultural labour for the farmers and ranchers in the valley; now there was no work to be had, and they were forced into poverty. Manzanar would not be resettled until March 1942, when it was used as a concentration camp to intern 10,000 Japanese Americans in the wake of the Pearl Harbour attack. 'We slept in the dust, we breathed the dust; and we ate the dust,' one internee recalled. 'Such abominable existence one could not forget, no matter how much we tried to be patient, understand the situation, and take it bravely.'[41]

A valley once called 'The Switzerland of the West' for its glacier-fed lushness was now a desert. A lake that once stretched over 110 square miles had dried up and mostly vanished by 1926. A few hypersaline brine pools remained along the western shore, colonised by tiny halophilic, salt-loving archea (a type of single-celled organism) that stained the water an unholy red. Much of the remainder of the lakebed was a bright, shining white – an eerie glow on the horizon for drivers on the 395 to Reno.

When the wind blew – and in the long, narrow Owens Valley it really does pick up – it became 'possibly the greatest or most intense human-disturbed dust source on earth,' writes Todd

Hinkley of the United States Geological Service.[42] It was certainly the largest source of dust in the United States, at between 900,000 and 8 million metric tons of total dust per year. During a major dust storm, 50 tons per second could come off the lake's surface, producing the highest concentrations of tiny PM10 dust ever recorded in the United States.[43, 44]

'There is every year three hundred thousand tons of dust and it goes someplace,' Harry Williams of the Bishop Paiute Tribe said in 1997 at a court hearing about air pollution. 'We are talking eighty years of three hundred thousand tons. I don't know what that adds up to, but it's landing someplace. It's landing on wetlands. It's landing on animals. Things are being affected. The Forest Service is telling you the oldest things in the world, it's up that high.'[45]

Dust plumes would rise high into the air and be blown north over the White Mountains, where they met the gnarled, twisted forms of the ancient bristlecone pine trees living in remote and airy fastness on the summits. These trees are astonishing beings – the oldest plants in the United States, up to 4,800 years old: trees that were saplings when the first of the Egyptian pyramids were being built, trees as ancient as the very concept of 'cities'. Drive up the long, rutted off-road track to visit them and it might seem as though they live wholly outside the runnings of the world, quite out of normal time. But the dust shrouds the trees in a powdery white veil, and its corrosive alkali salts cannot be doing them any good.

Down on the valley bottom, the dust did valley residents no favours either. 'Keeler fog,' they called it, after a town on the shores of what was once a lake. 'As a child, I believed it really was fog,' Karen Piper wrote in 2006. She grew up locally in Ridgecrest. 'It moved in like fog from over the mountains, blocking the sun. It hung in the air like fog even when there was no wind. There were days when the sun disappeared and it was hard to breathe – but no one ever explained to me what it was. No one knew, then, that it was dangerous.'[46]

'This dust, which is fine and white as flour, covers my home-town when the wind blows,' Piper said. 'It is sometimes white from the salt, sometimes pink from algae, and sometimes a dirty gray.'

What was blown up by the wind wasn't flour, of course, but salt. The bed of the Owens Lake had become a playa the size of San Francisco, a dry, deeply cracked surface of mud, sand and silt. Because they're formed by evaporation, playas are the flattest landforms on the planet, and it only takes a few centimetres of water to turn many kilometres of the lakebed into a shimmering mirror. These periodic inundations brought to the surface all the salts eroded over thousands of years as the Owens River rushed down from the Sierra Nevada. And when the water evaporated, it left behind a salt crust.

Salt crusts are strange. On the one hand, some of the dustiest places in the world are salt playa environments. Yet elsewhere – such as in parts of the former Aral Sea, in Uzbekistan, or in Tunisia or the Namib Desert – salt crusts form hard, tough surfaces that are highly resistant to wind erosion. So what exactly was going on at the dried-up Owens Lake that turned it into such an environmental hazard?

The first problem is that salt crystallises in what geologists call a 'displacive' manner – that is, it pushes other minerals apart, rather than filling in the gaps in between them. This both increases the amount of tiny particles in the soil by adding new salt crystals, and also pushes to the surface other sand or silt particles that are then free and available for wind erosion too. The mineralogy of exactly which types of salt are present also influences how dusty a salt playa will be. Different types of salts form different types of crystals, which are better or worse at clinging to each other when a strong wind blows. Unfortunately, the Owens Lake is made up of notably unclingy salts, such as thenardite and mira-bilite, which form delicate, pointed, prismatic crystals – and halite, which forms what geologists call a 'puffy' or 'hairy' salt crust that's essentially full of gaps. Finally, because Owens Lake occasionally refloods in heavy winter rain, the salts keep dissolving and only ever get the chance to regrow into tiny, individual particles before they're deluged and dissolved again.[47] All these factors add up to a playa surface rich in very small, weakly bonded salt and soil particles – which are easily provoked by a brisk breeze to become airborne dust.

This type of erosion is called saltation – from the Latin *saltus*, to leap or jump. Larger particles may only jump a short way before landing again, producing the characteristic low-lying haze in the valley. Smaller particles, however, are lifted higher and further by the wind – and are also a far more dangerous health hazard: so tiny that they slip past the body's defences and burrow deep into blood and bone.

During the twentieth century, Owens Lake emitted around 300,000 tons of PM10s per year, the tiny dust grains less than a hundredth of a millimetre across.[48] Thirty tons of this was arsenic and nine tons was cadmium, the Environmental Protection Agency (EPA) reported – and both were major carcinogenic hazards to residents' health.[49] Arsenic causes skin lesions, then skin cancer. Exposure in utero and in early childhood impedes cognitive development and increases deaths in young adulthood from lung and bladder cancers, heart attacks, lung disease and kidney failure. Cadmium, meanwhile, is toxic to the kidneys; weakens bones, causing osteoporosis and fractures; and harms the respiratory system as well. People in the Owens Valley were inhaling these heavy metals every day.

'I took an interest in Owens Lake because of eighteen years' worth of dust embedded in my lungs,' Karen Piper writes, continuing:

'My sister now has a form of lupus, and three other girls in my neighborhood were also diagnosed with this disease; two have since died. My childhood friend from church now has acute respiratory distress syndrome, as do four babies from the same church – they are currently all in the hospital. And

74

my mother's neighbor was just diagnosed with rheumatoid arthritis of the lungs and is on oxygen. I have suffered from bouts of pneumonia and asthma my entire life, as well as life-threatening allergies. The town is full of cancer, too, and my mother often calls to read me the obituaries. "So many people in their forties," she once commented.'[50]

The dust travelled. Fifty miles down the road, an emergency physician at Ridgecrest Community Hospital noted that attendance would rise tenfold during dust storms: 'When we see the white cloud headed down through the pass, the ER and doctors' offices fill up with people who suddenly got worse. It's pretty straightforward cause and effect.'[51] The dust travelled all the way to Los Angeles, a shadow following in the path of the Owens Valley's purloined water.[52] There's no getting away from consequences, much as the city has tried to deny them.

For a century this angered and upset Owens Valley residents, who had to live amid an environmental crisis Los Angeles sought to pretend wasn't happening. 'Year after year, we watch the grasses die back. We watch the shrubs move in. We watch the weeds come in,' said Sally Manning, environmental director of the Big Pine Paiute Tribe, in 2014. 'Years go by and nothing gets done.'[53]

Why? Because Los Angeles wanted the water. Its growth and progress depended on it.

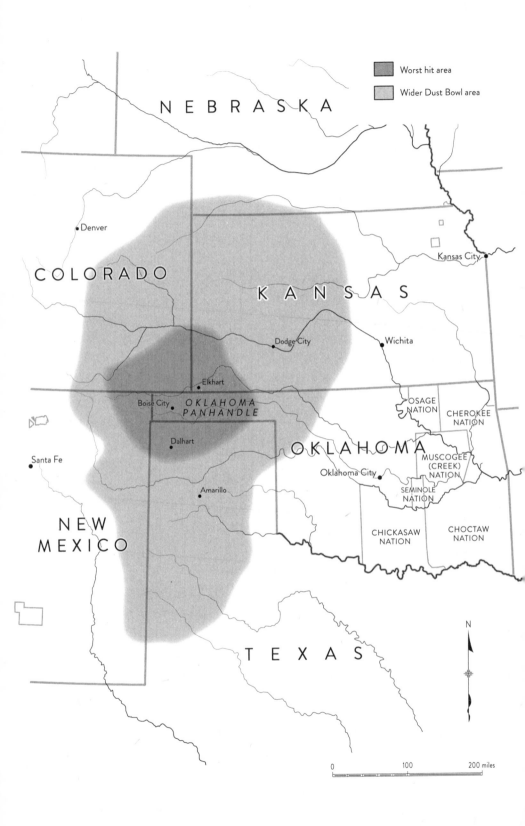

Chapter 3

Dust to Dust

On Sunday, 14 April 1935, a dust storm swept down through the Great Plains like an avalanche, smothering everything in its tracks.

March had seen dusters every day for thirty days straight; in Dodge City, Kansas, there'd been only thirteen dust-free days so far that year. Yet on that second Sunday in April, the morning dawned bright, sunny and clear. It was the best day of the year so far with temperatures up into the eighties: shirtsleeve weather. In the farms of the Oklahoma Panhandle, families opened up their front doors and breathed in deep. It was Palm Sunday, a week before Easter, and people hoped God was in a kindly mood.

They aired out their homes and started to clean. Wet sheets and blankets put up to catch the dust could be taken down, the tape and flour-paste strips sealing up windows and doors peeled off and windows thrown open wide. Dirt was swept and scooped out of the home by the bucket-load; roofs were shovelled; bedlinen and clothing got washed and hung up in the sun to dry. People went to church – the Methodist church in Guymon, Oklahoma held a 'rain service', the congregation praying for divine intervention to bring much-needed moisture. In Boise City they resumed plans for a rabbit drive, delayed a month by the dust storms. Elsewhere, families walked out to inspect their farms: the outhouses half-buried and ceilings falling in; new dunes nine and ten feet high, trapped by tumbleweeds piled up against the fences.

That same morning the sky turned purple and the winds rose, eight hundred miles to the north near Bismark, North Dakota. The temperature dropped 30 degrees Fahrenheit as the winds picked up and blew south and south-west, first 40 then 65 miles per hour as the storm raced over South Dakota, Nebraska, and into Kansas, picking up the dry, dry dirt from the exhausted land into a roiling mass of darkness two thousand feet high.

Thirty-two-year-old Ada Kearns went into the house, 'I believe it was around three or four o'clock,' she recalled to an interviewer from the Oklahoma Historical Society, many years later, 'and the radio was on, and it said "It's Dodge City, we're going off of the air."'[1] No time to say why: the storm carried so much static electricity that it was shorting-out electrical equipment and car engines. Barbed-wire fences visibly glowed with charge.

The stormfront rolled on southward, picking up more dirt and power as it went. The sky filled with birds racing ahead; the ground ran with jack rabbits. The dust storm rolled along the horizon, inexorable and terrifying – a wall of oblivion.

Logan Gregg was round a friend's house playing poker – a bit of a craze at the time. 'And someone got up and went out, and he said, "Come here" and that thing was coming down across this river. It just looked like it was rolling' – he made a gesture with both arms, to illustrate – 'and you think "Well that's the end of time" or something. By golly when it come over you didn't know whether there's going to be Jesus on the white cloud or the Devil on that black cloud – it just scared the heck out of you.'[2]

They all describe it the same way. 'What did that storm look like when it was approaching?' the interviewer asked Mrs Nellie Goodner Malone. 'It might just be the end of the world,' she said.[3]

That day became known as Black Sunday, and in newspaper

reports the day after the region was referred to for the first time as the 'Dust Bowl'.

***

The Dust Bowl was a crisis of drought, dirt and depression that hit the central United States in the 1930s. Over the course of a decade, 100 million acres of America's central Great Plains region – an expanse of land five hundred miles from north to south, and three hundred miles east to west – dried out and turned to powder. In 1935 alone, the year of the worst dust storms, 850,000,000 tons of topsoil blew clean away, to be scattered across the United States as far as Washington and New York.

Droughts are inevitable on the Plains: the rains fail as frequently as one year in four. A major drought between 1856 and 1865 contributed to the extinction of the buffalo – and Indian lifeways with them – as increasing settlement prevented the herds from reaching the last remaining living grassland in the river valleys. In the mid-nineteenth century, pioneer caravans reported experiencing dust storms where visibility dropped to no more than a hundred yards.[4] In Nebraska, the first wave of white homesteading was abandoned barely five years later as a seven-year drought hit in 1890 and farmers watched their wheat crops dry to a crisp.[5]

But the dust storms of the 1930s were something new: roiling black clouds, seething with static electricity, the air so thick with dirt you couldn't see your own hand in front of your face. These dust storms were not natural. Nature may have supplied the preconditions: the wind, the drought, the soil. The scale of the consequences may have been geological: a landscape transformed by deep drifts of dust; fields scraped bare down to bedrock. But the cause was humanmade. The Dust Bowl happened because white settlers arrived with ploughs and decided to 'break the plains'.

The Dust Bowl struck at the heart of the American continent, both spatially and symbolically. The counties at the centre of it – Cimarron, Texas and Beaver counties in the Oklahoma

Panhandle, a peninsula of land that sticks out of the north-west of the state much as its name suggests – were the last in the contiguous 48 states to be settled by white people, which took place between 1889 and 1907. Only a generation before disaster hit, this had been the frontier, with all that entails in the American mythos. That it should take so little time to all go up in dust might conceivably cast some doubt on the project of settlement in its entirety – yet this is the unthinkable in contemporaneous accounts of the Dust Bowl, and many more recent ones too. Most often the story is told as a moment of exception: a natural disaster, not a humanmade one; and a crisis safely contained to a specific region rather than, in truth, merely the most dramatic failure in a global crisis of soil exhaustion. This way, difficult questions about the sustainability of farming this land, then and now, can be evaded.

It's a story of tremendous loss. Approximately three million people were 'blown out, burned out, or starved out', in the words of one such emigré, Kansas wheat farmer, Lawrence Svobida. They escaped rural destitution by moving to towns and cities, or fleeing west to California.[6] It was one of the largest internal population movements in American history, as environmental disaster compounded with the Great Depression to hollow out what had been, for a few short decades, the centre of the American nation. The region has never recovered the population levels it saw in 1910. Beyond the statistics, what was lost might be unknowable. 'Dust pneumonia' is thought to have claimed perhaps seven thousand lives, but, as Svobida wrote, in 1940, as he looked back on the wreckage of the landscape and the life he had just left, 'Most of the grim tragedies of the Dust Bowl will never be recorded save in the hearts of near-friends and sorrowing relatives.'[7]

Not everyone wanted to remember. Don Hartwell's diary covering the later Dust Bowl period was nearly burnt by his widow, Verna, but rescued by a passing neighbour. Why would she destroy such an intimate memory of her husband, and a valuable historical record as well? Perhaps it reminded her of a time she just wanted to forget.[8]

***

Late December 2016, I headed down into the depths of the Grand Canyon with my friend Wayne Chambliss for a few days' hiking amid deep time. But the forecast turned: a storm was rolling in, and it seemed like a good idea to avoid it. Walking back up through 1.8 billion years of stratified time, we made another plan: let's drive to the epicentre of the Dust Bowl. No particular reason, just to go and see. To get a feel for the place. To contrast old and new. But mostly for the pleasure of movement itself: 1,500 miles on straight, unpeopled roads at 80 miles per hour, cars passing infrequently enough that drivers would flip up a finger or two in greeting.

And so the sun rose on 2017, over occasional oil derricks and flocks of wind turbines, whirling. We'd left Amarillo before dawn and headed north into the Llano Estacado, up into the panhandle of western Texas and then into Oklahoma. Small herds of black cattle roamed the grassland under a vast blue sky. For the first twenty years of white settlement, the southern plains were cattle country. The land was short-grass prairie – too high-altitude, too cold and too dry for longer grass to grow. It was once the southern range of the great American buffalo herd, and on that basis cattle ranching might have seemed sustainable on this dry turf, on the grounds of similar ecology. But even in the early days, it never was. Ben Kinchlow, a cowboy who drove cattle up to the railheads in Kansas and Nebraska in the 1870s, described a landscape so densely stocked that 'Goin' up the trail you never was out of sight of a herd. The trail was so worn, that the dust would be knee-deep to the cattle. You could ride right up to the rear of the cattle an' you couldn't see the cattle for the dust.'[9]

Today the cattle are still there, most fattening in intensive feedlots but some still roaming the plains. They drink from troughs fed by gleaming stainless-steel windmills twenty feet tall, pumping groundwater from the Ogallala aquifer below. Spindly centre-pivot irrigators rest gracefully at the side of fields like vast, mechanical dragonflies, a quarter or half a mile long, sustaining crops of sorghum and wheat. The animating economics of the place were unfamiliar to my European eyes: the houses and small-town main streets didn't look like wealth and yet the grain silos were huge, the trucks people drove were enormous and a newer model combine on the Guymon show lots cost half a million dollars. This was industrial agro-capitalism, a rural food factory.

Viewed from above, the landscape is a regimented grid of circles and squares, as constructed as Manhattan. Viewed from within, though, the world shrinks to a few degrees up and down from the horizon, as the road stretches away in a straight line until it vanishes. It's Montana that claims the title of 'Big Sky Country', but I have seen no skies bigger than here above these flatlands: a full one-eighty, a transparent but infinitely deep blue. You don't look all the way up in case you might fall.

The plains are a semi-arid landscape, the hostility of the desert only ever a few inches of rainfall away. In 1878, the geologist and explorer John Wesley Powell proposed that the defining climactic boundary in the United States was a line through the middle of the Great Plains that delimited the damp and humid east from the much drier west.[10] He was right. Close by the one-hundredth meridian (which runs up close by the Oklahoma–Texas Panhandle border, then up through Dodge City, Kansas) lies another line: the 20-inch isohyet. Imagine a contour line, but for rainfall. To the east of this line, the land receives over 20 inches of rain per year, with the result that farming is possible without irrigation, and a wide variety of crops can be grown. To the west, life is rather harder. In 1823, the US Army's Major Stephen Long mapped this land as the 'Great American Desert', his report describing it to be 'almost wholly unfit for cultivation, and of course, uninhabitable by a people depending upon agriculture for their subsistence'. For over fifty years until the settlers came, he was pretty much taken at his word.

Water here is described as 'ephemeral', a beautiful euphemism for drought. The Cimarron River only flows for a few months of the year, if that. Most of the time it is dry, sunken into the sand of its creek bed, and little good to man or beast. An early traveller on the Santa Fe Trail described it as the 'saltiest, most singular, and most abominable of all the villainous rivers of the prairies'.[11] Forests do not grow here: the subsoil does not hold enough moisture. The landscape is often flat to the horizon without even a single tree to punctuate the immensity. Only down in the creek beds might you find cottonwoods and willow trees, growing on sandbars along the waterline where the rivers strand their seeds.

***

It was water that opened the Plains to white settlement.

In the 1860s and 1870s, something curious happened: the climate shifted. It got wetter. Emigrants on the Oregon Trail began reporting that the land in western Nebraska, previously dry and yellow, had newly become green. 'Rain follows the plow,' the experts said, though no one seemed entirely clear on precisely how. Perhaps ploughing the soil for cultivation exposed the soil's moisture to the sky, or perhaps it was planting trees and shrubs that did it – or the steam from the trains, or the increased vibrations in the atmosphere created by all the new human activity. Whatever the cause, the boosterists were quite sure that the change was permanent: 'The raindrop never fails to fall and answer to the imploring power or prayer of labor,' claimed land speculator Charles Dana Wilber in 1881.[12] How convenient.

The Santa Fe Railroad printed a scientific-looking map showing the 20-inch isohyet moving west at eighteen miles per year, neatly following the new towns tied to the railroad. Land for thirty miles either side of the tracks was taken from Indian tribes and granted to the railroad companies by the federal government – an astonishing 175 million acres, one-tenth of the entire country

– which the railroads promptly turned round and started marketing for sale in the slums of East Coast cities and even back in Europe.[13]

Profit wasn't the only motive spurring America's westward expansion in the nineteenth century. 'Manifest destiny' was another, a vision 'to possess the whole of the continent which Providence has given us', as newspaperman John L. O'Sullivan put it, in a phrase that defined the age.[14] The 1862 Homesteads Act offered up 160-acre lots of public land for just a small registration fee, on the condition that settlers commit to making a go of working their plot. (Not an easy task.) After five years living on the land, people who might have previously had nothing could make permanent their own small stake in the American dream. This territorial expansion – this imperialisim and Indigenous dispossession – was championed as defining the nation. In 1890, historian Frederick Jackson Turner wrote that 'This perennial rebirth, this fluidity of American life, this expansion westward with its new opportunities, its continuous touch with the simplicity of primitive society, furnish the forces dominating American character.'[15]

Like the Owens Valley, then, the High Plains only seem to be a remote and peripheral place. Politically and sociologically, they occupied for many years the centre of the American national imagination. Marginal places – and the processes of their marginalisation – are in fact central to making the modern world.

***

From Missouri, Iowa and Illinois, by train, cart and on foot, people came.

Most of Oklahoma was at that time Indian reservation land – but not all. The territory had two anomalies, legal no man's lands – land which newcomers were clamouring to settle. At noon on 22 April 1889, an estimated 50,000 people lined up to claim homesteads in the two million acres of the Unassigned Lands made newly available for settlement. This was a land run, and it worked as the name suggests, a first-come-first-served dash by horse and buggy to claim a 160-acre 'quarter section' as your own. Oklahoma City sprang up (in tent form) literally overnight.

Over the next twelve years, all the Indian reservations in the state were forcibly sold to the federal government and their land opened for settlement. As the market price of cattle dropped, ranch owners subdivided their land to sell small sections to farmers as well. The Plains were now private property.

The Oklahoma Panhandle was one of the last places in America to be settled for a reason, and farm turnover was high on this dry and unpredictable land. Homesteaders weren't faultless stewards of the land, but they did at least see it as home, and a sense of care comes through in their diaries and oral history stories. As richer landowners bought out those who failed, however, the land was transformed into a pure investment vehicle. Plains geographer John C. Hudson explains how, 'because these areas were settled relatively late (especially after the magic date of 1890, which supposedly marked the closing of the frontier), early settlement within them bore the unmistakable imprint of early twentieth-century finance capitalism: large-scale holdings devoted to crop monoculture, high ratios of capital to labor in farming, factory like organization of production, and a general emphasis on the gigantic and the colossal in all things tangible.'[16] It was a recipe for environmental exploitation just waiting to happen.

Settlers were vastly ill-equipped for life in this arid land. They brought farming techniques and notions from their previous, more temperate climes – or consulted the bible for those newly moved west, Hardy W. Campbell's 1902 *Soil Culture Manual*, a guide to dryland farming. Preparing the hard-packed dry earth was the trick Campbell taught: loosening the earth by double-disking with a harrow, deep ploughing so that the wheat roots could spread. After each rain, a harrow should again pulverise the surface soil into 'dust mulch', to obstruct the upward capillary movement of subsoil water and save water wastage on weeds.

Then farmers should plough again straight after each harvest, and leave the soil exposed for months until it was time to plant again.[17]

Prairie grass had held this land together for thousands of years. In parts of Oklahoma the soil began to drift just four years after they broke the sod.[18]

***

Yet for a decade, little more, it seemed as though the settlement gamble might have paid off. When many years' hard labour started to come good for Caroline Henderson, a teacher and homesteader in Texas County, Oklahoma, 'It seemed as if at last our dreams were coming true.'[19]

The war that broke the fields of Flanders made Plains farmers' fortunes. Turkey's entrance into the First World War cut off Russian wheat exports, forcing European buyers to compete for the US crop. 'Plant more wheat! Wheat will win the war!' President Woodrow Wilson proclaimed, as prices more than doubled to around $2.10 per bushel – and plant they did, with vigour. Farmers in the Plains region ploughed under eleven million acres of prairie grasslands, and in 1919 the nation harvested 74 million acres of wheat, over a third more than the pre-war period.

With money in their pockets and banks falling over each other to lend more, Plains farmers went on a spending spree. Young men who'd grown up fetching hay for their fathers' carthorses could drive out of the agricultural showroom on a bright green John Deere Model D gasoline-powered tractor with glossy yellow wheels. They must have felt so proud. The new combined harvesting-threshing machines meant that bringing in an acre of wheat – a long, hard week's work by manual means – came down to less than three hours, as the flat, open lands of the Plains allowed for the most efficient farming in the country.

Wheat farming continued to expand throughout the 1920s: the climate was wet that decade, and yields were good, sometimes great. But it was all fuelled by speculation and debt. A farmer might make $8,000 from a good harvest – equivalent to around $100,000 today – but a combine harvester was $3,000, and some years the spring rains wouldn't come in time or a storm would batter your crop, and that year's income was gone. Farmers had borrowed heavily to fund expansion during the war, and their mortgages and rising tax rates pushed them into pursuing short-term gains, year after year. Continued monocropping wears out the soil, but, as Lawrence Svobida tells us, 'one good wheat crop sold at a good price might well yield returns equal to the profits to be earned in ten years of stock raising, and who was wise enough to see into the future?'[20]

It was cash crop agriculture that turned this land to dust, says environmental sociologist Hannah Holleman.[21] It is 'insatiable,' she says, 'very different in its social and ecological consequences from subsistence agriculture, or even farming by locals to supply local markets.' Plains farms were mortgaged, leveraged, financialised businesses – and investors demanded returns, and banks demanded debt repayments regardless of the vagaries of nature. And so 'fields are planted when they should rest, herds are expanded when they should be reduced,' and in this way, year by year, the land's productivity is exhausted and it starts to die.

By the end of the 1920s, three-quarters of winter wheat farms were mechanised – and all were tightly bound into the global economy. Put images of quaint family farming out of mind. Automated, financialised and globalised, Plains farmers were some of the most modern people in America.[22] Many were absentee landowners, living in towns and cities and hiring roving work crews to plough, sow and harvest.

As postwar wheat prices slid back down to $1.03 a bushel, these leveraged businesses were left with little leeway. Ann Marie Low – a young girl of about sixteen, living on her family's stock farm in North Dakota – wrote in her diary on 9 November 1929 how 'There seems to be quite a furor in the country over a big stock market crash that wiped a lot of people out. We are ahead of them. The hailstorm in July of 1928 and bank failure that fall wiped out a lot of people locally.'[23]

The next year, in 1930, land in the Dakotas started to blow. The year after, a drought hit – and crop prices plummeted to ruinous lows. In 1932, the dust storms moved south, to Kansas, eastern Colorado and the Oklahoma Panhandle.

The Dust Bowl had begun.

***

'The first dust storm I encountered was in the winter of '32–'33 . . . I would say early '33,' Mrs Shackleford told the Oklahoma Historical Society interviewer in 1984.[24]

'The farmers had the ground all tilled, and there weren't anything to hold the soil,' her husband, Elmer, added.

'And then it got such dry weather,' she continued. 'We always had wind out here to some extent, but then it just started up. Began way up in Colorado and just kept coming.'

'The scariest one was the first one that came rolling in,' said Mr Shackleford. 'The wind blew for probably a day or something. When it first came in it just looked like a big bunch of smoke on the horizon – it looked like the world was on fire over there. You see it rolling, the smoke rolling, just kind of like an oil well fire.'

'If you forget about the devastation and just look at the clouds, the formation of the dirt come boiling in, it was beautiful,' Esther Reiswig said, in a striking description to interviewer Joe L. Todd.[25] 'It was colourful: it had reds and yellows, and a dark blue colour': hues determined by the direction of the wind. The 'black blizzards' came from Kansas and the north – where the soil is rich and dark with organic matter. Red dust came from the Oklahoma Panhandle – where the soils contain a lot of iron. Ashy grey dust came from further west, from Colorado and New Mexico. Each colour carried its own distinctive, disconcerting scent, from sharp and peppery to greasy and nauseating.[26]

'It rolled like you had a bunch of balls, one on top of the other, and they were all rolling,' Esther Reiswig continued. 'They didn't all roll the same direction, it looked like they were stacked one on top of the other: this one's rolling here, this one's rolling here, this one over here is rolling this way.' She made a spinning motion with her hands, to illustrate. 'And then it was just straight up! Almost perpendicular. Maybe a thousand feet. And the birds couldn't get over it, or they didn't: the birds flew in front of it, just like there was a net behind them, pushing them, and the birds all tried to stay ahead of it.'

Not that anyone could stand and watch: when you saw the birds fleeing then you too needed to run if you valued your life. Being caught outside without shelter could be deadly. After a storm on 15 March 1935, a seven-year-old boy was found suffocated in a dust drift.[27]

'When that thing hit, sometimes there was real hard wind in it, sometimes there wasn't, it just moved in silently,' Esther Reiswig said. 'And it would be just as dark as pitch.'[28]

Uncanny colour and darkness, ungodly heat and dirt: this was the sensory world on the Plains for days, weeks, seasons at a time.

'There are days when for hours at a time we cannot see the windmill fifty feet from the kitchen door,' Caroline Henderson wrote in a 1935 letter to the Secretary of Agriculture. 'There are days when for briefer periods one cannot distinguish the windows from the solid wall because of the solid blackness of the raging storm. Only in some inferno-like dream could anyone visualize the terrifying lurid red light overspreading the sky when portions of Texas are "on the air".'[29]

The fact they were electrical storms can only have added to the terror. Dusters bring not only thunder and lightning, but also generate large amounts of static electricity. When soil particles knock together in the air, the smaller, lighter dust grains 'steal' electrons from larger particles of sand – then get lifted away from them, creating an electrical field. Researchers have measured field strengths of over 100,000 volts per metre in desert dust storms and twisting, tornado-like dust devils – generating sufficient electrical force to pick up even more particles, creating even dustier conditions than the wind alone might explain.[30]

Juanita Wells was only seven when Black Sunday struck. 'I remember my dad at night would go out and look for the cows,' she recalled years later. 'He could follow the fence line because it was just like an electric fence, he'd see a light on that fence from the static in the air.'[31]

The electricity would scorch the tips of the young wheat shoots, and that could take out a crop – if the heat or the drought or the plagues of grasshoppers hadn't got to them first. Each year was a gamble: will we have a harvest? As the decade progressed, and failure stacked on top of failure, farmers were gambling on increasingly borrowed money.

***

'How did you keep the dust out of your house?' the oral history interviewers always asked.

'You didn't,' people responded.

The dirt was so tiny it got in everywhere. 'You take some flour that you make bread out of and rub it in your fingers: blow dirt was that thin, that fine,' as Esther Reiswig put it. 'It's just like God had sifted it all out.'

The dust crept in through every crack and crevice in these hand-built wooden houses. It drifted under the doors and weaselled its way in round the window frames. Dust worked its way under roof tiles and slipped down through the rafters to lie in a layer an inch or more deep all over the floors, and on tables, chairs and beds. Dust got inside the cupboards, dirtying the plates and glassware, which had to be washed before every meal; there was so much of it in the air it'd settle on the surface of a stew cooking on the stovetop, leaving an off-putting dirty skin. Serving dinner to the family required a carefully timed choreography of cleaning and wiping, rinsing and washing – and still you'd feel little pieces of grit crunch between your teeth as you chewed. 'When you went to bed at night, why you turned the pillow over,' Arthur Leonard recalled.[32]

It was hard even to breathe when the dirt was on the air. 'I can remember Mom always took a wet sheet and stretched it across the top of the bed,' Juanita Wells remembered, years later, 'and we all slept, just me and my brother, and Mom and Dad; we'd all sleep in the same bed so we could breathe at night.'[33]

Women did everything they could think of to keep the dust out of their homes. 'We'd tape the windows,' Mrs Shackleford said, 'or you could put some kind of paper on the outside of your windows. Later you could buy this heavy plastic and put that on the outside of your windows. But there was no way to keep it out completely.' Her voice was emphatic. 'No way.'

'It'd come down through the ceiling rafters,' her husband added.

'There's no way you could keep it all out. You just had to clean.'[34]

This was something of the horror of it: the fact there was no escape. 'Cleaning up after dust storms has gone on year after year now,' young Ann Marie Low wrote in her diary on 1 August 1936. 'I'm getting awfully tired of it. The dust will probably blow again tomorrow.'[35]

It was called the Dust Blues, that sinking sense of despair and powerlessness against the unrelenting wind.

***

The dusters kept coming.

After just a year, the fences were choked with tumbleweeds, and dust piled up behind them until even the top wire was buried. 'Over much of this area the wind and eroding sand have obliterated even the traces of cultivation,' Caroline Henderson wrote in July 1935. 'Pastures have changed to barren wastes and doorways around humble little homes have become scenes of dusty desolation. Small buildings have almost been buried' – you have likely seen some of the famous photographs.[36] Dunes formed out where there had been fields. Temperatures touched record highs: in the summer of 1934, Nebraska hit 118°F (48°C); in the summer of

1936, over 4,500 people died from excess heat. A swarm of hungry grasshoppers – as had plagued Kansas before in 1873 and 1919 – descended again, eating what little remained of farmers' crops and having a go at pretty much everything else too, from the leaves on the trees to linen on the washing line.[37]

Down in Cimarron County, in the very west of the Panhandle, the average harvest in the 1930s was just 0.9 bushels an acre – not even enough to provide seed for next year's crop. The gales were a problem, and pests were a menace – but most of all it just didn't rain, and so crops shriveled in the fields or didn't even sprout. This land on what was meant to be the 20-inch rainfall line was barely getting half that: Cimarron received just 11.65 inches of rain a year between 1931 and 1936, only an inch and change away from desert.[38]

Some of the cows made it through the first three years on what little fodder there was to feed them – tumbleweeds, often, their teeth and gums worn to stumps by chewing on the dirt that coated everything. But they were skin and bone, and that skin was worn to sores by the blowing sand and their eyes were blinded by the dust. Farmers would find their livestock suffocated after a storm, sometimes, and if you cut them open you found their bodies were heavy with mud.

People hung on by any means they had: 'We just tried to stick it out and then it got so bad,' Juanita Wells recalled. 'I can remember Dad going out and smoking out the rabbits. They were really bad, and they were eating everything that they could get their teeth into. And I can remember we'd go out and dig a hole where they were at and smoke 'em out, then we'd catch them, then we'd eat 'em. Ya know that was 'cos there wasn't much to eat any more by then.'[39]

But the dust got inside everything – and everyone.

After days working his fields in the blowing dirt, trying to hold on to what soil he could, Lawrence Svobida reported how he didn't sleep that night, 'for the dust I had labored in all day began to show its effects on my system. My head ached, my stomach was upset, and my lungs were oppressed and felt as if they must contain a ton of fine dirt.'[40] The day after, his instinct was to remain in bed – but there was another gale blowing, and once it lessened up he headed back out to retrieve the broken-down tractor he had abandoned in the height of the winds the day before, and kept 'listing' up his land. Listing was a practice of ploughing ridges into the fields to disrupt and slow the windflow over them, reducing how much dust the wind could pick up. Svobida was fighting to save his last three-quarter sections of wheat. 'If the engine was laboring against an intolerable load,' he wrote, 'so was I. Dust was coming into my system through my mouth as well as my nostrils, and I was choking, wheezing, and gasping for air. I was hovering on the verge of unconsciousness, but fought off these spells, with the thought that if I fell [from the tractor] it would mean almost certain death . . . I held to the thought that this land had to be stopped from blowing, and I kept driving on and on.'[41]

In the end Svobida was bedridden for two weeks, time he could little afford to spend away from his farm. In hindsight he reckoned that had the dust picked up any worse that night, he probably wouldn't have made it. They called it 'dust pneumonia' – scientifically known as silicosis and close cousin of the 'black lung' that affects coalminers.

As we have seen in London and Los Angeles, dust inhalation does nobody any good. In 1935, the Kansas Board of Health analysed the air their citizens were breathing. Their research found that just five big storms deposited 4.7 tons of dirt on each acre of land – and as we have seen, this dust was super fine. Consequently tiny particles could slip through the airways, deep into the lungs themselves – and the Health Board knew that silica, the main mineral in soil, was 'as much a body poison as lead'.[42] The immune system reacted to it as an invader, attacking it and causing inflammation that developed into areas of scarred, hardened lung tissue. You coughed, you choked, you struggled to breathe. The Red Cross anded out dust masks and general stores sold out of goggles.[43] But the dust storms kept on coming, and people's lungs filled up with mud.

In Ford County, Kansas, one-third of all deaths in 1935 were from pneumonia; in Morton County, thirty-one people died from pneumonia that year, compared to an average of just one or two.[44] Respiratory disease wasn't the only killer: nurses reported a rise in child malnutrition, and the infant mortality rate in Kansas rose by almost a third.[45] The dirt was pathogen-free, containing only fungal spores and innocuous soil bacteria – but nonetheless a particularly bad measles outbreak seemed to coincide with the worst dust storms in the spring of '35, for reasons unclear.[46]

Laurence Svobida mentions a friend who went to the doctor and found his bronchial tube from his airway to his lungs was completely filled with dust. The doctor worked on his chest, and Svobida reported the day after, 'my friend coughed up several pieces of solid dirt, each three to four inches long and as big as a lead pencil'. The doctor had found the same affliction in no fewer than 367 of his patients that season.[47]

***

A phenomenal amount of dirt went flying that decade.

In one dust storm alone, on 12 May 1934, 350 million tons of dirt became airborne, and dust filled the skies of the entire nation east of the Rockies in a belt 1,800 miles wide. A pleasant spring day in Chicago turned chilly as dust blocked the sun; in New York, streetlights had to beturned on mid-afternoon because of the darkness. Head of the Soil Conservation Service Hugh Bennett reckoned that 'more than 100 tons of dust per square mile' fell in some areas.[48] Snow ploughs were called out to clear roads in Indiana.[49]

Dust fall is a public danger and an enormous insult to housekeepers everywhere. Dust erosion is something else, though. We might think of it as geoengineering: planetary-scale human intervention in environmental processes and the Earth's natural systems. The Anthropocene is not just a matter of climate change but terraforming. The dust storms were the size of volcanic eruptions. Six hundred and fifty million tons of topsoil were lost in just two storms in 1935 – for comparison, Iceland's Eyjafjallajökull volcanic eruption in April 2010, which stopped European flights for nine days, produced 380 million tonnes of airborne tephra and fine ash.[50] In 1938 alone, 850 million tonnes of topsoil was lost to erosion, more dirt than washed down the Mississippi that year.[51]

By 1938, the Soil Conservation Service claimed that 80 per cent of the land in the southern plains had lost soil to wind erosion. Ten million acres of land lost at least five inches of topsoil – and 'In our part of the Dust Bowl, only some five inches of topsoil is really fertile,' Svobida noted mournfully.[52] The good bits blew away first. Texas dust deposited five hundred miles away in Iowa contained ten times as much organic matter and nine times as much nitrogen – the key building blocks of fertility – as the soil left behind. Farmers were left with sterile fields of sand dunes, costing the region $400 million a year in lost agricultural productivity – alongside inexpressible losses to nature, biodiversity, habitats and wildlife, never measured and only privately mourned.[53]

To watch five inches of soil blow away from your farm is to watch thousands of years of geological labour disappear into dust. Topsoil is precious because it forms incredibly slowly: an inch of soil takes perhaps 500 to 1,000 years to form; a lifetime's careful husbandry of a farm might produce just a couple of millimetres. You start to understand why Lawrence Svobida would risk his health to ridge his fields and try to hold on to the dirt when a storm blew up.

Svobida described how he had seen 'fields where an oil pipeline, originally two feet underground, is exposed above the surface. There is a rock in a neighbor's field, which used occasionally to be struck by the plow, because it lay beneath the surface. That rock now stands four feet above the surface, and the land is still blowing with every wind.'[54]

Large areas were scrubbed bare of soil entirely, right down to hardpan. Things long-buried reemerged: items lost on the trail by the Spanish conquistadors and in pioneer days, and thousands of Indian arrowheads, which, in the absence of anything better to harvest, people collected and sold.[55]

As the land went flying, the Plains were held together through government programmes – loath as many of these proud pioneers were to acknowledge it.

When President Franklin D. Roosevelt took office on 4 March 1933, he took the reins of a shattered country. Twenty-five per cent of Americans were unemployed, with millions more on meagre part-time wages. Banks had crashed, manufacturing output had plummeted and farm income had halved since 1929. People were desperate and demanded immediate action. In response Roosevelt launched one of the most productive hundred-day periods in US political history.

The 1933 Agricultural Adjustment Act paid farmers subsidies for leaving land fallow in a bid to reduce American farming's crisis of overproduction – with the intent that prices might thereby recover to sustainable levels. That same year the Civilian Conservation Corps was set up, employing young men aged 18 to 25 on deliberately labour-intensive projects such as conservation work and road building. They received a wage of $30 a month (equivalent to $600 now) along with bed and board, clothing and healthcare – $25 of which had to be sent home directly to their parents or wives, money which kept many a family farm together. Just one year into the Dust Bowl, in 1933, the Dalhart Texan newspaper reported that one-third of their population was being supported by charity or government relief work – and 90 per cent of farmers were taking on debt in the hope of having a chance at a better year next year.[56]

When the May 1934 dust storm swept an unfathomable 350 million tons of topsoil east – blotting out the Empire State Building in New York and making Washingtonians choke and cough – the crisis on the southern Plains finally got Roosevelt's full attention. On 9 June, Congress granted $525 million in dedicated drought relief: $275 million went to cattle ranchers as emergency feed loans and buying stock; and $125 million to other farmers in public works jobs and income support. Congress offered seed loans, so farmers who'd lost a harvest the year before could borrow and still plant again the next autumn or spring.[57] Where markets had failed – many cattle were unsellable due to their scrawny condition and low prices that didn't even cover the cost of transport – the government stepped in, paying $14 to $20 a head for animals that were sometimes simply shot and buried, too thin even to be worth processing for emergency food aid.[58]

By 1936, over two million farmers were receiving government relief cheques, including between a third and a half of farming households in southwestern Kansas.[59] But the pioneer mythos was strong on the southern Plains, and people chafed at being seen as weak or their independence lacking. Caroline Henderson describes concern about 'a widespread and serious degradation of moral character' as a result of government handouts. Work programmes may not have been the most economically efficient solution, she thought, 'But in our national economy manhood must be considered as well as money,' and so they were preferable to direct handouts.[60] Lawrence Svobida writes of his pride in himself and the minority of his neighbours who held to a 'policy of independence'.[61]

These individualist principles sometimes only worsened the problem. In Beaver County, Oklahoma, the main cash benefit was from the wheat acreage control programme, Henderson tells us, a scheme in which farmers were given money to set aside and not plough part of their land each year. Yet they found this almost immoral, she tells us, that they should have to restrict their production in this way, and take money that was not from the efforts of their own labour. The practice of leaving land fallow dates back over 2,500 years in both China and the Middle East; the Old Testament Book of Leviticus orders the Israelites to observe a 'Sabbath of the Land' every seventh year, in which they should let the land rest from crop production. Yet it seems that on the Plains in the 1930s, the Protestant work ethic had overridden this age-old lesson: instead, capitalist productivity was all.

Hugh Hammond Bennett could see the insanity of this. In 1933 he was appointed head of a new federal agency, the Soil Erosion Service. Bennett was a good choice: he understood the environmental crisis on the Plains better than almost anyone in the country. Back in 1909 the US Bureau of Soils had asserted, 'The soil is the one indestructible asset that the nation possesses. It is the one resource that cannot be exhausted; that cannot be used up.'[62] Bennett's response: 'I didn't know so much costly misinformation could be put into one brief sentence.'[63]

In Bennett's mind, soil erosion was 'a national menace'. The fertile loam soils of the Plains had been built up through millions of years of geological processes, as rivers washed silt down from the Rocky Mountains and wind carried particles of volcanic rocks to form crumbly loess in Nebraska and Eastern Colorado. It was all held together by a dense web of buffalo grass roots – and as soon as that went, so did the soils. Bennett knew his history. He knew that the Plains had started blowing 'shortly after cultivation was introduced'.[64] And he knew this was serious: it threatened the viability of farming life on the Plains.

Bennett also knew that erosion problems could be largely resolved by some simple changes in farming techniques. Growing up in North Carolina, he'd seen his father digging terraces on his family's cotton plantation before seeding. Why, young Hugh asked? 'To keep the land from washing away,' his father said. In 1905 Bennett had visited a hillside in Virginia where heavy rains had sluiced off inches of topsoil on the cultivated slopes, while the woodlands next door were spared. It was an epiphany, and he spent the 1920s running research studies to find out how to save the soil.

In 1934 Bennett set up operations just north of Dalhart, Texas, under the title of 'Operation Dust Bowl' and set out to demonstrate the difference that could be made by better agricultural practice. Terracing and contour ploughing going across the slope (not up and down) would hold more rainwater on the plants and prevent run-off stealing away soil. Cover crops could hold the dirt in place against wind erosion – and strip-cropping with rows of tall, quick-growing sorghum would help break the wind as well. Drought-resistant grain varieties, meanwhile, might stand the best chance of germinating at all in these dry and difficult years. When another dust storm swept the United States capital in 1935, Bennett won a further $125 million from Congress to extend this work. It funded tree planting programmes to construct shelterbelts that would block the relentless wind: across the Plains as a whole, an astonishing 217 million trees were planted (many by Civilian Conservation Corps workers) by 1942. Bennett also championed Soil Conservation Districts, local bodies in which farmers agreed to farm the land as a single ecological unit. Dust is no respecter of fences or farm boundaries and plenty of responsible farmers were tormented by dirt blowing in from absentee neighbours' fields. Collectively, though, they might make a difference that as individuals they never could.

Day by day, though, Dust Bowl farmers describe fighting the dust storms with the same tools that were causing them: the plough. Robert Howard, then a boy of twelve living on his family's homestead in Goltry, Oklahoma, recalls his father getting on a 'lister' – a type of plough that throws soil either side of a central furrow – and going out to see what time he'd finish so he could get the feed in for the horses. 'I'd be, oh, sixty, seventy feet from him sometimes and I could just barely see them, the dirt was really blowing.'[65]

It was slow, hard, Sisyphean labour – farmers would be coughing and choking, eyes streaming in the dusty air, as they ploughed on through storms, trying to stop their land from blowing – and do right by their neighbours too. But it was often futile. Lawrence Svobida up in Kansas describes how the 'furrows soon filled level with blow dirt, and the winds continued worse than before'. In hindsight, he wasn't sure it was a good idea at all: 'without accompanying moisture the treatment further broke down the soil structure and made it more susceptible to wind erosion', not less.[66] He almost killed himself to ridge up his fields regardless. It was a matter of honour: other men might let their fields blow over their neighbours' crops. He would not.

Every historian notes the distinctive psychology of the Plains wheat farmer: a near unshakeable optimism that next year the rains would come. Lawrence Svoboda's memoir is an astonishing study in psychological toughness and resilience: one man out farming all alone in big country, with barely any neighbours from one horizon to the next, watching his crops get battered by wind and hail, 'drouth' and 'hoppers'. He describes how season after season he planted two, sometimes three, crops. 'I had worked incessantly to gain a harvest, or to keep my land from blowing, and no effort of mine had proved fruitful.' When 'the long-awaited, desperately needed rain' did not arrive until too late, and he knew that his crop was 'irrevocably gone', he writes of how he 'experienced a deathly feeling which, I hope, can affect a man only once in a lifetime . . . I had reached "the death of utter despair"', adding: 'The scars of that year are with me still.'

His spirit was broken – but then, even mid-sentence, he turns and says, 'the world did not come to an end and, before the first of September I had taken advantage of a slight dip in the market to buy seed wheat at $1.04 a bushel . . . By the first of October I was out in the fields again – drilling wheat.'[67]

***

'July 20. I wonder if in the next 500 years – or the next 1,000 years, there will be a summer when rain will fall in Inavale. Certainly not as long as I live will the curse of drought be lifted from this country.'

    – 1938 diary of Don Hartwell, a Nebraska farmer[68]

The Dust Bowl is perhaps one of the more comforting disasters in this book – because it did at least come to an end. Some part of the credit goes to the Soil Conservation Service and related programmes. The changes in farming practices they promoted, such as contour ploughing and crop rotation, spread quickly, with nearly 40,000 farmers participating by May 1936 and 5.5 million acres under soil-preserving cultivation.[69] By 1938, despite the dust storms not letting up in the slightest, soil blowing had been reduced by up to 65 per cent.[70]

That summer, Lawrence Svobida asked himself a by now familiar question: 'Should I try once more or get out while I still had a shirt to my name? Again I chose to fight,' he wrote.[71]

The course of the next year demonstrates the steel nerves Plains farmers had to possess. That summer Svobida prepared the ground and planted wheat again in September. Then much to his disappointment, 'no more moisture fell during the balance of the year' and his crop started to die. Rain did arrive in January 1939, reviving some of the plants, only for them to be threatened by high gales the following month; they, nonetheless, persevered. 'Two more rains fell in March, and by the time the first of April rolled round, my wheat was a foot and a half high,' he wrote. 'This was the best wheat I ever had out so early in the growing season . . . but, as usual, the prospect soon proved to be an illusion, since no more life-giving moisture fell, and my crop burned up.'

Seven crops had failed in eight years, and Svobida finally had nothing left. 'With my financial resources at last exhausted and my health seriously, if not permanently impaired, I am at last ready to admit defeat and leave the Dust Bowl forever. With youth and ambition ground into the very dust itself, I can only drift with the tide.' He was not the only one.

The 'Dirty Thirties', as they were named, saw over three million people forced to leave the Plains.[72] The central Dust Bowl counties saw their populations drop by a third as people just walked out on their broken land and their broken lives – ten thousand farmhouses left empty on the High Plains; nine million acres of land left to the tumbleweeds and sand dunes.[73] They took no more than might fit in a car or a trailer: mattresses, the odd piece of furniture, kitchen items; the few items of hard-worn clothing they had left. As Dorothea Lange famously photographed, many simply walked for hundreds and hundreds of miles, their lives in a handcart.

But not everyone left. Unlike those a little further east, many farmers in the eye of the Dust Bowl did own their land, and government aid proved sufficient for them to hang on during these lean years in the hope of better times to come. Most migrants here didn't go far, moving to nearby towns or to the next county over.

Nonetheless, places like Elkhart, Kansas, were left almost ghost towns, and the population of the Plains has not recovered to this day.

***

In the end, only one thing could stop the dust: water.

'I can remember, out when we was still out on the farm, when the first rain came,' Juanita Wells recalled almost sixty years later. 'Mom had some garden out there. She was out there raking around in it and it started raining: that was the best smell you've ever smelled, you know, we could smell that rain.'

'You still remember that?' the interviewer asks.

'Yeah, I can, I really can.'[74]

Juanita would have been about twelve at the time.

The rains returned in the autumn of 1939, though rainfall remained below average until 1941. As the US mobilised once more for war, the Plains began to ripple again with the promise of wealth. Farms were much larger now, as smaller operators had been bought out at bargain prices – and farmers had money to invest in new technologies: electric and diesel-powered water pumps that could draw up water from hundreds of feet below, and centre-pivot irrigators to distribute it across new, circular fields, some a full mile in diameter. In this arid and most unpredictable of climates, they had found the holy grail: artificial rain. The High Plains were transformed into a circular patchwork of deep saturated green. The 1950s saw another run of drought years, but the dust never returned.

It came at a cost. Each pump and pivot were drawing on the Ogallala Aquifer, one of the world's largest groundwater supplies stretching the length of the Great Plains, from South Dakota and Wyoming in the north, down to the far west side of Texas. Its source: ancient, geologic 'paleowater' that collected beneath the prairie over millions of years, as rivers ran down from the Rocky Mountains and drained into the clay, sand and gravel below. One hundred and seventy thousand square miles – and 170,000 wells. It seemed a free and abundant source: farmers (who used 94 per cent of the water) upped production and turned the Plains into the grain basket of America, their water usage subsidised by tax breaks.

Yet the Ogallala is essentially a closed basin: it receives less than an inch per year of freshwater recharge,while in some places four to six feet of water per year was being pumped out.[75] As a result, water levels in all but the deepest parts of the aquifer have fallen catastrophically: over 100 feet in parts of Kansas, Oklahoma and Texas. On many farms, the wells are already running dry.

Political ecologists Matthew R. Sanderson and R. Scott Frey describe this as a 'metabolic rift in the hydrological cycle'.[76] It's a powerful phrase – and an analytically useful one. The concept originates with Karl Marx who, in *Capital*, wrote of the once simple 'metabolic interaction between man and the earth' where early cities largely fed their populations from the bioregion around them, and recycled the results of that metabolism (that is, the city dwellers' shit) back to this same land as fertiliser. Yet with the rise of capitalism and world markets, that balanced ecological exchange no longer took place. Cities filled with their own waste and pollution (as we saw in Chapter 1) – and the farmland beyond them was 'robbed' of its nutrients, which were no longer returned and recycled. The metabolic rift is that gap, that debt borrowed and never repaid – and the resulting rupture. Imperial Britain responded to increasing anxieties about declining soil fertility at home by importing guano from Peru and sodium nitrate from Chile – even dissolving coprolites (fossilised shit) in sulphuric acid to use as a source of essential nitrogen.

Once again, we borrowed from the deep past to sustain our needs in the present. This is never sustainable. 'All progress in capitalist agriculture is a progress in the art, not only of robbing the worker, but of robbing the soil; all progress in increasing the fertility of the soil for a given time is a progress toward ruining the more long-lasting sources of that fertility,' Marx wrote. In this way, capitalist production only progresses 'by simultaneously undermining the original sources of all wealth—the soil and the worker.'[77] We might add to that list: the soil and the worker *and the water*.

Farmers in the region have long known they were borrowing from borrowing from long ago and that this could not last. Lindsay Wise reported for the *Kansas City Star* in 2015 on how local farms were trying all kinds of solutions, from more efficient irrigation techniques to satellite weather tracking and genetically engineered drought-tolerant crops, even considering pacts with their neighbours to reduce water withdrawals. But drought in the early 2010s saw one farm get less than seven inches of rain for four full years in a row – so dislike it as he might, that farmer kept pumping.[78]

Wise described how the farmers talk of 'mining water': this is an extraction economy as naked as if they were pumping oil. Given that much of the grain this water grows is used to fatten methane-emitting cattle, it's a mining economy with quite the global warming impact, as well. But not necessarily for that much longer. Across the southern Plains, the water is running out. In 2004, Donald Worster imagined mournfully how like 'all mining economies, its end will inevitably be ghost towns, abandoned houses, derelict farms, and bankrupt businesses. Then, when there is little or no water left to grow crops, when the cycle has come full circle, will the dust begin to blow? It always has in the past.'[79]

***

The High Plains landscape took a little while to understand. I didn't see the number of abandoned 1930s homesteads I had expected – not until we started to look better. Thickets of a few trees marked where houses had once stood – trees planted to provide a little shelter against the high winds that pummel these seas of grass. One thing was abundantly clear in this big open land: when a duster rolled in, there would have been nowhere to hide.

Once my eyes were attuned, I started to see how the ghosts of its history already littered this landscape. 'Historic Marker' signs dot the roadside, and we pulled over to one to find out more. It signposted a mere metre of rusty metal fencing: the drift fence. Barbed wire was one of three technologies that facil-itated white colonisation of the American West (the others being the windmill, powering the water pumps that sustained life in this barren land; and the revolver, which took it). Used for fenc-ing, barbed wire ended open-range herding and the cowboy way of life. In the fierce winter of 1886, it killed 200,000 head of cattle as herds were trapped against the fences and could not escape south to shelter.

Other signs pointed off the roadway to the old Santa Fe trail and the Cimarron Cutoff, a hundred-mile shortcut for wagon trains and traders heading west in the early 1800s, before the railways were built; a route for chancers in a hurry willing to risk the dry featurelessness of this arid land to save ten days over the mountain trail. We found ourselves tracing this line back home, having driven in on I-40, overlaying and replacing the old Route 66, the escape route of John Steinbeck's Joad family as they fled dust and dispossession in eastern Oklahoma. The Trail of Tears led here too, a death sentence for ten thousand

Cherokee, Creek and Chocktaw as they were forced from their homes and land by American expansion. There are no innocent paths in the American West: they all echo with displacement and forced migration.

Heading back into New Mexico with many miles yet to go, we flicked on the radio. The car filled with the crackle of drum dances on 90.5 FM, the radio station of the Jicarilla Apache tribe, before the signal was lost under a haze of Christian preaching. Electromagnetic stratigraphy. A hawk fought a raven, the birds turning cartwheels above the road. We picked up 102.5 Coyote FM, which played 'Dust in the Wind' by Kansas, followed by Queen's 'Another One Bites the Dust'. Wayne – who is, among other things, a poet – talked of 'pulverised narrative: the black snow that falls on the white page'. I laughed. Our journeys always ended up uncanny with synchronicity.

If modernity is haunted, the Plains today might be doubly so – haunted once by the ghosts of a farming prosperity now lost, and once again by the buffalo and the Indigenous world that was annihilated to make that profit possible. The dust itself is a revenant, threatening return.

Scientists fear a sequel to the Dust Bowl may be inevitable.

In 2020, University of Utah researchers reported that atmospheric dust levels were rising across the Great Plains by as much as 5 per cent per year.[80] Though agricultural practices have improved substantially since the 1930s – for example, farmers now plough a lot less, if at all – more land has been brought into cultivation, spurred by rapidly rising demand for biofuels since the late 2000s. Much of this land is marginal, highly sensitive to drought and soil erosion: if it weren't, it'd have already been in cultivation. As such, the scientists report: 'If the Great Plains becomes drier, a possibility under climate change scenarios, then all the pieces are in place for a repeat of the Dust Bowl that devastated the Midwest in the 1930s.' They reel off a list of health consequences: meningococcal meningitis; Valley Fever; a potential fourfold increase in hospitalisations for cardiovascular and respiratory illnesses by the end of the century.[81] It would be another environmental – and human – catastrophe.

Today, something far more severe than a regular dry spell is gripping the inland of America. The region is now in mega-drought, defined as a multi-decade period of severely low rainfall This is more than the years without rain that sparked the Dust Bowl: it's the hardest drought seen since the late 1500s, and the second worst since 800 CE. The epicentre is further west, in New Mexico, Arizona, Nevada and Utah – but drought ravages the Plains nonetheless in Colorado and Nebraska, and claws at the edges of Oklahoma and Kansas. 'This appears to be just the beginning of a more extreme trend toward megadrought as global warming continues,' the researchers write. It's 'a drought bigger than what modern society has seen'.[82]

Another multi-year drought of Dust Bowl scale could deplete US wheat stores by 94 per cent and produce cascading shocks through the world's food system, agricultural modellers have forecast.[83] Rising temperatures are also a major threat in their own right. By 2050, five in every six summers on the Plains are expected to see average highs even hotter than the freak summer of 1936 – which, even with normal rainfall, will hammer the maize crop and again produce losses of Dust Bowl severity.[84]

So the days may be numbered for a high-modern model of agriculture on the Plains. This might have been the lesson of the Dust Bowl itself, but for seventy years agribusinesses borrowed from the deep past to delay the inevitable – and for a while, it seemed to work. But dust is patient. Eventually the bill for a century of environmental domination comes due.

***

For thousands of years, people lived very lightly on the great grasslands of the Plains. Evidence of early Paleo-Indian cultures is sparse, often only flint points and arrowheads; as small, nomadic bands of hunter-gatherers, they left little trace.

In the early 1600s, Britain established thirteen colonies along the eastern coast of the United States. The population doubled each generation as English settlers were joined by Germans, Scots and Irish Protestants seeking religious freedom, and convicts and enslaved Africans brought over by force. As the colonisers pushed inland, bringing deadly waves of epidemic disease with them, a domino effect was set in motion across fully half the continent as tribal communities in their path fled or were forced westwards, and in turn displaced other tribes. The Cheyenne, originally from the Great Lakes region, were pushed westward into the Dakotas, then further south, into Oklahoma – where they allied with the Arapaho, and pushed the Kiowa south beyond them. Meanwhile, Comanches had migrated from Wyoming to the southernmost plains around 1700, stealing the Apaches' horses (which they'd obtained from the Spanish) and taking over their buffalo hunting grounds.

From this upheaval emerged a striking warrior culture, nomadic once more but now fleet-footed and far-ranging. Mounted riders could keep pace with the buffalo herds, first hunting with bow and arrow, then guns. The horse made the Plains fully habitable: they enabled Plains Indians to cover the vast range required to achieve a secure supply of food and resources in this shifting, drought-prone land. The abundance of buffalo – twenty-five or thirty million of the great beasts – provided an ease of subsidence that enabled a nomadic luxury: clothing richly ornamented with beads, fringing and porcupine quill embroidery; bodies decorated with tattoos; the course of the year marked by elaborate ceremonial gatherings, such as the gruelling ordeal of the Sun Dance. Horses were, of course, inexpressibly valuable – resulting in incessant raiding and warfare between rival tribes. This brave, spectacular Plains Indian culture came to define the image of the Indian for a white American audience and beyond.[85]

Ultimately, this was a society not of excess but avowed ecological equilibrium, guided by a deep reciprocity in which, as Kiowa novelist N. Scott Momaday describes: 'man invests himself in the landscape and at the same time incorporates the landscape into his own most fundamental experience'.[86] To the Indians, the Plains environment in which they lived was not a 'thing': it wasn't property you owned or a resource to be exploited – but rather something alive that they were materially and spiritually 'in relations' with – relationships of interdependence and mutual care. The population stabilised at around 10,000 people, or just one person per 31 square miles – a figure far lower than still-sparse Cimarron County today (two people per square mile) – and a way of life that was not wholly settled in one place, but could roam with the buffalo herds, the seasons and the climate.

'They did so not because they were especially righteous,' Donald Worster writes. 'More important to their adaptiveness was their assumption of complete dependence: their unwillingness to consider that any other relationship with the grassland might be possible.'[87]

***

'Where land-utilization practices are firmly established and have become the basis of the country's economy, the adoption of a new land-utilization programme conforming to the limits imposed by the natural environment, may well involve a social and political revolution.'
 – Graham Vernon Jacks & Robert Orr Whyte, 1939.

Can a metabolic rift ever heal? Karl Marx thought it could.

Capitalist market forces could never be the solution: the imperative is to extract advantage and profit, producing environmental short termism, exhaustion and crisis. (A survey of history 1862 to date indicates he had a point.) But Marx was a modern man, and he did not advocate a return to some kind of 'state of nature' or hunter-gatherer subsistence economy. Instead, he imagined a future in which we might 'govern the human metabolism with nature in a rational way' by means of 'collective control'.[88] Or as we might say now, Land Back.

109

'Land Back' is a movement for returning Indigenous land to Indigenous hands. It means recognising that America's native peoples were forced off their homelands illegally, and seeking to resolve that injustice. It's a recognition that the present way of governing these lands is not working and is not sustainable – and a route to trying something different.

Given how America took the land in the first place, some may have anxieties that Land Back could be a programme of revenge, entailing forced evictions of multi-generational farmers and so on. It's not. In fact, most land transfers to date have been achieved through that most patriotic of means: commercially. In 2021, for example, the Passamaquoddy tribe in Maine raised $355,000 with the support of conservation charities and purchased Pine Island, land that had been officially granted to the tribe in 1794 then snatched back in 1820. Environmental groups and tribes are now often collaborating, as we shall see in Chapter 9, as conservationists recognise that traditional Indigenous land management techniques are effective at protecting and rebuilding biodiversity.

Across the Plains region today there is a lot of under-occupied land. Many rural counties have seen their populations decrease 60 per cent or more since the Dust Bowl years. Farmers are getting older and their children may not wish to take their place – and where the aquifer has dried up, they face a choice between low-yield dryland farming and abandonment. Some counties are now effectively 'frontier territory', with fewer than two people per square mile. From here it is not so far to imagine going just one step further back, to the way this land was before colonisation and homesteading chopped it up into private property.

There's talk of restoring the Buffalo Commons.

In 1987, land use planners Frank and Deborah Popper published a paper aptly titled 'The Great Plains: From Dust to Dust', reviewing the boom-to-bust history of settlement in the region. They sought to inspire a new path of large-scale, generations-long land restoration to return sterile monoculture fields and cattle feedlots back into the grassland they once were, with ecological corridors connecting wildland reserves, allowing native prairie wildlife to roam freely over their natural ranges. It is an ecosystem capped by an iconic herbivore, the Plains buffalo (or bison), the nibbling and trampling and pooing of which keeps the whole structure in balance.

It is a vision of a landscape once again whole in itself, that can live – indeed, flourish – without nitrogen fertilisers made in factories or water drawn up from deep time. It is a vision of a landscape once again held together by a deep web of grass roots and fungal mycelium: a landscape proofed, as best it can be, from strong winds and blowing dust. In this way, the Great Plains Restoration Council hopes, they might 'simultaneously restore the health and sustainability' of their communities, 'so that both land and people may prosper for a very long time'.[89]

It proceeds by fits and starts, however. These rural Plains counties are the Republican heartland – Indigenous reservations in South Dakota, as bright, anomalous spots of blue in a sea of deep red – and unsurprisingly many have been suspicious of the 'Buffalo Commons' idea. They distrusted federal involvement, the Poppers observed, and 'intensely disliked the commons portion of the metaphor, associating it with collectivism and lack of choice'.[90] Above all, people hated hearing that their home region was declining, and were angry at suggestions that depopulation might in any way be a good thing; at one point in the 1990s, the Poppers were getting death threats.

But slowly minds are changing.

Interior Secretary Deb Haaland (who is herself Indigenous and a citizen of Laguna Pueblo) speaks of a vision to 'restore grassland ecosystems, strengthen rural economies dependent on grassland health and provide for the return of bison to Tribally owned and ancestral lands.' The Interior Department currently manages 11,000 bison on public lands, and has transferred 20,000 more to tribes and tribal organisations over the past twenty years, under the aegis of the InterTribal Buffalo Council.[91] In 2021, the

US Department of Agriculture announced a record four million acres signed up in the Grassland Conservation Reserve Program, a scheme that subsidises farmers to remove 'environmentally sensitive' land from production and instead restore native prairie species.[92]

Slowly, the idea that the High Plains is not land meant for capitalism is taking root. Five inches of topsoil cannot sustain endless year-in, year-out growth.

Such a thing can't be sustained anywhere, but that's a matter for another book.

Chapter 4
Cleanliness and Control

We have not yet spoken of the intimacy of dust.

I have written about places you perhaps already know and had wondered about: you may walk London's still sometimes soot-stained streets – or perhaps you have seen the movie *Chinatown* and its depiction of the Los Angeles 'water wars'. I have talked about how its tiny particles infiltrate our bodies with every breath – all of us without exception, but some violently more so than others. I have talked about how this does nobody any good at all.

But this only starts to grasp the *presentness* of dust: its absolute relentless ubiquity. In every lungful of air, on every surface we touch, it is inescapable, and in that way monstrous, a constant reminder of the friction between us and the world. A reminder, of course, of our mortality. Dust is not in fact mostly skin, as people like to think: we are not so important on the surface of this planet – and we are far from unique in our decay. Everything around us is falling apart: our clothes, our carpets and chairs and duvet covers; the brakes and tyres of our cars (and buses and trains and bicycles); the surface of the road. Brick and concrete, tile and plasterboard, the soil itself. Everything we build, all the achievements of human civilisations past and yet to come – all the joy and beauty of nature which doesn't accomplish anything and is all the better for it – it all meets the same end, given time. Dust is a reminder that, given a long-enough view, the world tends in but one direction: towards the formlessness from whence it came. Entropy is scrupulously fair, it does not discriminate – but neither does it grant reprieve, that is the horror of it.

Dust is not only about death, but about this monstrous indifference: it marks the absence not only of life, but the prospect of ever mattering. In the Book of Genesis, God cautions Adam against pride: 'for dust you are, and unto dust you will return'. On Ash Wednesday, Christians press ashes into their brows in the shape of a cross as a reminder that the world is impermanent and to trust only in God.

I look at the fluff that has gathered on my desk and it is like gazing into the void.

Yet the void is, for most of us, something of an abstract concern – whereas dust is terribly material. It has the impudence to require something of us. We are here to talk about housework, that most Sisyphean of tasks – and of all the chores of material reproduction, none is more futile and maddening than dusting. Why? Because dust is remarkably hard to truly get rid of – this is, as you will see, the leitmotif of this chapter, for all the moderns might try. Each faff of the feather duster merely transfers your energy to these tiny particles – which merrily fly into the air, float for a few moments, then gently settle down, distributing themselves right back over the surfaces you just wiped. But you're smart, you say? You use a dampened cloth, or some fancy microfibre invention with electrostatic fibres that you saw on an infomercial? No matter: the very pressure of your hand abrades both cloth and surface alike, leaving a trail of microscopic destruction in its wake. You will never, ever get anything perfectly clean: it cannot be done.

Given such frustrations, what exactly are we to do? You may question the nature – and question the *function* – of dirt itself and go down a decade-long rabbit hole of strange and desiccated obsession of which this book is the final product (and hopefully exorcism). Or you may declare dust an abomination, an offence against health, hygiene and rational order. The modernist answer is to seek to destroy it. (Good luck with that, Monsieur Le Corbusier.)

This chapter explores modernity's invention of cleanliness in the home, in architecture and urban planning, and in the laboratory: its attempts to dominate the indoor environment, as well as the outside. In the simultaneous necessity and futility of eradicating dust, modernity might come to be revealed as something impossibly utopian.

***

People used not to give a damn about dust.

The notion of the 'housewife' has existed since about 1200, the indoor equivalent to the smallholding farmer's 'husbandry' of crops and animals. (It first sees light as 'husewif' in *Sawles Warde*, an early Middle English religious allegory.)[1] Yet in the medieval and early modern household in Europe, 'housework' as we currently understand it – dusting, sweeping, scrubbing and so on – was far from the housewife's primary duty. She was busy with household production: milking cows, weaving cloth, making soap and candles and taking goods to market. Cleaning took place on both a weekly cycle – laundry day being a major ordeal in the years before piped water, and as an annual event, as each spring a year's worth of soot and smut from the household fires was washed off the walls and replaced with a fresh coat of limewash.

It's tempting to see the medieval world as filthy – but historians insist this is unfair. We project dirt backwards in order to have a narrative of progress. 'The past must be seen as garbage,' John Scanlan argues, so that today we might be clean.[2] Nonetheless the case can be made that dirt has not always been so morally freighted: once upon a time, a stain was just a stain, and not a black mark on your character. Suellen Hoy describes the world of American farm families in the early-mid 1800s who 'still held dirt in too high regard. Not only did it give life, in their view, but it was also a sure sign of plain living, honest toil, and physical fitness.'[3]

Being clean was a matter of gentility and politeness: people made the effort to wash and put on their best clothes for church on Sundays, but otherwise believed that to be 'dressed up' when there was work to do was just 'puttin' on airs'. The reformer Catherine Beecher, instructing on 'domestic economy' in 1841, tended towards demanding neatness and order rather than calling for everything to be sparklingly, spotlessly, impractically *clean*.[4]

Yet the industrial city of the nineteenth century saw dirt become a crisis. A new intensity of production created a new intensity of poverty, as people crowded into slums to serve as a labour force for factories. The population of London doubled between 1801 and 1841; some northern English cities such as Leeds nearly tripled in size. Systems of sanitation that had operated with ad hoc effectiveness in the past – latrines, cesspits and waste collected by 'nightsoil men', who sold it to farmers as fertiliser – crumbled under the pressure of this new human density. After a day on the Thames on 7 July 1855, the scientist Michael Faraday wrote a letter to *The Times*, describing how the river had become a 'fermenting sewer' where 'Near the bridges the feculence rolled up in clouds so dense that they were visible at the surface.'[5]

In both the UK and US, the urban poor often lived in tenements, cheap brick-built apartment blocks four or five storeys high. The conditions in which they lived were horrific. Charles Dickens described 'Wretched houses with broken windows patched with rags and paper: every room let out to a different family, and in many instances to two or even three . . . filth everywhere—a gutter before the houses and a drain behind—clothes drying and slops emptying, from the windows.'[6] Sanitary inspectors made similar reports of conditions in New York City, with recent immigrants to the city living packed in damp cellars downtown, unable to escape the stench of the city's 173 slaughterhouses. The city's population of 100,000 horses produced over a thousand tonnes of manure a day – London, meanwhile, had three times that number, prompting *The Times* to forecast, in 1894, that 'in 50 years, every street in London will be buried under nine feet of manure'.[7]

These dense, unsanitary cities were ripe for the spread of infectious disease. Five cholera pandemics swept the world during the 1800s, killing 55,000 in the UK in 1832, and carrying away many survivors of the Irish Famine, already weakened by starvation and fever, in 1849. The US lost 150,000 lives to cholera in these two waves; Germany, Mexico, Japan, and more, faced similar such tolls. Mosquito-borne yellow fever travelled steamboat routes from New Orleans, and ravaged Philadelphia and New York. The American journalist H.L. Mencken described Baltimore as smelling 'like a billion pole cats'.[8]

It was time to clean up the city.

***

Curiously, the first wave of sanitary reform took place *before* scientists understood the hygiene problem as we know it today.

Rigorous social surveys, such as Edwin Chadwick's 1839–43 report on the dire housing conditions of the British working classes, and John Snow's mapping of Soho's 1854 cholera outbreak established a clear and causal relationship between environmental filth and infectious diseases. At the time the vector of disease was usually thought to be 'miasma', a malignant atmosphere arising from rotting vegetation and foul water, which affected those with a weak constitution or other bodily predisposition. Dr Snow was (rightly) a sceptic of this 'bad air' theory, but a more accurate 'germ theory' of disease didn't get properly established in the West until the 1860s, from the work of Louis Pasteur and Joseph Lister.

The idea that microbes might cause disease was far from new. Roman scholar Marcus Varro had warned in 36 BCE of swamp-dwelling 'minute creatures which cannot be seen by the eyes'.[9] The Persian scholar of medicine Ibn Sina (Avicenna) had said much the same in his 1025 book, *The Canon of Medicine*, which was used as a textbook in Europe up until the seventeenth century. Yet the idea didn't seem to stick. Instead, the cholera bacterium had to be discovered multiple times between 1854 and 1883 before medical institutions started to do something with this information.

Fortunately, this causal uncertainty didn't stand in the way of effective solutions to the problem. Summer 1858 saw a 'Great Stink' in London, as the sewage-filled Thames festered in the heat. Parliament – operating out of a building next to the river – duly signed up to civil engineer Joseph Bazalgette's proposals to build a hundred-mile main sewer and thousands of miles of connecting pipes to enclose many of the tributary rivers that flowed into the Thames and keep effluent out of the river. It worked, eliminating cholera from the water supply and substantially reducing typhus and typhoid. Florence Nightingale, newly returned from the Crimean War, set about demonstrating how cleanliness, fresh air and sanitation could save lives. The filthiest housing – the 'rookeries' of Saffron Hill in Farringdon and St Giles, north of Covent Garden – was simply demolished. Thousands of people, the city's very poorest – casual labourers and costermongers, thieves and sex workers, waifs and strays – were made homeless. 'Some of the reformers who pioneered slum clearance thought a proportion of the inhabitants could be saved, if they were removed from the disorderly environment they lived in, though others were wedded to their immoral ways and were just unredeemable,' the working-class history blog Past Tense reports.[10] It was not just a physical scouring but social cleansing. The dirt of the city threatened not just health, but moral and social disorder.

This mid-century clean-up focused on infrastructure: water, waste and housing. Dust less often entered the picture. It frustrated only the experts, such as household servants. Robert Roberts' *Guide for Butlers and Household Staff* (1827) complained at length of the effects of softer coal: 'Your clothes are smutted – your flesh begrimed – your furniture dirtied – your walls blackened . . . I say nothing . . . of those globules of pure greasy black in the shape of *pollywogs* [tadpoles] that go sailing around the room.' Hard anthracite produces a much better class of dust, he found: 'light and fine . . . clean and pure, and easily dislodged from the mantel-piece and furniture, without leaving the least trace of its presence'.[11]

Florence Nightingale doubted that creating a clean environment was quite that easy. In her 1860 book, *Notes on Nursing*, she bemoaned 'the present system of dusting' as merely ineffectual 'flapping'. She believed that 'To dust, as it is now practised, truly means to distribute dust more equally over a room,' with the result that, 'from the time a room begins to be a room up to the time when it ceases to be one, no one atom of dust ever actually leaves its precincts'.

Nightingale knew that dust was anything but harmless: the particles flaking from your green wallpaper contained arsenic; your fireplace distributed coal dust about the place. Instead, rooms should be designed with the fewest possible surfaces that would store dust: no inaccessible ledges or sills, and no wallpaper or carpet, and furniture should be made of materials able to withstand the frequent contact of a damp cloth without flinching. 'The best wall for a sick-room or ward that could be made is pure white non-absorbent cement or glass, or glazed tiles, if they were made sightly enough,' she wrote.[12] Ceilings should be high and windows open to permit crucial ventilation.

Nightingale may have greatly influenced hospital design, but this stark, even proto-modernist outlook did not catch on in the nineteenth-century home. Quite the opposite: people's homes filled up with *stuff*.

***

Clutter is a contemporary problem. One of the striking differences between the pre-industrial world and today is how few things people owned. When every item has to be handmade – when a shirt or dress did not just have to be handsewn, but its fabric handspun – then possessions are both expensive and time-consuming to acquire, and so most people owned very few. Homes were relatively spartan, their surfaces rougher: before the coming of machine manufacturing, perfect flatness and shine were luxury commodities. As such, if dust fell on a rough wood or dirt floor, you perhaps might not much see it or mind. For dust to become widely perceptible as a problem – for dust to become a type of *dirt* – it turns out that the objects on which it lands are just as important as the particles themselves.

It requires the invention of consumerism.

Historians describe a 'transformation of desire' that took place in the seventeenth and eighteenth centuries, as colonialism and global trade delivered tea, coffee and Indian cotton to Europe, and, under a nascent capitalism, people learnt how to want novelty and luxury. Late Ming China enjoyed what Chinese historians call the 'sprouts of capitalism' around this time too, with a profusion of silks and lacquerware, porcelain and books. Yet it took the Industrial Revolution and the growth of mass manufacturing for the middle-class home to really start to fill with a profusion of newly affordable possessions: china and cutlery, curtains and carpets, glass and mirrors. Brought into smoky firelit parlours, these fine goods quickly lost their sparkle: the literary critic and philosopher Walter Benjamin describes 'plush as dust collector', the new upholstered, padded style of furniture gathering soot.[13]

These new possessions required cleaning – yet the newly prosperous middle classes aspired to mark their status by not having their women do any of it. Housework was for another class: 'domestics', that is, servants.

By the close of the nineteenth century, one in eight of all girls and women over the age of ten worked in domestic service – an astounding one in every three young women aged between fifteen and twenty.[14] Much of the housemaid's day would be taken up with battling dust and soot, silt and grit, beginning at dawn – or well before it, in winter – sweeping out the ash from the hearth and lighting the fires so that the household might rise to a warm house. Before breakfast, she should sweep the hallway and the front step; later in the morning she might deep clean one room in particular.

'A fair-sized room takes fully three hours to clean properly' if it is to last a fortnight between cleanings, the 1889 *The Art of Housekeeping* instructs – one of many household manuals teaching the new middle-class how to manage their staff. Its dictats for dust management are meticulous. 'All the ornaments have to be removed to another room or packed on a central table and covered with a clean dust-sheet, like the chairs, which should be rubbed and brushed before turning out of the room . . . then comes the sweeping, after which the housemaid will make time to run down for her "lunch" of bread and cheese and glass of — let us hope not beer.' (Let the poor girl have a pint!) 'The dust takes half an hour to settle, and during this process, a few ornaments can be washed or metal goods polished. Then comes the tidying and the beeswaxing, and last the dusting.' The mistress of the house was instructed to 'peep at corners, high shelves, &c.' to ensure the housemaid had done a proper job.[15] Dust as a means of micromanagement. Dust as a mode of class war.

So that the dust wouldn't settle on newly cleaned surfaces, some housework experts advised working top down around the room. Others educated on how carpets must be sprinkled with damp tea leaves before sweeping, to prevent the dust from rising – though one guide, cognizant of tea stains, recommends 'fresh cut grass' instead: 'It is better than tea leaves for preventing dust, and gives the carpets a very bright fresh look.'[16] (Bright fresh green chlorophyll stains, perhaps.)

A middle-class house of a dozen rooms 'above stairs' might thereby occupy one housemaid full time, each room deep cleaned once a fortnight, with Sunday mornings off for church if she were lucky.

***

Yet all this cleaning wasn't enough. By 1890, waterborne diseases such as typhoid and cholera may have been much less common in London than a generation previously, thanks to Bazalgette's grand Byzantine 'cathedrals of sewage' and other infrastructural improvements. Yet tuberculosis remained a menace and infant mortality stubbornly high: in both Europe and America, one in every seven infants died before their first birthdays; in some cities, nearer one in three.[17] This time around, civil engineering could not solve the problem, because the leading causes of death were bacterial infections transmitted directly from person to person in coughs and sneezes, spit and saliva. They required a much more intimate mode of intervention: the creation of 'personal hygiene'.

People had of course washed themselves before this time. Some cultures had been great fans of bathing: both ancient Rome and early Islamic cities developed elaborate plumbing systems and leisurely sauna routines. There was even a Greek cult dedicated to Hygieia, goddess of cleanliness, health and sanitation, which spread following the Plague of Athens in 430 BC. The word 'hygiene', though – the idea of cleanliness in order to prevent disease – was, despite these classical origins, barely used in English prior to the mid-1800s.[18] It took until the 1880s and 1890s for the germ theory of disease to be fully accepted within Western medicine, and a few years more to infect the imaginations of sanitarians and social reformers. Yet once the concept was accepted, it proliferated wildly.

Medical historian Nancy Tomes calls it the 'gospel of germs'.[19] Its evangelism took two forms. First, an array of newspaper and magazine articles, household manuals and Women's Institute lectures to persuade middle-class women to up their housekeeping standards through fear and norm-setting. Second, a kind of 'philanthropic maternalism', as a class of do-gooders swept into the houses of the poor, basketfuls of cleaning supplies under one arm and educational pamphlets in the other, in order to instruct the less fortunate on how to wash behind their ears, cover their sneezes and stop swapping saliva through shared drinking cups and toothbrushes. Health visitors urged 'the importance of cleanliness, thrift and temperance on all occasions.'[20]

The message was simple: germs were public enemy number one, and this was war. Harriette Plunkett, writing in *American Kitchen* magazine, Feb 1899, commented:

'The whole science of bacteriology has been developed within these last wonderful twenty years, and the sanitarian must study that; ... a woman must read and study and observe, she must mark and inwardly digest, and then she must rise up with all her womanly might, and translate her knowledge into action, – absolute, aggressive, ceaseless action, against the kingdoms of dust and filth and dampness and bacteria.'[21]

Microbes were a tremendous locus of anxiety. Remember that we are talking about an age before antibiotics: penicillin wasn't in medical use until 1941. You cannot see germs, you don't know where they are and they can potentially kill you: what could be more terrifying?

Dust became a focus at this time due to its particular association with TB. 'About one in every seven of all the people who die are prematurely carried off by tuberculosis, and a large proportion of these through dust-poisoning, which if we choose we can largely prevent,' reported the American pathologist Theophilus Mitchell Prudden in his 1890 book, *Dust and its Dangers*.[22] Scientists had found it was possible to cultivate the tubercle bacterium from ordinary house dust, and the theory of fomite transmission – the idea that objects and materials could harbour active, infectious germs for long periods of time – made household surfaces into a deadly hazard. 'Only later did further investigations reveal that the germs cultured from dust had little infective power,' Nancy Tomes notes, but, of course, by then it was far too late to stop the panic. Consequently, 'Women were taught to assume that where there was dust, there were germs, and to act accordingly' – that is, to clean as if their lives depended on it.[23]

Mr Prudden's 'sanitary preachments' were enthusiastically taken up by a new type of woman: the domestic scientist. Home economics programmes in schools and colleges had taught girls cooking, sewing and childcare since the mid-1800s, but in the early twentieth century the rhetoric took a sudden shift. The world was changing, consumer goods becoming increasingly affordable and abundant, and electrification on the cusp of revolutionising daily life. Women must be trained for this new world and schooled as modern consumers. Their needs must be advocated to businesses and government. And household management should be elevated to the status of a profession, one as rigorous, rational and intellectually fulfilling as managing a factory.

In 1906, a young Boston woman named Christine Campbell graduated from Northwestern University and married the business executive J. George Frederick. Giving up her work as a teacher and settling into family life with four small children, she found herself facing 'a daily struggle to " get ahead " of household drudgery': the cleaning and the cookery, the laundry and the mending. 'It was a continuous conflict to do justice to all the house-work tasks, and yet find enough time for the children,' she later wrote, blaming herself for what seemed to be 'my lack of ability to manage my household problem'.[24]

Meanwhile, Christine's husband was coming home from his work fired up with talk of 'scientific management' and the new 'efficiency' ideas, as advocated by Frederick W. Taylor and others. Tired as she was, she listened into his conversations with friends, learning that their object was to 'short-cut and reduce work to such a system that the shop or office or any business would be managed with less effort, less waste, and even at a lower cost'. What if this could be applied to her own work in the home, she wondered? It would be difficult: tasks conducted by housewives were of a far greater variety than those carried out by factory production line workers. Nonetheless, the energetic Mrs Frederick quickly set up a laboratory at her home in Greenlawn, New York, to conduct experiments on her domestic labour, at each step streamlining and simplifying, scheduling and standardising – and above all developing a rigorous 'efficiency attitude' turn of mind. What was the ideal layout of sink, stove and refrigerator that would minimise motion? What was the optimum sequence of gestures for doing the washing up? Our kitchen worktops are 90cm high (go and measure them!) because Christine Frederick told us so a century ago. Approached analytically enough, she believed, household problems 'became objects of keen mental interest – quite the same as the tasks of the business and industrial world which men tackle with zest and results'.[25]

In this way, housework was no longer 'drudgery or degrading' – not least because Christine Frederick was no longer a housewife, but a 'household efficiency engineer'. She would come to run a fair-sized business as an author and magazine editor, selling correspondence courses and promoting products for brands under the guise of 'consumer education'. Ninety-odd years before Instagram was invented, this enterprising woman defined the template for the contemporary 'influencer'. But the main reason I include her in this story is because, with her vigour for research and quantification, rationality, optimisation and control, Christine Frederick is the epitome of twentieth-century high modernism just as much as any architect.

The manuals and magazine articles that Frederick and women like her wrote are unflinchingly techno-bureaucratic in their language, with titles such as *Household Engineering: Scientific Management in the Home* and *Running the Home Like a Factory*. Progress, it would seem, required transforming the domestic environment from something soft, comfortable and cosy to something much harder (and ideally wipe clean). There is little room for sentiment in Frederick's approach, only the rational use of time, rigorous scheduling and firm 'self-discipline and control of the will'.[26] David Harvey notes that, some years before the architect Le Corbusier's famous claim that a 'house is a machine for living in', that it was these women who imagined the home as 'nothing more than a factory for the production of happiness'; a place that should be 'equipped, accordingly, with machinery'.[27]

Daily life was transforming in front of people's eyes at a truly staggering rate. In Europe and America, the twentieth century dawned on a domestic landscape powered by wood, coal and elbow grease alone. Only a third of US households had running water; fewer than one in five, flush toilets. Then within a couple of decades, the rhythms of daily life were writ anew. In her 1934 autobiography, American novelist Edith Wharton observed: 'I was born into a world in which telephones, motors, electric light, central heating (except by hot-air furnaces), X-rays, cinemas, radium, airplanes and wireless telegraphy were not only unknown but still mostly unforeseen.' By the time she came to set down her memoirs, these once-startling novelties were commonplace.[28]

Yet for all that these technologies might seem to make the world bigger – the day longer; the span of your daily movement freer and further; the news you are aware of that much *newer* – the effects they had on women's lives were often quite the opposite. Far from liberating women from domestic labour, these technologies only made more of it. Brighter light meant dust and dirt were now more visible, and so women had to clean more thoroughly and more frequently to remove them. Clothes were no longer dirty when they had marks on but had to be laundered after a day or two; one's children, likewise. Furthermore, middle-class women discovered – to some disquiet – that they now had to do these cleaning tasks themselves. In 1927, the novelist J.B. Priestley declared domestic service 'as obsolete as the horse' in an era of motor cars, as young working-class women were taking up new jobs in shops and clerical work in preference to the backbreaking live-in labour of the house-maid.[29] Doing their own hands-on Hoovering for the first time, middle-class women risked becoming in some sense downwardly mobile through their new responsibilities for manual labour. No wonder the domestic science movement sought to professional-ise it.

The technologies themselves didn't produce this change in social attitudes to cleanliness and dirt, of course – it had to be advertised, evangelised and taught. Public health officials, house-hold equipment manufacturers and the domestic science move-ment worked hand-in-hand to do precisely that. The modern template for public health communication was developed during the 'antituberculosis crusade' of the early 1900s, historian Nancy Tomes tells us. 'Borrowing overtly and enthusiastically from the new advertising culture of the 1900s, antituberculosis workers turned out posters, slogans and other forms of propaganda to "sell" their message of protection against the ubiquitous tubercle bacillus' – approaches that 'with surprisingly little modification remain in use today'.[30] (Very true: during the coronavirus lock-downs of 2020 and 2021, public health billboards repeating the mantra 'HANDS. FACE. SPACE' were for a time some of the only advertising I would see on my daily tedious walks around north London.) Then as now, commercial advertising served to hammer these lessons home through establishing not just new consumer habits, but new social norms: new ways of being a person in the world.

Science – or the appearance of it – was central to these appeals to buy these new, modern consumer products and technologies. The Standard Oil Company invited readers to send off for Theophilus Prudden's pseudo-scholarly treatise, *Dust And Its Dangers*, to persuade them at wearisome length of the need for floor varnish.[31] Margaret Horsfield explores Hoover's sales tactics in her 1997 history of housework, *Biting The Dust*. One of their adverts presented an 'Analysis of Rug Dirt' featuring a highly magnified diagram of a carpet ecosystem containing no less than five types of different filth, from 'visible debris and dust' to 'pocketed furrow dirt', for the housewife to fret over. Household manuals and home economics courses would subsequently instruct women in a litany of different vacuuming motions to tackle each one.[32]

In her role as a Hoover brand advocate, Christine Frederick also testified to the superiority of their vacuum cleaners for being 'based in fact on a very superior principle of hygiene: freedom from dust and germs, and ultimate disease'. She promoted vacuum cleaners on the grounds that they were the efficient option: they actually removed dust, not just scattered it around a bit as a feather duster or cloth might. You can imagine her in an advert, eyes earnestly turned to camera as she entreats, 'What housewife could refuse to accept such an offer, and even skimp and save if need be, to purchase the device that would bring more health into her home?'[33] Little has changed in a century: these days, Proctor & Gamble sponsor a glamorous blonde by name of Mrs Hinch to evangelise for Dettol from her spotless Essex abode.

Alongside the health angle, advertising also operated through emotive and social narratives claiming to help middle-class women navigate their gendered roles in a fast-changing social world. During the interwar years, vacuum cleaners were sold as solutions for the 'servant problem'. A 1937 advertisement in *Woman* magazine declared, 'I've just got a maid at fourpence a day! Tireless workers who never need afternoons off or evenings out.'[34] After the Second World War, advertising narratives took on a highly conservative stance on ideal domestic femininity. In the US, Vienna-trained psychologist Ernst Dichter applied the insights of Freudian psychoanalysis to consumer marketing, declaring 'buying is only the climax of a complex relationship, based to a large extent on the woman's yearning to know how to be a more attractive woman, a better housewife, a superior mother.'[35] The Eureka Roto-Matic Swivel-Top Cleaner duly promised women 'everything your heart desires' to be a perfect suburban housewife: a cleaner that is 'So quiet! Baby sleeps, neighbors chat, nerves stay serene.'[36] Notice how fragile the housewife's emotional equilibrium is thought to be.

Yet though the vacuum cleaner was sold as a labour-saving device, in reality it did no such thing. Christine Frederick knew this perfectly well: her Taylorist time-and-motion studies measured it. She reported in *Household Engineering* that the vacuum cleaner offered 'little if any' time saving, 'the chief advantage being only in the absence of the scattering of dust, and in the somewhat greater thoroughness of the work', such that cleaning might need to be done less frequently.[37] But even this minor advantage was elusive: instead, cleanliness standards rose to fill the time available. The *Ladies Home Journal* noted in 1930: 'Because we housewives of today have the tools to reach it, we dig every day after the dust that grandmother left to a spring cataclysm.'[38]

Author and activist Barbara Ehrenreich observes acidly that the emphasis on domestic perfection was no accident. It served a highly conservative social purpose:

'Housework as we know it is not something ordained by the limits of the human immune system. It was invented, in fact, around the turn of the century, for the precise purpose of giving middle-class women something to do. Once food processing and garment manufacture moved out of the home and into the factories, middle-class homemakers found themselves staring uneasily into the void. Should they join the suffragists? Go out in the work world and compete with the men? 'Too many women,' editorialised the *Ladies' Home Journal* in 1911, 'are dangerously idle.' Enter the domestic-science experts, a group of ladies who, if ever there is a feminist hell, will be tortured eternally with feather dusters. These were women who made careers out of telling other women that they couldn't have careers because housework was a big enough job in itself.'[39]

'Digging every day after the dust' would go on to be a full-time occupation for a generation of women in post-Second World War America, as a renewed gender conservativism made deep incursions into their lives. After years of threat and anxiety, the late 1940s and the 1950s were a time of 'family fever' and a resurgence of domesticity. Men and women of all classes, races and education levels married more, married younger and had more children than any other time in the twentieth century, producing the fabled 'Baby Boom'.[40]

During and between the world wars, women had gained independence as they'd gone to work and filled crucial jobs while men were away abroad – but afterwards this trend sharply reversed. In 1945, 37 per cent of American women were in paid employment, but just two years later only 30 per cent were. Magazines and media drove home the idea that a woman's place was in the home, taking care of her family and taking pride in the housewife role. Many could afford not to take paid work: rising productivity and a strong labour movement meant that men in working-class manufacturing jobs could support their families on a single salary, perhaps even comfortably. The federal government was investing heavily in building new housing and the suburbs expanded as millions of immigrants – Irish, Greeks, Italians, Poles – moved out from the inner cities to enjoy a new, middle-class lifestyle. They become assimilated into whiteness; Black Americans and people of colour, meanwhile, were explicitly excluded from these new suburbs through discriminatory bylaws, realtors and mortgage lending practices.

We're not speaking of all American women here – but it's a very large number, nonetheless. A new consumer demographic for a new generation of housework manuals to instruct in the attitudes, behaviours and anxieties proper to the housewifely role. *The ABC of Good Housekeeping* (1949) directs the housewife's schedule from 7 a.m. to 7 p.m. every day. Dust work begins at 9.30 a.m., when she is to dust and tidy the bedrooms, before moving on to sweeping and dusting the lounge, dining room, landing and stairs at 10.15. Between 11.30 and 12.30 and 3–4 p.m., she is assigned 'special weekly duties', which means more dusting on four days out of six as a particular room is deep cleaned. In addition, all floor types need sweeping, dusting or carpet sweeping each day, with carpets vacuumed weekly; furniture needs daily dusting and 'rubbing up', and wall surfaces require weekly dusting.

Do you dust your walls every week? I have spent more time thinking about dust than all but a handful of people on this planet, yet it has never even crossed my mind. Cobwebs do not form that quickly: this is an instruction to pursue invisible dust. It is an instruction toward obsession.

The manuals never really explain *why* dusting had to be done – and done so frequently – the turn-of-the-century TB fear is no longer mentioned. It takes 1950s' satirist Elinor Goulding Smith to say it out loud: 'Dust is probably the housekeeper's worst enemy and must be fought tooth and nail'.[41] Cleanliness is, as ever, about respectability: 'You never know when a dear and trusted friend or relative may drop in and run a white-gloved finger along a skirting-board behind the couch, and how would you feel then?' But there's also a deeper anxiety, a fear of invasion: dust 'comes in through doors and windows, ventilators, children's pockets, and it comes out of mattresses, sofa cushions, last year's overcoats, books and cupboard floors.' It's relentless. Historian Elaine Tyler May, in her book about the impact of the Cold War on American families, described how the stability of the suburban home symbolised meaning and security at a time of deep geopolitical threat.[42] Dust reveals that the home's sanctity is a fiction: it threatens its status as a haven from the outside world. The battle against it looks profoundly trivial, but subconsciously the stakes are existential.

In her 1963 polemic *The Feminine Mystique*, Betty Friedan describes how 'millions of women lived their lives in the image of those pretty pictures of the American suburban housewife, kissing their husbands good-bye in front of the picture window, depositing their stationwagonsful of children at school, and smiling as they ran the new electric waxer over the spotless kitchen floor'.[43] Their lives were reduced to servitude, she argues, their own ambitions and interests put aside in favour of their family's needs. Friedan called it 'a problem that has no name', a sickness of the soul caused by a life filled with inane tasks and profoundly constrained horizons. Think of Betty Draper, the perfect blonde housewife in the television drama *Mad Men*, her hands going numb with suppressed psychosomatic rage when she has to do the washing up and other domestic chores. Seeing the day stretch away into emptiness, she picks up a BB gun and goes out into the garden to shoot the neighbour's pigeons for daring to exercise the airy freedom she lacks.[44]

Critics argue that Friedan overstated the plight of the desperate housewife. All the women she wasn't writing about – the women of colour and working-class women; the single mothers, lesbians and singletons – had their own struggles too, many far more materially urgent than boredom. Still, there's something vividly symbolic in this cultural moment: the perfect white suburban housewife, driven mad by a smudge of dirt.

***

The postwar American suburb is the site of one moment of 'peak cleanliness' in twentieth-century history – 1920s France is another. If we are to talk of modernity, we must consider its grand architect, Charles-Édouard Jeanneret – better known as Le Corbusier – who sought to turn the home into a 'machine for living in'.

Born in Switzerland in 1887, Le Corbusier pioneered the 'International Style' of modern architecture, characterised by its rectilinear and repetitive forms, rejection of ornament and colour, and use of concrete, steel and glass. At every scale, from his designs for furniture and private homes up to his masterplans for cities such as Paris and Chandigarh, Le Corbusier was a utopian, one of the century's deepest believers in the rational perfectibility of human life. As such, his work carries within it an abject horror of dust as its antithesis: this formless symbol of decay and disorder.

In *Toward An Architecture* (1923), Le Corbusier argued 'Modern life demands, and is waiting for, a new kind of plan, both for the house and the city.' Since the turn of the century, technology had transformed the texture of everyday life: electricity, telephones and cars had arrived in half of American households; aircraft were fighting dogfights in the skies and crossing the Atlantic. It had been fully sixty years since French poet Charles Baudelaire had coined the term '*modernité*' to describe the flitting, fleeting experience of urban metropolitan life – and yet design, to Le Corbusier's mind, had not remotely kept up. London's street plan was Victorian, and between the Hausmann boulevards, much of Paris's was older still. The 1925 Exposition of Decorative Arts in Paris was a festival of chintz and chinoiserie, as art deco revived classical styles and exoticised the Orient.

Le Corbusier was repulsed by the dirty air and disorder of Paris's city centre, where 'apartment buildings pile up, all crammed together, and narrow streets interweave, full of noise, gasoline stench, and dust, and where the floors are completely open to inhaling this filth'.[45] There had to be a better way to design a city. A better architecture, one that more truly expressed the spirit of the age to which it belonged – an architecture better suited for living in it. Only a new architecture could cure the sicknesses (both bodily and social) of the industrial city – nay, even fend off revolution. 'Space and light and order. Those are the things that men need just as much as they need bread or a place to sleep,' Le Corbusier wrote. A purity of architectural form and material would produce rationality, harmony, beauty.

What form should it take? In his 1925 Plan Voisin, the architect sketches a vision of a new Paris of 'widely-spaced crystal towers which soar higher than any pinnacle on earth', their glass facades 'flashing in summer sunshine, softly gleaming under grey winter skies, magically glittering at nightfall'. The buildings are huge, and yet 'seem to float in the air without anchorage to the ground', standing raised up above it on sculpted concrete pylons. He envisioned a high-rise architecture, with towers 600 feet tall – the street abandoned to machines, and people raised high above them to look down on the world and its motion as remote and distant spectacle. 'When night intervenes the passage of cars along the autostrada traces luminous traces that are like the tails of meteors lashing across the summer heavens.'[46]

The new city was to be a parkland city, only five to ten per cent of its land given over to building footprints – the rest for 'speedways, car parks and open spaces'. Somehow, despite this car-centricity, 'The air is clear and pure; there is hardly any noise.' Through fresh air, sunshine, and pleasant, leafy garden surroundings, Le Corbusier sought a harmony of man, architecture and nature – a kind of physical and moral hygiene.

He may have insisted on the importance of fresh air and nature – but that did not mean the air should be natural. The problem is, of course, nature is unpredictable. The seasons are first too cold, then too hot, then too damp and generally unhealthful. It is rational, then, to improve on nature. To control and perfect it. 'Let's give the lung the constant which is the prerequisite of its functioning: exact air,' he wrote in 1933. 'Let's manufacture exact air: filters, driers, humidifiers, disinfectors. Machines of childish simplicity . . . Send exact air into men's lungs, at home, at the factory, at the office, at the club and the auditorium: ventilators, machines so often used, but so often used badly!'[47]

Perhaps the first air-conditioned building was the Great Hall of Westminster Abbey in London, when, in 1620, Cornelius Drebbel 'turned summer into winter' for King James I through the chemical reaction of saltpetre (the explosive part of gunpowder) in water.[48] Florence Nightingale also advocated for good ventilation as the 'very first canon of nursing . . . keep the air he breathes as pure as the external air, without chilling him.' Her 1863 book, *Notes on Hospitals*, called for open windows as the best solution – with air shafts and mechanical ventilation units as back-up in poorer weather.[49] Years later, in 1902, the first modern electrical air conditioning system was installed in the Sackett-Wilhelms Lithographing & Publishing Company building in New York, controlling both temperature and humidity to keep the paper rolling smoothly on the presses.

Although not the inventor of air conditioning, Le Corbusier was among the pioneers seeking to find ways to incorporate nineteenth-century refrigeration technologies into building-scale solutions. He proposed two solutions: '*respiration exacte*', a mechanical air-conditioning system providing ventilation at a constant temperature of 18°C; and '*mur neutralisant*', a double-glazed wall with conditioned air circulating within the cavity, to moderate heat exchange between inside and outside. In this way, a system of architecture built on the foundational principle of fresh air as a remedy to the choked and polluted early industrial city turned out, in the end, to desire its total exclusion. He wrote in *The Radiant City*:

'From then on, we will witness the inauguration of a new era: buildings will be altogether hermetically closed, the use of air in the rooms being provided for by the closed air circuits mentioned above. Windows will no longer be needed on the facades; consequently neither dust nor flies nor mosquitoes will enter the houses; nor will noise. Experiments on the absorption of sound, especially in the buildings of steel and reinforced concrete, must be pursued!'

Here Le Corbusier makes the central point of this chapter explicit: dust is a threat that must be expelled from the modern world. It must be whitewashed out.

'Whitewash is extremely moral,' he wrote in 1925. Where Christine Frederick shilled for Hoover, Le Corbusier proselytised for Ripolin, a brand of glossy enamel paint. He fantasised about it as a top–down decree. 'Imagine the results of the Law of Ripolin. Every citizen is required to replace his hangings, his damasks, his wall-papers, his stencils, with a plain coat of white ripolin. *His home* is made clean. There are no more dirty, dark corners. *Everything is shown as it is.*'[50]

My friend Douglas Murphy, an architecture writer and lecturer, reminded me that an obsession with hygiene and cleanliness is not itself unreasonable: Le Corbusier was writing in a pre-antibiotic era when TB was rampant.[51] Nonetheless, his obsession exceeds these practicalities to become a moral principle of progress and perfectibility. From whitewash 'comes inner cleanness, for the course adopted leads to refusal to allow anything at all which is not correct, authorised, intended, desired, thought-out . . . Once you have put ripolin on your walls you will be *master of yourself.*'[52]

This is an architecture of domination. Le Corbusier believes in the absolute authority of the top-down planner – in the drive to obliterate dust, he allows no space for the messy reality of what it means to inhabit a place. There is no room in this architecture for bodies – our wonderful sweaty, disintegrating, vulnerable bodies – or any subjectivity beyond the architect's.

Not even the buildings themselves could live up to this ideal: they notoriously leaked. Eugénie Savoye, co-commissioner (alongside her husband) of the famed Villa Savoye, wrote repeated letters to Le Corbusier complaining: 'It is raining in the hall, it's raining on the ramp and the wall of the garage is absolutely soaked . . . it's still raining in my bathroom, which floods in bad weather, as the water comes in through the skylight. The gardener's walls are also wet through.' In 1937 she wrote again, declaring the house 'uninhabitable'.[53] The clean rationality of high modernism was, in fact, impossible, defeated by the tiny yet relentless forces of dust, rising damp and dripping water. As Douglas Murphy writes, 'the history of modern architecture is inextricably tied to failure and ruin'.[54]

***

After the Second World War the quest to expel dust from the world only intensified.

In 1960, Willis Whitfield, a physicist at Sandia National Laboratories, invented the cleanroom, the cleanest place on earth.

'I thought about dust particles,' Whitfield told *Time* magazine in 1962. 'Where are these rascals generated? Where do they go?'[55]

The problem is that the world is much dustier than you might realise. The quantity of particles in the air is typically described in micrograms or millionths of a gram per cubic metre – and the numbers here seem small, for example, decent outdoor air has just 10 µg/m³ of PM2.5s. This doesn't seem much – but dust is, of course, very very small. Count it mote by mote, and a cubic metre of typical room air might contain 35 million particles over 0.5 µm in size.[56] When you're trying to build semiconductors, or package high-spec photographic film – or analyse just one hundred particles of interstellar dust – 35 million tiny particles are a very big problem indeed. You need to make an environment not just a little bit cleaner, but whole orders of magnitude less dusty.

The problem is that the dust you can see in a shaft of sunlight is just the tip of a (tiny) iceberg. The limit of visibility is about 10 micrometres, or a hundredth of a millimetre. 'Big dust' this size or larger just falls to the floor in a few seconds, from which you might straightforwardly hoover it up. 'Little dust', the stuff under one micrometre in size, is more difficult. It makes up the vast majority of all the particulates present, and it's annoying in two particular ways.

First, it turns out to be sticky. A 1961 review paper notes dust's 'great tenacity' at hanging on to things.[57] Turn the surface upside down? Doesn't work: gravity can't overcome the electrostatic adhesion forces. Vigorously blowing on the surface similarly dislodges only a small proportion of particles. Instead, getting tiny dust off a surface requires more physical force, which is to say scrubbing. But scrubbing also generates the great paradox of dust: in trying to get rid of it you risk only creating more of it, through the abrasion of the surface you're cleaning.

The second problem with tiny dust is that it's floaty. It's too small and light to 'sediment out' under gravity, so instead of falling to the floor, it surfs around the room indefinitely, riding the air currents.

What Willis Whitfield realised is that you'd essentially have to hoover the air itself. He designed a small, sealed room where the movement of air could be fully controlled. Air would enter through a bank of ultrafine filters, 99.97 per cent efficient in removing particles greater than 0.3 μm in size – and be completely replaced ten times every minute, so no matter what the scientist might be doing at the workbench, air pollution couldn't build up. Air would flow through the room at a speed of about one mile per hour, gently sweeping particles from their rest and encouraging them to fall to the floor – which was covered with grating, so they could be safely sucked away. When the first cleanroom was tested the dust counters went to nearly zero. 'We thought they were broken,' Whitfield said in a 1993 interview. They'd gone down from a million dust particles per cubic foot, to just seven hundred and fifty. Sandia Lab claimed that 'cigarette smoke blown in one side comes out the other as clean air'.[58]

In 1964, Whitfield received US Patent no. 3,158,457 for his 'Ultra Clean Room'. Ultra-cleanliness would subsequently be formalised in the ISO 14644 standard, which defines just how clean different grades of cleanroom must be.[59] This reveals a curious facet of dustiness: it's highly non-linear. Dust instead proves to be a thing like earthquakes, sound, or light intensity that has to be measured on a logarithmic scale.

Room air is ISO-9 – or much worse, if you have been cooking toast or burning incense, filling the air with carbon soot. Spacecraft are assembled in ISO-8 and ISO-7 cleanrooms that are 10x or 100x cleaner respectively. At the NASA Jet Propulsion Lab (JPL) in Pasadena, California, engineers and technicians have to gown up in full head-to-toe 'bunny suits' to ensure they do not shed dead skin or bring in foreign microbes. Strict 'planetary protection' rules seek to ensure that space missions do not contaminate the rest of our solar system (or the great beyond) with earthly microbes, and so humidity is closely monitored and exposed surfaces are regularly swabbed, cultured and even DNA-sequenced to check for more lively passengers.[60]

More specialist missions require even cleaner environments. The *Genesis* mission (2001–4) sought to collect the solar wind, the continuous flow of charged particles from the sun which permeates the entire solar system. For 850 days, polished collector arrays of gold, silicon, sapphire, and diamond-like carbon were exposed to the sun's outpourings, capturing the atoms that could reveal the precise composition of the Sun – and the origins of our solar system. All was nearly lost when, after three years' voyaging, the capsule's parachute failed to open on return to Earth, and it crashed into a remote part of Utah at 300 kilometres per hour.[61] But its rare and precious cargo was recovered – to be analysed in ISO-4 labs at JPL, five orders of magnitude or 100,000x cleaner than room air. In this environment there are zero stray particles over 5µg in size, and just 10,000 particles 0.1ug in size per cubic metre.

Yet there are cleaner places still. ISO-1 cleanrooms exist, though they seem surely impossible: just ten particles 0.1µm in size per cubic metre of air, and zero larger than that. Not even the coronavirus might get in through the HEPA filters: at 0.05 to 0.14µm in size, it could well be too big. Working in these strange, rarified spaces must feel a little bit like preparing to go on a spacewalk. Scientists engage in strict rituals of 'air showers', 'scrubbing in' and careful costumery as they cross the threshold into these new, pristine worlds. Some anxiety must be ever-present: these ISO-1 cleanrooms are where the finest sensors are made for detecting life on Mars. One error might change interplanetary history.

Sandia National Laboratories was government funded, and so the cleanroom patent was shared freely with manufacturers and hospitals. As a result, the technology quickly spread. Within a few years, $50 billion of laminar-flow cleanrooms were built worldwide, enabling developments in industries ranging from semiconductor manufacturing and space science, to biotechnology, nanotechnology and healthcare. The latest iPhones run on five-nanometer transistors, the size of just ten large atoms – and to build them, Apple's chipmaker, Taiwan Semiconductor Manufacturing, must build cleanrooms over 160,000 square metres in expanse.[62] In this way, Willis Whitfield's musings about 'rascal' dust particles have given us the technological world we know today. The cleanroom should be understood as a critical enabling infrastructure of modernity.

'Humans are the dirtiest things that enter a cleanroom,' comments Victor Mora, a technical supervisor at JPL, but dirtier technologies come out.[63] It turns out, the main reason to build a better cleanroom was to build a better nuke. In 1959, components in nuclear weapons – especially mechanical switching parts – were becoming smaller, and dust particles were preventing engineers from achieving the quality needed. The National Laboratory where Whitfield worked was managed and operated by the Sandia Corporation, a wholly owned subsidiary of Lockheed Martin, the world's largest defence contractor. The Sandia Labs are just one locus in a nuclear archipelago of military–industrial sites across the desert southwest that I'll return to in Chapter 6. As such, the 'industrial hygiene' of the cleanroom connects in uncomfortably few steps to much larger geopolitical issues of borders and defence, nuclear contamination and fear of invasion – that is, the hygiene and biopolitics of the nation-state.

Think about it too long, and dust starts raising some big, political questions.

***

When I'm stressed or overwhelmed, or when I feel a lack of control in my life, I feel a compulsive – and very common – need to clean my flat. It's a desire to restore order and equilibrium in my environment, in the hope that some sympathetic magic or metonymic sleight of hand might restore such order to my agitated mind, too.

The same belief exists at a societal level.

You may know the famous definition of anthropologist Mary Douglas, that 'dirt is matter out of place' – perhaps you have been waiting for its appearance in this book. What's significant in this insight is that she's not just talking about the material ordering of things, but societal order. 'I believe that some pollutions are used as analogies for expressing a general view of the social order,' she writes.[64] The history of anxieties about dust and domestic hygiene prove this at every turn.

The interventions of both the mid-nineteenth century sanitary reformers and the 'domestic science mania' at the turn of the century were often wildly class prejudiced in tone. Adrian Forty, an emeritus professor of architectural history at University College London, connects 'the fetish for hygiene' with 'bourgeois fears of losing social and political authority'. As he writes, 'anxieties about pollution arise when the external boundaries of a society are threatened.'[65] Urbanisation and industrialisation was a massive upheaval to the established social order, creating a new urban working class who took part in protests, marches, labour strikes – and in France and Germany, revolutions – in order to win the rights to a fair wage, better working conditions, and the vote. Hygiene became a means of differentiating the 'good', 'respectable' poor from the lumpen rabble. Those who followed middle-class public health worker diktats might perhaps get rehoused when their slum was razed to the ground to cleanse the city; those who were less compliant were evicted. The least fortunate literally grovelled on London's dustheaps – the great refuse piles, one just south of King's Cross – scratching a living from the dregs of what everyone else threw out.

Similarly, the middle classes had to make such a big show of household cleanliness in order to mark themselves as elevated above those who did 'dirty work' – a status often new, and not always altogether certain. Mrs Beeton's famous *Book of Household Management* (1861) bemoans how 'domestics no longer know their place; that the introduction of cheap silks and cottons, and, still more recently, those ambiguous "materials" and tweeds, have removed the landmarks between the mistress and her maid, between the master and his man'.[66] She wrote her instruction manual for the benefit of the upwardly mobile middle classes, new to employing servants and rather unsure of how to do it. Enforcing unreasonable (but aspirational) upper-class standards of house-keeping – a gloved finger run pointedly over picture frames and mantelpieces in search of errant dust – was how the new and uncertain mistress might continually test her housemaid's diligence and assert her authority over a young woman not as dissimilar to herself as she might like.

Dust – or preferably its absence – continued to signify status and respectability for working-class communities in the UK throughout the mid-twentieth century. Women in inner-city terraced homes would scrub the front step as a daily or weekly ritual, buffing it up nice with red polish or scouring it with a donkey stone until it shone. The street outside also would be swept to keep down dust and dirt (important in industrial areas), and a bucket of soapy scrubbing water sluiced over the pavement once done, to clean that up too. 'It showed you were house-proud,' said eighty-five-year-old Margaret Halton to the *Lancashire Telegraph* in 1997. 'You could tell who was clean and who wasn't just by looking at their doorstep.'[67] On a forum about 1950s' Nottingham life, another woman notes nostalgically: 'As well as keeping the front step clean, they were always washing the net curtains in the front parlour. No use having a clean step and mucky nets!'[68] In these narrow streets and tight-knit communities, all eyes were on you. Maintaining pristine cleanliness among difficult conditions was how you showed you had pride.

At first, I thought this chapter was about how the expulsion of dust was a tiny but intrinsic part of the creation of modernity – the making of the new through the conquest of disease and dirt, cleanliness as a doctrine of control down to a microscopic level. As I wrote, however, 'whiteness' kept turning up alongside the dirt – not only as the visual image of the ideal, hygienically spotless home environment but also in the whiteness of the archetypal 1950s' housewife, living in the suburbs from which Black families were systematically excluded by redlining and racial covenant. In

London, you'd have to shut your eyes not to notice that the majority of people doing dust work, cleaning homes and offices, are people of colour, often Latin American and Black. The history of twentieth-century cleanliness is, thus, a history not only of the making of gender and class distinctions, but racialised inequalities too.

Cleanliness is rarely just cleanliness, a practical, functional process of vacuuming the carpets and washing your hands with soap. It is rarely neutral but rather always marked, always burdened with additional social meaning and significance. It's not only that dirt is morally contaminating and social stigma produces conceptions of 'dirtiness' – the relation goes both ways.

Hygiene proves to have this curiously slippery quality too, as though everyone still subconsciously believes the old maxim that 'cleanliness is next to godliness' such that the benefits of cleanliness overflow into perceptions of moral probity and social status. The early twentieth-century fixation with domestic hygiene ran alongside a 'social hygiene' movement equally dedicated to public health by means of stamping out 'venereal diseases' (STDs), sex work, drug use and other such 'vice'. The original *Boy Scouts Handbook* of 1911 makes this plain: 'A scout is clean. He keeps clean in body and thought, stands for clean speech, clean sport, clean habits, and travels with a clean crowd.'[69] The social hygienists went well beyond this into overt eugenics: women's rights and family planning advocate Marie Stopes argued for sterilisation of the 'hopelessly rotten and racially diseased'. A language of contamination and racial hygiene structured the greatest crimes in European history. Roma and Traveller communities are often referred to as 'dirty' to this day.

We may seem to have travelled some distance from my dusty desk – but, as Mary Douglas writes, 'Reflection on dirt involves reflection on the relation of order to disorder, being to non-being, form to formlessness, life to death.'[70] The supposedly self-evident virtue of cleanliness becomes muddied when we recognise how frequently it is used as a means of creating categories of person: the virtuous citizen vs. the marginalised, the scapegoat, the abject outsider. Women, in particular, are disciplined through words such as 'slut', 'slattern' and 'sloven' that link sexual immorality to matters of dirtiness and carelessness – the 'good woman' still far too synonymous with the careful, 'clean' housewife.

Don't get me wrong: please do continue to hoover. Dust mites trigger asthma, and the endocrine-disrupting fire-retardant chemicals released by your sofa as it ages are unlikely to do anyone any good. The sanitary reform do-gooders of the nineteenth century did genuinely do good for public health; those proselytising the 'gospel of germs' saved lives. But can we ever strip dirt of its moral horror? That is the question.

It is tempting to argue for some kind of 'wabi sabi' approach to dust, modelled on the path Japanese aesthetics suggests for living with the similar inevitabilities of imperfection, decay and time. A gentle philosophy of living with dust. Given the relentless ubiquity of our subject, this might seem the only sane course of action.

The geographer Alastair Bonnett has written on Le Corbusier's loathing of dust and, like me, believes it goes beyond a functional concern with hygiene. He suggests that dust's challenge to Le Corbusier is ultimately one of 'hiddenness': it poses a challenge to his regime of visibility, to the 'total visual and intellectual availability of the urban environment' that the modern city is meant to deliver. By contrast: 'To live with dust, to admit that it will continue to fall, continue to settle, suggests an ability, a desire, to live with the irrational, with history' – with what Le Corbusier called 'the darkness your eyes cannot penetrate', a darkness he abhorred. Gleaming, pristine surfaces are, Bonnet suggests, utopian: that is, they have no place in this world.[71] Even the Class ISO-1 cleanroom has ten particles of dust. There might be better things to do with our lives than fight it absolutely.

Even Mary Douglas, ur-theorist of dirt and the unclean, turns out to agree. In the concluding chapter of *Purity and Danger*, she writes that we must finally consider 'how dirt, which is normally destructive, sometimes becomes creative'. The attitude to waste and rejecta goes through two stages, she writes. 'First they are recognisably out of place, a threat to good order, and so are regarded as objectionable and vigorously brushed away.' The problem is that dirt at this stage still has identity: 'they can be seen to be unwanted bits of whatever it was they came from, hair or food or wrapping', and, as such, it's dangerous, because it's boundary breaking. But, given time, this changes: 'a long process of pulverizing, dissolving and rotting awaits any physical things that have been recognised as dirt. In the end, all identity is gone. The origin of the various bits and pieces is lost and they have entered into the mass of common rubbish.' Once identity has vanished, this material transgresses no longer. It regains a clear place to belong to: rubbish in the rubbish heap – and dust, gently limning the form of each and every thing around us.[72]

Douglas observes how water gains such great symbolic powers of renewal not just from its life-giving properties, but the fact it dissolves – 'breaking up all forms, doing away with the past', and offering a new start.[73] Similarly, through its return to this state of non-differentiation and formlessness, dirt becomes as 'apt [a] symbol of beginning and of growth as it is of decay.' In the last, it is disgusting no more: 'Even the bones of buried kings rouse little awe, and the thought that the air is full of the dust of corpses of bygone races has no power to move.'

We are talking, in the last, of life. And, '[a]s life is in the body it cannot be rejected outright' – however hard modernity might try.

Dust is simultaneously a symbol of time, and decay and death – and also the residue of life. Its meaning is never black or white but grey, and somewhat fuzzy. Living with dust – as we must – is a slow lesson in embracing contradiction: to clean, but not identify with cleanliness; to respect the material need for hygiene while distrusting it profoundly as a social metaphor.

As Mary Douglas shows (if we read her closely), dust isn't always a threat, a transgressor 'out of place'. It does not always require modernist mastery and domination (which is, as I have tried to show in both Christine Frederick's and Le Corbusier's language, often really an unwinnable fight against the *self*, our unruly ill-discipline and bodily porosity).

Mastery over dust is of course a microcosm of seeking mastery over nature. At every scale, from the personal to the planetary, this keeps failing – as this book repeatedly shows. We might then attempt other, less hierarchical approaches: an ecology of coexistence, even collaboration.

Aralsk

Shevchenko Bay

Syr Darya

U s t y u r t   P l a t e a u

**ARAL
SEA**

1 9 6 0   s h o r e l i n e

Moynaq

Q A R A Q A L P A Q S T A N

Nukus

Amu Darya

K y z y l k u m   D e s e r t

T U R K M E N I S T A N

N

0   50   100 miles

Chapter 5

The Vanished Sea

It's midnight on the dance floor when the DJ drops a track that makes my heart skip.

I'm in Moynaq, a tumbleweed town of 13,500 people in remote western Uzbekistan – and the site of one of the worst human-made environmental disasters of the twentieth century: the drying up of the Aral Sea, once the fourth largest lake in the world.

Here, for two nights in late August 2019, at a memorial to the vanished sea, world-class techno pulses from an unexpectedly capable sound system out into the empty desert night. Green lasers trace a sky thick with stars as, around me, hundreds of people go wild, dancing like they have never heard this before – because most of them never have.

This is Stihia, a word which means 'an inevitable force of nature'. It's a two-day techno festival curated by lawyer Otabek Suleimanov and the Fragment crew, nurturers of a nascent techno culture in Tashkent, as Uzbekistan begins to open up culturally and politically since the death in 2016 of the repressive President Karimov.

The festival is a public event. A couple of hundred people have come in from outside – mostly Uzbeks from Tashkent, but also a fair-sized group of techno-heads from Moscow. People have travelled from across Central Asia: feminist activist Arina hitchhiked 1,200 miles from Almaty, Kazakhstan; a German health NGO worker is here from Bishkek in Kyrgyzstan and French travellers have come from Azerbaijan by ferry, over the Caspian Sea. Me and journalist Tom Faber, reporting for *Resident Advisor* magazine, are both from London. A handful of Americans; DJs from Japan, San Francisco and Berlin. And then there are the two thousand or more local Qaraqalpaq people, who have left their homes after dinner and walked up the road to see what is going on.

The stage and sound system stand on a tarmac lot at the end of the road where Moynaq meets the new Aralkum Desert, a scrubland of low thorny bushes and drifting sands. All along the street local people have set up stalls selling beers, soft drinks, sunflower seed snacks and sweets; mobile phone credit, phone accessories, little plastic tchotchkes for children. Braziers and kitchens do a fast trade in grilled lamb kebabs and Uzbekistan's ubiquitous *plov*, oily and delicious. The woman providing catering at the yurt campsite apparently made as much money during the three days of the festival as she normally did in a year.

A yurt camp stands on a cliff, overlooking a dozen rusty fishing vessels marooned on the sand below. These boats are the memorial; they're testament to what has been lost.

As the sun goes down and the cloudless blue sky turns transparent gold then dusky pink, orange and violet, a crowd builds. They walk past the police-manned roadblocks, women in floral-print dresses carrying toddlers on their hips, small children running around in amazement at the music, the lights, the strangers, everything. Teenage girls, hair and makeup carefully done, pose like girls of their age everywhere, in twos and threes, arms outstretched, shooting selfies. As the music picks up, a teenage boy in an AK-47 DEFEND UZB t-shirt and skinny calf-length jeans is one of the first to throw shapes on the dance floor, soon joined by a guy in his twenties with bleached blond hair and a traditional ikat-patterned T-shirt. The crowd builds. An old man in a flat cap, eyes crinkled by decades of sun, stands in the middle of the dance floor, a half-smile on his face, as the party really starts to kick in.

What motivates Stihia, organiser Otabek Suleimanov tells me on a Skype call a couple of weeks before the festival, is a belief that techno music can change the world. 'My first encounter with live electronic music was back in the 1990s when I was doing my law degree in Sheffield,' he says, when he went to the era-defining superclub Gatecrasher. 'That was listening to proper Detroit techno, and that was that was really inspiring because Jeff Mills said that the techno music is not about the dance music. It's about the futuristic statement. And that's something that I completely share.'

To Otabek, the dusty desert wasteland that is the Aralkum – the Aral sands that stand where the sea used to be – is kind of a Mars-scape, toxic and inhospitable to life. The health crisis brought on by the drying sea is profound. In the late 1980s, two-thirds of the population were suffering health problems as environmental conditions collided with poverty and undernourishment.[1] Deaths from acute respiratory infections were three times higher than in the USSR as a whole, as dust and sandstorms carried salts and toxins deep into people's lungs. The poisoned water supply saw liver cancer double, and gallstone and kidney stone diseases at six to ten times the typical rates. Children were born with rickets, anaemia and stunted growth. Despite a major World Health Organization initiative in this region around the turn of the millennium, substantial health inequalities and deprivation remain.

A visionary and sci-fi dreamer, Otabek wants to be a part of rehabilitating this ruined landscape. 'The big inspiration came from the Martian Chronicles of Ray Bradbury,' he tells me, and goes on to describe the story: an astronaut sent to Mars, who starts planting trees to address the low oxygen level, and doesn't stop until the last seedling. 'All of a sudden, overnight, because of the magical properties of the Martian soil, the entire forest emerged. So we think this is also similar to our very ambitious and difficult task.'

He's not kidding about the ambition: alongside Stihia, Otabek talks of a programme, Six Billion Trees, designed to support the afforestation of the dried-up bed of the Aral Sea.

'It's called Six Billion because this is the exact number of trees you need to cover the entire Aral Sea. We literally calculated it. There are about four million hectares of Aralkum, and the specialist experts say you can grow about fifteen hundred trees in one hectare. So you just simply multiply it and you have six billion trees. It's a very, very funny figure' – because of its audacity, I think he means – 'but, you know, this is a target that we have.'

It's a little unclear to me on this call whether this is actually Otabek's plan, or really the Uzbek government's target for afforestation work that's already underway. But this blurring of boundaries between a clearly well-connected international lawyer and the state bureaucracy seems illustrative of how things seem to operate around here.

Tree planting has been on the table for a long time – the southern part of the Aral has been dry for nearly forty years now – but little happened until a big push this year with 'a thousand trucks digging the bed of the Aral Sea', Otabek tells me.

We had driven past a couple of dozen tractors parked up at the northern end of Moynaq (a town that is essentially a seven-kilometre main road). It's August and this is the off-season, too dry and dusty to dig – but last winter they were out on the former seabed, planting saxaul trees, a short, shrub-like tree native to the Central Asian deserts. Saxauls are 'halophytes', salt-fiends able to live, even thrive, in this saline ground. Their roots fix the soil against strong winds, one tree able to prevent ten tons of earth from becoming dust – and they start the process by which life returns to barren sand.[2]

The 40,000 square kilometres of Aralkum Desert is the largest site of primary ecological succession on earth, the closest thing to a blank slate that this planet has ever seen in human history.

Can this be the first major landscape to be terraformed as Elon Musk dreams of doing on Mars? More to the point, should it be?

Nothing will bring the water back. Not now. But can anything stop the dust?

And can a rave possibly do anything meaningful for this remote and ravaged place, the 'last stop on the road', as Aleksandr, a Tashkent techno DJ and one of the other Stihia organisers called it, as we stood in blazing sunshine on the steps of the Saviinsky Art Museum in Nukus, after it was all over? Or are some places too peripheral, and some dreams too unlikely?

What happened here, at the Aral Sea in the mid-twentieth century, is one of the biggest environmental disasters in human history. It gets called the 'slow' or 'silent' Chernobyl: a human-made catastrophe sixty years in the making, yet one nobody seemed to notice until years after the water had already gone. A dream of modernity and scientific progress turned to toxic dust, with human consequences that echo on for generations. And it's the harbinger of countless more water crises to come, as the world heats up this century and water becomes a resource that's fought over.

Yet, in the West, no one seems to know anything about it.

***

I was in Moynaq because I wanted to see a disappearing sea – and the dust that is left when the water is gone.

The Aral Sea was once the fourth largest lake in the world, but half a century ago it almost entirely vanished. What was once a bustling fishing port – and (as the town museum proudly claims) the biggest canned fish exporter in the entire Soviet Union – is now beached: economically, politically and quite literally. Moynaq once stood on a peninsula, surrounded on three sides by water. These days you can stand on a cliff edge and look out over desert as far as the Kazakh border.

We drove into town at lunchtime on a Thursday, a day before the festival started, and I knew for all the wrong reasons that I was in the right place. People sat chatting in the shade by the side of the road, their entire faces hidden from the dust. Bandanas over mouths and noses; sunglasses over eyes; white scarves wrapped over their faces entirely. The entire town was a building site, piles of earth lying by the side of the street; finer cement powder getting picked up by the wind as it waited to be mixed into concrete.

President Shavkat Mirziyoyev had visited the region that week to check on the progress of the major development push he had instigated since his election in 2016. Qaraqalpaqstan, long the forgotten corner of the country, was finally gaining investment in housing, infrastructure and jobs, to make up for the devastation of the past half-century – and so everything in Moynaq was under construction. The wind blew and the dust rose. The windscreen of the car was thick with it. The main street might have been tarmacked but the rest of the town was still a grid of dirt roads; the new buildings being put up along the main street might have been cinder block and render, but out at the edges of town people were still building with mud brick construction. We turned the car and watched a dust devil twist and waver at the end of the street, a spectre hundreds of feet tall, a sandy stain against the sky. 'A tornado,' my guide, Sarsenbay, called it. The earth was alive, and rising up.

The dust lifted into the air was construction cement, sand, salt, pesticides and fertilisers – the residue of all the rain and irrigation and run-off of six Central Asian nations, all concentrated here, at the very end of the river. Once airborne, the winds can carry it as far as Greenland and Japan.[3]

I had come in search of exactly this, yet it was daunting. The town felt unreadable: buildings set back from the road with few visible signs or any evidence of what was open; the streets unwalkable, 38°C and swirling with construction dust.

Previously on my journey I'd walked miles every day, crisscrossing Tashkent and the historic Silk Road city of Samarqand to get a feel for the place I was in. I'd walked through high-walled traditional *makhalla* neighbourhoods, glimpsing courtyards and mulberry trees and pools through open gates – and drifted through the newer high-rise Russian districts. In Nukus, the capital of this Qaraqalpaq region of western Uzbekistan, I'd felt somewhat more out of place as I'd zigzagged between banks, trying to find a foreign exchange bureau (no language shared with anyone I met) to get some dollars to pay my desert driver and guide-slash-translator. Walking back to my hostel with $700 in my wallet, I knew I was carrying six months' wages; had it been a bundle of Uzbek soum and not US dollars, it'd have been a brick six inches thick.

Now here, in Moynaq, I was beyond any place or landscape I'd previously known. I checked into my hotel room – the building poorly constructed, like almost everything else in town. The corridor opposite was cut off with plastic sheeting, that side of the hotel still under construction; splatters of paint lay on the unfinished bathroom tiling, and the ornately patterned blue striped wallpaper ran out halfway round the room to be replaced by another, subtly different pattern. But the shower was good and hot, the twin beds thin but clean and I had some biscuits in my bag if I couldn't find food. So I holed up there – read, wrote, edited photographs – and talked myself into staying for the techno festival. Even though it was why I was there, I was nervous about what was ahead. I repeated the mantra of a good friend of mine: 'It doesn't have to be fun to be fun.' Even if the festival was awkward and out of place, even a Fyre Festival disaster, I was here to tell that story.

***

Central Asia is a vast landmass – four million square kilometres of mountains, riverine valleys and desert, just under half the size of China or the United States. It spans from the shores of the Caspian Sea to the Taklamakan desert in Xinjiang (or East Turkestan), in China – and from the fringes of Siberia to the Iranian and Afghan borders. It is at the very heart of the Asian continent, and Uzbekistan is one of only two countries in the world to be 'doubly landlocked', that is, surrounded only by other landlocked countries, requiring at least two national borders to be crossed to reach a true coastline. (The other's Liechtenstein.) Given this, you can start to imagine the importance of the Aral Sea and the rivers that used to feed it.

The Amu Darya (Amu River) lies to the south. The Ancient Greeks and Romans knew it as the Oxus; a thousand years later it was called Jayḥūn by the Arabs, after one of the rivers in the Garden of Eden. Its tributaries rise in the great mountain ranges of the Tien Shan and Pamir Alai, before irrigating the deserts of Turkmenistan and blooming into a delta at the Uzbek border and the historic Silk Road city of Khiva. From here, the Amu Darya flows 220 miles northwest, to feed the Aral Sea from the south at the once coastal town of Moynaq. (Another river, the Syr Darya, feeds the sea from the north, in Kazakhstan.) The Aral – meaning 'island' – was named for the countless thousand tiny islands that stood in the ever-shifting silt.

Viewed from above – from an aeroplane, or the heavenly overwatch of a satellite – this region is truly an oasis, a deep green fractal ink blot spilling out among pale fawn desert sands. Water brings life. It still does, to the delta region: the river still flows, all the way to Moynaq but no further. It feeds the cotton and the rice paddies and the melon plants, generously and copiously – Uzbek cotton needs 1.0 to 1.2 metres of water per year to grow.[4]

Imagine fields flooded with water to waist height, and you'll have got the gist: simple, inefficient flood and furrow irrigation has been used here for thousands of years. Moscow saw no reason to invest in anything more effective, such as underground drip pipes or centre-pivot sprinklers as are used by cotton farms in the deserts of America: the labour to dig new canals was cheap or even free if staffed by gulag prisoners. And this was purely an economic extraction zone. Yet average rainfall in Nukus is just 109 millimetres per year, a tenth of what the cotton needs. As a result, up to 62 cubic kilometres of water a year – enough to submerge the entirety of Greater London under 40 metres of water – was extracted from the Amu Darya River and from deep groundwater reserves, and diverted through a network of count-less canals to feed the crops.[5] Under the 40°C summer sun, it evap-orates – leaving behind its silt load, its pollutants and its salt.

The Aral is an endorheic basin – as are many of the places in this book, such as the Salton Sea and Owens Lake in California. Endorheic basins are low-lying depressions where water flows in from rivers and streams, yet no water flows out to reach the ocean. Instead, the rivers spread out into wide deltas, forming lakes and marshes – where their water simply vanishes, evaporating under the hot continental sun. The Okavango Delta in Botswana is perhaps the best known and most beautiful, as its wide grassy plains flood each year and provide a rich habitat for wildlife rang-ing from hippos and crocodiles to lions, giraffes and strange, semi-aquatic antelopes. Yet most endorheic basins are much drearier, and much dustier. Rivers transport eroded soil and sedi-ment downstream, but in endorheic basins that material doesn't reach the ocean and is instead deposited on the land. This silt can be very fertile – it's why the Amu delta region has been an oasis for thousands of years. But these sedimentary particles are also very tiny, and easily caught by the wind, which makes them a fine and ferocious source of dust.

As a kid in the mid-nineties I pored over atlases to discover what lay east of the Mediterranean: first the Black Sea, then the Caspian Sea, then the Aral, a blob of baby blue in the middle of a continent. That was how the world was. Yet by the time I was learning of its existence, the water had already gone.

***

The Aral Sea started drying up in the 1960s, and the town of Moynaq on its southern coast was marooned in the dust and dunes of the newly formed Aralkum ('Aral Sands') by the early 1980s. But the groundwork was laid a century earlier, in the Russian conquest of the region then known as Turkestan. Armies came from the north, capturing the Khanate of Khiva on the southern side (still governed by descendants of Genghis Khan) in 1873. The Aral Sea was now the largest lake in the Russian empire.

In 1882, the Russian climatology pioneer Aleksandr I. Voeikov called for its destruction. 'The existence of the Aral Sea within its present limits is evidence of our backwardness,' he wrote in his geographical report on the rivers of Russia, and to him it illustrated 'our inability to make use of such amounts of flowing water and fertile silt, which the Amu and Syr rivers carry'.[6] The Amu Darya might be allowed to flow to the Aral Sea in winter, when it was not needed for irrigation, but in summer it was needed to feed cotton fields and rice paddies, the better to sustain the empire and free it from dependence on US imports. Voeikov recognised that the water level in the Aral Sea would drop as a result – but, though he was a leading figure in nineteenth-century scholarly debates about the influence of human activity on climate, he did not address what the impact of a falling sea level might be. Had he done so, the world map might be drawn differently today.

The expansion of irrigation that followed may have been relatively haphazard and low intensity, yet it's significant because both the vision and the geopolitical dynamics established in the late nineteenth century would carry through into Soviet rule despite very different systems of government. Whether tsarist or socialist, it was all imperialism: Central Asia would never be on equivalent footing to the core in Moscow and St Petersburg. Instead, its vast distance and cultural difference made it more akin to a colony, like French North Africa or British India. The Aral Sea region was doubly peripheral, not only subordinate in relation to Moscow but a sparsely inhabited backwater with respect to Tashkent as well. As such, like the Owens Valley in relation to Los Angeles, it was seen only as a land to be exploited.

Following the Russian Civil War, America cut its exports and Russia caught cotton fever as it scrambled to make up production. Restoration of production in Turkestan was a priority, and the Central Committee ordered that cotton should be the dominant crop in all state and collective farms that could grow it.[7] Stalin continued these policies, with each of his five-year plans setting increasing targets – and in just over a decade cotton production more than tripled. By 1940, 'Lenin's concept of "obligatory crop rotation" has disappeared forever from the vocabulary of the obedient Uzbek agrarian,' the economist Vladimir Zuev writes, 'and nobody thinks of water: how much it is needed, how it can and must be used, how to irrigate and for what purpose'. Orchards and vineyards were uprooted and pastureland destroyed so that the best soil might be turned over to cotton production. Irrigation channels were dug with no thought for minimising evaporation loss or draining the used water away from fields. The land became waterlogged, saturated with the polluted fertiliser- and pesticide-ridden run-off from six Central Asian nations upstream. The soil – and the Aral itself – started to fill with salt.

The devastation of the Second World War was followed by further disaster in the Soviet Union, as wartime underinvestment in agriculture combined with massive droughts saw over a million people die of famine from Ukraine to Kazakhstan. In response, Stalin declared a Great Plan for the Transformation of Nature.

The environment would no longer stand in the way of progress. 'We cannot resign ourselves to the fact that the water abundant Amudarya River carries its waters to the Aral Sea without any use,' declared Uzbek premier Yusupov, echoing Moscow doctrine and the words of climatologist A.I. Voeikov sixty years before. Rivers would become productive machines, with dams generating hydroelectric power, and canals and reservoirs irrigating crops to greater yields. Seventy thousand kilometres of forest shelterbelts would be planted to protect European farms from feared drying winds from Central Asia, much as Roosevelt had done in America a decade before to help the country recover from the Dust Bowl. Yet the Stalinist motivation was more romantically Promethean than it was technocratic, argues environmental historian Stephen Brain. The aim was not to restore landscapes to their natural form, or an imagined primeval state – but to create 'an improved nature.'[8]

Work started on one of the largest irrigation canals in the world: the Main Turkmen Canal, diverting water from the Amu Darya along the ancient, dried-up channel of the Uzboy river, along which it had twice before flowed into the Caspian Sea. Like most 'Great Construction Projects of Communism', the canal would be built on the back of forced gulag labour and displace tens of thousands of villagers as it was built. But no matter, Soviet ingenuity would make the Karakum Desert bloom. Water could open up huge new tracts of land to agriculture, expanding cotton production and enabling shipping. Vast amounts were needed: the canal would carry around fifteen cubic kilometres of water a year and divert fully a quarter of the Amu Darya's flow – yet the structure was unlined. Not only would a huge proportion of this water evaporate from the surface, as it travelled for 1,400 kilometres under hot desert sun – yet more water would be lost into the sand, through seepage. It was an Ozymandian folly from the start.

Yet it seemed for a moment like it might work. As in the US Dust Bowl disaster of the 1930s, the vast environmental damage of this drive for growth was for a while concealed by a few years' favourable weather conditions. During the 1950s, there was more water available in the Amu Darya catchment than average – and increased irrigation use meant less water evaporated from reservoirs. As a result, the Aral Sea's water level stayed where it was while state cotton targets strove ever upwards.

Uzbek First Secretary Yusupov pledged an oath to Stalin that he would deliver annual production of 2.4 million tons of cotton by 1953. Khrushchev succeeded Stalin and demanded three million tons a year by 1960. The Uzbek SSR Academy of Sciences estimated that twenty million hectares of land was potentially irrigable, but only four million had been irrigated so far. What potential! More land was put to the pump and duly a four million-ton cotton harvest was achieved in 1964, the year Krushchev was deposed.

New president, same plan. Krushchev's successor Brezhnev declared that 'Golden hands create white gold,' and set Uzbekistan the target of producing five to six million tons of raw cotton by any means possible. In 1977, peak cotton yield was achieved at 5.7 million tons, and the landscape had been transformed into a monoculture. All the land that could possibly be used to grow cotton was growing cotton. Land that quite obviously could *not* be used to grow cotton was nonetheless irrigated too: land that didn't drain, land that was full of salt; soils that were high in gypsum, a mineral that cotton doesn't tolerate. It all drained the river unnecessarily as top-down targets ran roughshod over on-the-ground realities. People living and farming in mud-built *qishlaq* villages, traditional wintering places from semi-nomadic days, were evicted to work on larger central farms, and a way of life was lost forever. And each September, hundreds of thousands of people – particularly students and children, their smaller hands thought to be more dexterous and adept at cleanly picking the 'bolls' of cotton from each plant – would be sent out into the field for the backbreaking task of picking their bodyweight or more in cotton each day, a forced labour practice that hasn't yet fully ended.

Brezhnev's six-million-ton target would never be achieved, because it was impossible.

The water level in the Aral had started dropping in 1960. It dropped an average of 22 centimetres per year that decade, as more water evaporated from this large, shallow sea than was allowed to keep flowing down the rivers feeding it. During the 1970s, as irrigation radically expanded, the sea level fell 58 centimetres a year, rising to 76 centimetres a year in the 1980s – the shoreline retreating markedly every season, hundreds of thousands of hectares of seabed exposed each year to the open air and to the wind.

From the late 1970s, no water at all reached the sea from the Syr Darya in the north, and the 1980s saw inflow from the Amu Darya become irregular as well. In three years, the river petered out entirely into mudflats and reedbeds before it ever entered the sea. The Aral broke in two, into a small northern sea and a larger southern Aral. They have never rejoined, only dividing further, the southern sea splitting into two lobes like a pair of lungs. The easternmost lobe dried up entirely in 2009 and now fluctuates seasonally, no longer a sea but only a shimmering, saline puddle. Ninety per cent of the fourth largest lake in the world has evaporated into thin air. Socialism in the Soviet Union had achieved the same as capitalism in the High Plains of America: it had terraformed the planet at geological scale, a vast disaster written in trillions of tiny grains of dust. I write of modernity to have a word for the hubris that made this possible.

Tom Bissell, a US Peace Corps volunteer and journalist, spent time in the Qaraqalpaqstan region in the late 1990s, back when the sea was more than just the memory of old men. He tells the story of how fisherman in Moynaq 'chased the sea' as it retreated from the town, digging canals out to reach the ever further waters so that they might steer their boats out to fish, and make a living. In 1986, the last of the native fish in the Aral died and the boats were left where they stood to rust.

'I found I did not want to contemplate the long, difficult walk back to town,' he writes, 'that these brave, deluded men must have taken, the day they realized everything was over.'

***

I came to Uzbekistan to see the absence of a sea.

I packed a 25-litre rucksack with William Atkins' book on deserts and Robert Byron's *The Road to Oxiana*, a dress and two pairs of false eyelashes for the festival, before flying into the capital, Tashkent. I spent a few days in Samarqand, agape at azure, domed mosques and glittering, golden mausoleums, the legacy of fourteenth-century conqueror Timur – Sword of Islam, self-styled heir to Genghis Khan and last of the nomadic conquerors of the Eurasian steppe. Then I caught the night train west, from Samarqand to Nukus, the main city in the Qaraqalpaq region.

We board the train at 10 p.m. and I find my place: coach 6, *mecto* 24. It's the top bunk: good! I'm sharing the cabin with two

Uzbek men who are both deeply asleep while a phone alarm hammers away on the table unnoticed. I make my bed with the pillow and cotton sleeping bag provided and drift quickly into unconsciousness as the train chugs steadily westward into the night.

I wake around 6.30 in the morning to a rich green landscape. The train line crisscrosses the Amu Darya River, the land around it lushly fertile. The houses are rectangular and flat roofed, of mud-brick construction, roofing timbers sticking out of the mud plasterwork or more modern and topped with corrugated steel. Wonky wooden verandas are built from the available local timber, thatched with rushes from the river and provide animal stabling near the home or shade structures in the middle of fields as midday respite for labour. Each house has its own gardens: tall stands of corn, orchards of fruit trees and nut bushes. Small paddy fields of rice – all at human scale: from the window I see donkey carts but little in the way of tractors or mechanisation. Each farmstead ringed in tall vertical poplar trees as boundaries and windbreaks.

My cabin mates are having breakfast and offer me bread – at first I am shy, but I know this is a hospitable culture so I come down and join them. I am soon glad for their offers of green tea with sugar and coffee (also sweet) – and share the curly poppy seed-topped cinnamon buns I bought at Samarqand station. Between their few words of English and my fewer of Russian, we endeavour to have a conversation. They establish that I'm going to visit Nukus, Moynaq and the Aral Sea, that I am a tourist and when asked my profession by my interlocutor – who's a machinist in Kungirot, the town at the end of the line – I pause for a second. How do I describe 'internet research', a job difficult enough to comprehend in much of the UK let alone here? I decide on the much easier, generic 'business'. 'Ah, bizniz,' they reply, nodding.

What they're most interested in are my family relationships: am I married? Who are my parents? Do I have any brothers or sisters? I share a photo of my father, picking blackberries in the Sussex countryside a few weeks before, and am introduced to all their children, in turn.

After breakfast, I stand in the train corridor gazing out the window, interrupted by half a dozen hawkers selling all kinds of things: food, snacks, electronics and China-made wrist watches, jewellery for wives and daughters back home. Soldiers stand about in desert camo fatigues. A small girl with pigtails smiles at me and then disappears into her family's compartment, suddenly shy.

At some point around Kartaube, the landscape switches from riverine bounty to semi-desert: we have entered the Kyzl-Kum, the red sands to the east of the Aral. Short, scrubby bushes stand with feet of bare soil between each. Dried-up flower spikes from the spring bloom look like upturned bottle brushes. A herd of black and white sheep graze in the distance, a feral horse stands alone by a telephone pole. The scrub is crossed by dry stream beds, the occasional arroyo carving canyons down into the sand. I am in the desert again. It feels good to be back.

In Nukus, I arrange a car and driver to take me north to see what remains of the Aral Sea – and what is left in the places where the water is gone. This is surprisingly easy, because the first stir-rings of tourism are arriving in this region. Most visitors just take a car or a bus into Moynaq to see the abandoned ships and the Aral Sea Museum, returning the same day – but some have been asking to go further, to see the sea itself, and so an enterprising man called Tazabay has set up an operation to deliver this. White Toyota Land Cruisers are parked outside the Besqala guest house he manages in Nukus to take visitors the three hundred kilome-tres north to the yurt camp he's established overlooking what remains of the western sea. I ask to see as much as he can possibly show me, and he's only too happy to oblige.

I am curious as to how this region can be brought into the international register of tourism – and to discover who other than me might want to see this difficult landscape, too. Can it be a history tour? But it lacks any of the architectural crowd pleasers of the Silk Road cities, boasting only a few crumbling *kala* fortresses and a lone caravanserai. The Sudochie wetlands in the Amu delta might offer scope for eco-tourism. Located on the Central Asian flyway for migratory birds, each spring they become a biodiversity hotspot with over two hundred species at the lake, from flocks of flamingos to rarities such as the Dalmatian Pelican and the slender-billed curlew. But really it's the difficulty of the place that's the draw. As I talk to travellers at the guest house, I find they have come, as much as anything, for the sake that 'here' is rarely travelled to at all.

My guide to the Aralkum is a man called Sarsenbay. He is a local Qaraqalpaq, 'a country boy', as he puts it – born in a small town about seventy kilometres from Nukus. We zoom in on Google Maps and he shows me his mother's home; he has bought a plot of land out here too, where he hopes to build a house. Sarsenbay studied English at the university here in hopes of becoming a teacher, but that didn't work out – so instead he works as a guide during the main tourist seasons in the spring, summer and autumn. The work can be intermittent and he wants more for himself: he thinks he might set up his own tourism agency, he says, and to this end is carefully observant of me throughout our trip. His manners are impeccably solicitous: do I have enough water? Am I too hot or too cold? Each time I take out my camera, he asks if I want to stop to take pictures, and when I photograph something, he photographs it too. Evidently this sight is something tourists will be interested in, and he should put it on his newly purchased travel website. When I use an English word he's unfamiliar with, he asks me to define it and we practise it as we go: five new words in three days. I am unused to being the subject of someone so attentive – but of course, I am observing back.

Starting from Nukus, we drive to a mausoleum, an ancient fort – and then up to the Ustyurt Plateau, our route north, to find what remains of the sea. The Ustyurt Plateau is an austere 200,000 square kilometres of chalk and limestone – a landscape of clay and khaki, fawn and buff. No roads go up this western side of the Aral Sea: instead we drive at a brisk eighty kilometres per hour on dirt tracks, the Toyota Hilux truck kicking up a plume of dust hundreds of feet long as we go. The tracks braid in and out of each other, yet Max, the driver, knows where to go unerringly, no map, compass or GPS needed.

As we drive, birds rise from the low, dun-coloured vegetation on top of the plateau, flurries of life and alarm. They too are mostly dun-coloured, because everything here is dun-coloured: the plants, caught in some desiccated moment between life and death; the bird life, the ground, the earth that rises into the air, the old Kazakh grave sites that show where trade routes used to run, connecting with the Silk Road further south at the now abandoned, waterless city of Konye-Urgench. Only the sky above is blue, a clear, high desert blue that is, of course, cloudless. A ragged cliff edge rends and collapses down to what was once the sea's edge.

And then, finally, I see it.

It is late afternoon, the light golden, shadows growing longer over the scrublands and dunes running down to the water's edge, mineral deposits staining the soil cream and ochre. The sea is bright, a supersaturated Persian blue, the colour of lapis and tiles. It looks almost unreal, like a child's drawing of what a sea should be. I can see the shore on the other side. The western Aral – all that's left of the sea in Uzbekistan – is only about twenty kilometres wide.

Close up, the Aral Sea smells like something sick. It's catastrophically saline but there's none of the briny ozone freshness you associate with the ocean. Instead, down by the water, the air is heavy and smells strange, semi-edible and nearly sweet, like stir-fried noodles left too long in the sun. The depths are contaminated with hydrogen sulphide: rotten eggs. I stand barefoot, calf-deep in the fine, sticky mud of its shoreline, and find myself too repelled by it to swim.

The sea is a mirror. It changes with the light, reflecting the sky

– from roseate pink at dawn to mirrored silver-white minutes later. During the morning it becomes the exact same blue as the sky so that there is no boundary visible between the two, and it is as though the whole sea has vanished. Sarsenbay says he's seen it black, during storms.

Given the chance to go down to the water again the next morning, I declined. I came to Uzbekistan to understand what was once a sea, but found that easier to do from a distance, from a vantage point where the sea wasn't just a sad, stinking, dead place – but also somewhere golden eagles soar on thermals off the edge of the plateau, hunting for mice and gophers in the saxauls and scrubby, salt-loving succulents reclaiming the land metre by metre, as the water withdraws.

Despite the apparent ruined minimalism of this landscape, it is alive, even vibrant. Sarsenbay warns of wild boars when we walk down into the canyonlands where the Ustyurt Plateau crumbles at its edge – and wild dogs, when I walk down to the shore of the Aral Sea as the sun rises. Unseen birds pip-pip-pip and call in the valley below. A dragonfly helicopters around, glistening metallic green, its dual set of wings whirring like rotors. Butterflies flit and sunbathe: blue, tortoiseshell, black and white swallowtail. A vivid orange creature feeds on the second wave of late summer flowers. Each spring the deserts in Uzbekistan turn crimson with the waxy parachute-silk petals of poppies and, for a few days, the Kyzylkum red sands live doubly up to their name. It must be spectacular. For now, I see only remnants, seed pods and possibilities – the holes and burrows that promise unseen mammalian life, busy sleeping the heat of the day away, safe in the cool of the ground. Not such environmental riches as Sarsenbay talks of – the lost Caspian tiger and the deer his ancestors used to hunt. But still, this is by no means a dead place, for all the region is littered with ancient fortresses and lost cities, reminders of the relentlessness of sand and time.

We stopped at Gyaur-Kala, a fourth-century BCE mud-brick Zoroastrian stronghold on top of a hill, melting back into the rock, and the Mizdahkan cemetery on the hilltop opposite. In Zoroastrian tradition, Gayomard – the first man and the father of humankind – was buried here, and his mausoleum is a clock of vast, geologic scale. The mud-brick walls of a ruined building rise eight to ten metres high, ragged at the top, the ceiling caved in, though the stacked brick arches that once supported it still stand high in each corner. Cracks run down its walls. This old ruin loses one brick a year, Sarsenbay told me, and in this way counts out time until the end of the world.

***

'The most critical issue is the dust,' Otabek had said, the unrelenting, salty dust. 'It lays down in all neighbouring countries: in western Kazakhstan, in Qaraqalpaqstan, in Turkmenistan; it also goes to the Caspian countries.'

It gets everywhere.

In 1975, Russian cosmonauts reported massive dust storms scouring the newly exposed seabed in the southeastern part of the Aral Sea. The water level had dropped five metres by that time, exposing sandy sediments that had been deposited over thousands of years. With no plant life to stabilise these sediments, this virgin surface was vulnerable to the wind. It blows hard in this part of the world, straight down from Siberian steppe, and under its relentless pressure a new landscape was created, an aeolian landscape: one shaped by the action of the air. The sand blows into crescent dunes called barkhans which wander, constantly shifting.[9] Smaller particles get blown higher up into the air, where they can travel for hundreds of kilometres. The more the sea shrank, the more land was exposed – and the worse the dust grew.

Somewhere between 43 million and 140 million tons of dust are blown off the Aral sands each year.[10] Winds of seventy kmph or more produce the worst dust storms on the northern Kazakh side: one monitoring station called Ujaly, fortunately uninhabited, saw dust storms for 84 days a year. In the south, Moynaq would be battered by eleven storms a year and the city of Nukus twenty.[11]

Once a month or more, people would have to race indoors as a towering tsunami of dirt would crash over the city, a cloud blocking out the sun and turning the air to choking, corrosive sandpaper.

The wind doesn't only pick up sand particles, of course, but everything which lies on the Aral seabed. Back when it was still a sea, 'The Aral contained an estimated 10 billion metric tons of dissolved salts,' geographer Peter Micklin reported.[12] The water was a toxic soup of not only the familiar sodium chloride sea salt but also sulphates and carbonates – the herbicides, pesticides and fertilisers that had run off water-logged cotton fields for decades, and gathered in the sump of the Aral to stew. When the sea dried out, all those toxins remained, as dust.

People call this salt dust the 'dry tears of the Aral'. The salt is blown by the wind and deposited by the rain, and it rises up from waterlogged ground to leave the soil crusted white, 'salinised', dead. It stops plants from bringing life to the new sands, and it blows south over the fields and gardens of the Amu Darya delta to poison the cotton and rice and fodder growing there, too. The salt harms the animals who eat these plants, and it makes power lines short out and catch fire.

Needless to say, it's not good for people either.

***

When the land gets poisoned, people do too.

Seventy per cent of women of maternal age living in this region in the 1970s and '80s were anaemic, and they faced three times the risk of dying in childbirth than women in the rest of the USSR. Their infants faced multiple assaults on their health: babies were born with congenital abnormalities, under threat from infectious and parasitic diseases such as typhoid and viral hepatitis, alongside chronic bronchitis and respiratory illnesses brought on by breathing dust day in and day out. Children grew up malnourished, anaemic, with rickets and stunted physical development: they lived in an oasis, but pasturelands and hayfields that used to feed cattle for meat and milk have been turned over to cotton production and the region could no longer feed itself. Protein deficiency weakened people's immune systems, and increased their vulnerability to water and airborne diseases.[13]

Over time, everything that has been sprayed onto the cotton fields ends up finding its way into people's bodies. And so much has been sprayed onto the cotton fields. Monocrop agriculture requires massive applications of weedkillers and pesticides, as these unbalanced ecosystems lack both natural resistance to disease and the insect life that helps keep diseases in check. Crop rotation was abandoned so that cotton could grow year on year to meet Moscow's ever-increasing targets – exhausting the soil, and requiring vast applications of fertilisers, ten or fifteen times the supposed USSR limit. And flood irrigation methods wash all these chemicals off the fields – meaning more must be applied to keep production up. DDT, butyphos, hexachloran, Lindane: the clever chemical inventions of twentieth-century agro-innovation are highly stable and persist long in the landscape (and the body), well after the Soviet regime had gone.

In adulthood this can produce cancer: of the liver, of the oesophagus and digestive system. Nephrolithiasis and kidney diseases are six times normal rates, as people's bodies try to filter out the pollutants they drink and breathe – but fail. Analysis of people's hair find high levels of heavy metals such as lead, cadmium and molybdenum, which suppress the body's ability to make blood cells, retard the development of the central nervous system and produce lesions of the kidneys, bones and skin.

The Soviet Union falls and the sea vanishes, but this does nothing to stop the dust, or the health crisis. Fully three in every four people in the region are sick. The 'white gold' of cotton had instead given the region the 'white death', and in the early 2000s the rate of tuberculosis in Moynaq is one of the highest in the world.[14] Médecins Sans Frontières move in, an organisation who normally only operate in warzones. People move out.

Sixty thousand jobs in the fishing industry were lost with the disappearing waters, as a 44,000-ton fishing catch in the 1950s collapsed to zero. Former fishing villages were abandoned, and Moynaq's population fell from thirty thousand to around ten thousand. But moving wasn't easy: it required an intra-national visa. Those with money or connections tried to bribe their way into getting one; Russians who'd moved to the region repatriated themselves back home. Many Qaraqalpaqs headed north and slipped quietly across the border into Kazakhstan, a culture linguistically and culturally more familiar than Uzbek Tashkent. Besides, Aralsk still had a sea and there might be a chance to continue the life they had known.

'I worry for the future of my country,' Sarsenbay told me, concerned that Qaraqalpaqstan was losing its identity. Young people wanted to leave, ideally to study in England; law and international relations the courses of choice, to open up government positions on their return back home. As a young man, he'd once hoped to study abroad himself, but someone better connected got the state permit instead. Instead, here he was in Nukus, reading old Qaraqalpaq books of stories and mythologies and showing travellers the mausoleums and gravesites of his ancestors.

Like everyone, he was glad that the new president was bringing investment and modernisation to the region, that Qaraqalpaqstan was no longer being left behind. Yet he did not seem to have faith this was for the good. 'There used to be a good life here', he told me, a life of fishing and hunting deer; the climate was softer, moderated by the waters, the summers less searing, the winds less vicious. Modernisation was, on the other hand, a mixed blessing. Mobile phones had arrived in Moynaq a few years before, and people were 'getting lazy', he thought: they'd just call or text, and would no longer visit each other in their homes.

Sarsenbay had never known the sea: it had been gone for a decade by the time he was born. 'Old people are sad,' he told me, for the way of life that had gone. But there was a melancholia to him, too. He had also lost something irreplaceable.

***

The landscape remains an extractive one – except now it's not water that's pumped, but gas.

We drive into Moynaq from the northwest, across what was once the sea. An off-road descent a hundred metres down the crumbling cliff face of the Ustyurt plateau, then two hours and perhaps one hundred kilometres across the cracked clay of the old seabed. It's a saxaul forest now: tough, salt-proof shrubs as far as the eye can see, alongside low domes of glossy-leafed, ground creepers. We get out of the car, and it is utterly silent. Seashells crunch underfoot, pristine, as though they were deposited by the tide that morning. I try to stop my breath, so I can hear nothing better.

In the distance is a gas plant, a vast new alien industrial steelscape dropped into the desert where the water used to be. The state oil and gas company Uzbekneftegaz provides 15 per cent of GDP and wants to become the next Shell or BP; Uzbekistan is the fourteenth largest gas producer in the world. Sixty-two billion cubic metres of gas were processed in 2014, most of it used domestically – but the ambition is to export. Up on the plateau we had crisscrossed newly laid pipelines hauling the gas out to Kazakhstan and Russia. Another gas plant stood a hundred kilometres from any sign or squiggle on the map – yet dozens or hundreds of people lived here, in the company village providing labour for the plant. Concrete slab apartments stood amid a few trees, a playground, a school. I tried to imagine being a child growing up in this de facto work camp on this flat, featureless plain. Camels roamed, one piece of the nomadic past surviving stubbornly on.

This is the new Silk Road, a northern, overland fork of China's 'Belt and Road Initiative'. As I write, the China National Petroleum Corporation has just signed a partnership to integrate Uzbekistan's transmission into the Central Asia pipeline that runs from Turkmenistan through to Xinjiang. Yet China must compete with Korea for access to this fossil wealth: the 'Uz-Kor' gas chemical complex on the plateau, a partnership with three Korean producers, is a $3.9 billion development seeking to capture more of the value-add from the hydrocarbon manufacturing chain. Instead of the natural gas being shipped abroad for processing, polyethylene and polypropylene plastics are now being manufactured in Uzbekistan. This is what modernisation means here today, as one environmental disaster gives way to another. It brings jobs to a region that badly needs them – but the money stays where the money always stays, back in the metropole.

I'm told to stop taking photographs: the infrastructure in this country is a matter of state security.

Meanwhile, word is that the president wants to redevelop Moynaq into a Central Asian Las Vegas. It seems hard to credit, but I keep asking the Tashkent contingent at the festival and they keep insisting it's true. It's difficult to imagine this dusty, ravaged place as a tourist attraction – and yet on a small scale it is one already. The attraction it sells is disaster.

Some years ago, ten fishing boats were transported from where they had beached as the waters receded, and arranged in a kind of park beneath the sail-shaped Aral Memorial. They are wholly rusted out now, engines long gone and sold for scrap. One has lost most of its hull and stands like the skeleton of a great whale, all metal ribs and memory. They are listed in all the guidebooks to Central Asia; they are why international tourists come to Moynaq, the only reason. The rusted metal glows copper and orange in the early evening light and makes for quite the photograph.

It is a strange playground. There's little in the way of signage to guide or instruct any particular kind of response. People climb on and into the boats, which would seem disrespectful except that stairs have been added to a couple of the boats for this very purpose. Festivalgoers sit up there for hours, drinking beers, talking about everything and nothing, as travellers do.

Most of all, everyone shoots selfies and posts them to Instagram, local people and travellers alike. The light is golden and very flattering – and it gets branded as narcissism. In 2019, *Calvert Journal* reported on the trend as 'another example of disaster tourism's self-indulgence'.[15] But I'm not so sure it is.

At that time, photography at Chernobyl itself was also under examination, as images of an influencer posing artfully in a hazmat suit and thong underwear circulated to considerable condemnation. 'It seemed wrong to reformat horror in an aesthetically pleasing way, then hope people engaged with it by tapping a Like button,' internet culture reporter Taylor Lorenz wrote, analysing the controversy. 'Part of the reason people get so angry at images like [these] . . . is because of a perceived gap between where attention should be aimed (the tragedy of Chernobyl) and what viewers interpret to be the focus of the photo (a person posing for the camera).'[16]

Yet putting oneself in the photograph is how pictures are taken these days. People take selfies at the Aral Sea memorial to show that they were there, and to say to others that this is a place of significance. It might be diluted as a method of bearing witness, but it is nonetheless a means of seeing, not turning away. The world has turned its back on the Aral Sea disaster: it's old news now, and the region doesn't trouble the international press with more than a handful of stories a year. If a self-portrait on a friend's Instagram feed is how people get to learn about what happened at this place, I see little to fault with that.

***

At the yurt camp on the Aral Sea, I get talking to another guide, a man in his mid-twenties named Ahmed. Born and raised in Moynaq, Ahmed works part of the year for Médecins Sans Frontières as an assistant – using guiding for an income boost during the European holiday season in the summer. What's MSF doing these days, I ask, curious. The TB programme of twenty years ago has come to an end, he explains, and now theirs is more of a backstage role, supporting the Ministry of Health with training and monitoring. He tells a hair-raising story about local standards of medical waste disposal.

The next morning, at breakfast, I ask him what people in Moynaq thought of the Stihia festival. His answer is careful, measured. Some people were not happy about it, he says. They did not like the focus being put on the bad things that had happened in this place, and they were not sure it was right to be 'cheerful' in this way. They feared the festival could be 'making fun' of the Aral Sea disaster: it seemed contradictory, even mocking to be having a party at the memorial for a catastrophe.

Other people, Ahmed adds, think it's more positive that there are 'festivities' in this place. They like that it's drawing attention and visitors to the region. He talks about another festival that happened in Moynaq some years before, a food festival featuring ninety-nine different fish dishes, celebrating the heritage and culture of the region. He and Sarsenbay both say they preferred this kind of event.

Lounging on this *tapchan* overlooking the Aral Sea, I sit a little with the discomfort of being here for an event that doesn't necessarily sit well with locals. Journalism might justify this in a way that tourism cannot – whatever the line is between the two that I'm doing.

But of course, techno is alien to a place like this – Otabek, the festival organiser, might argue that is actually something of the point. It seeks to be the sound of the future, and the new is always strange.

Besides, everyone here does want modernisation for their region. The open question is the form that modernity should take.

On Saturday afternoon I get a new roommate: Mama Snake, a DJ from Copenhagen. We spend the afternoon talking nineteen to the dozen. We're the same age, with similar politics and priorities, both of us steering a line between professional and creative lives (she is also a doctor, alongside being a DJ).* Beers in hand,

* Her pseudonym gives her the space to live these two lives, so it's the name I'll use for her here, too.

walking around the festival site as the sun slips into the sand, we mourn what feels like a decline in club culture in London and beyond. Venues closing, licencing cracking down. A generation's cultural horizons shrinking to Uber and Netflix and Deliveroo: services that cocoon us from the city and from each other. None of the twenty-somethings I used to work with seem to know or care about the things a dance floor can do. The magic it can work.

We're being a pair of early thirties Rave Grandmas: 'Kids today don't know their history! They don't know what the Summer of Love meant!' We talk of community and resistance and liberation. Mama Snake says she now tends to prefer playing in developing countries, places with new and emergent electronic music scenes, places where what she does isn't just another Saturday night but seems to really mean something to the people who're there. She's come in from another event in Azerbaijan.

And then it's time for her set. I am centre front on the dance floor, the better to immerse myself in the crowd. The energy is a little wild in the way of something young and coltish: everyone here is, in one way or another, testing the edges of what they understand to be possible. A thousand local young people who may never have seen or heard anything like this before. A nascent techno scene testing the boundaries of what dance music is allowed to do in Uzbekistan. A country where this festival could not have happened even three years previously. Can girls wear bikini tops and dance with foreigners and drink beer in public? Might this be OK for the first time?

It's mostly young men here at the front of the dance floor, teens and twenties, a young masculine energy only moments away from becoming a mosh pit. Teenage girls stand further back in groups. Everybody takes selfies on big new Chinese Android phones with wide-angle lenses that can fit three and four people into the frame. They're keen to invite the stranger into these shots; I oblige. Outside, at the edge of the dance floor light, navy blue-uniformed police stand implacable, a line of dominion merging into the desert night.

And now there is this extraordinary sound ringing out, resonant as a gong, into the Aral night. Mama Snake has just taken over the decks, an industrial blonde in a black T-shirt dress and combat boots, standing on a low stage riser in front of the crowd. My god, what is this? What she is playing is immense. Endless organ chords sustain under an ululating woman's voice, drenched in reverb, a track I half-recognise yet cannot name, as much Arabic as Western, as if the soundtrack to an as-yet-unwritten sequel of *Bladerunner* or *Dune*. It is strange and audacious and only keeps getting bigger. Mama Snake is teasing the crowd, playing with us and leaving us hanging for minutes, the whole track an eternity.

Me and the Tashkent techno-head next to me are unable to believe what we're hearing on this incredible sound system, in this extraordinary place. We just look at each other and mouth 'Oh my god, oh my god, oh my god!' Up on the decks in front of us, Mama Snake is wearing a grin as wide as her face.

And then – finally – she drops the beat and the whole place detonates.

After her set is over, I go backstage and babble incoherently with delight at her and Otabek – still somehow crisp in his preppy blue Tommy Hilfiger shirt – as DJ Sodeyama begins to wind down the night.

The final track played is a special one, a gift from Eduard Artemiev, father of twentieth-century electronic music in Russia and composer of the soundtracks to many of Tarkovsky's movies, including the haunting and apt *Stalker* (1979). A film set in 'The Zone,' a landscape of industrial ruin and Soviet over-reach – and a place where an event has occurred, where nothing has been quite the same since. A landscape that remains despoiled – though, in places, nature is beginning to reclaim it.

The music stops, and we mill about, joyous and delighted, on the road, underneath a sky glittering with countless stars. Billy passes around a bottle of three-dollar vodka and I drink deep from it, the liquid spilling down my chin and evaporating cool in the night air.

Then a shout goes up and we rush towards it. On the side of the road, big ten-feet wooden letters spelling S E A have been lit on fire. They go up fast in flames that dance high, fractal and evanescent, vanishing out into darkness as the plywood quickly burns down, leaving the 2x4s scaffolding the structure silhouetted.

It doesn't seem to mean anything as such. It's pure spectacle. But this doesn't seem to matter. We stand together, faces illuminated in the heat, and the sparks, and the light.

Sunday lunchtime I hitched a ride with the DJs back to Nukus. The driver is hours late. 'Moynaq!' we say, as if in explanation. Nothing quite works as planned here. Not yet.

Phone calls are made to Otabek, who makes more calls, and eventually a car turns up, a big 4WD SUV to handle the battered old Soviet road back to the city. It's a bumpy ride. We hang on to the grab handles as the car swerves round crater-size potholes and sometimes veers off-road entirely. And then a little way outside the town we see the first one: a dust devil, and then another, and then the biggest yet, a swirling column of sand out in the saxaul scrub roaming and twisting a kilometre or more into the sky before diffusing into haze. A small herd of brown cows pass in front of it, their hooves kicking up more dust from this shattered soil.

What a sign, what a symbol for my departure from this place. There is little in nature more uncanny, I think, and then correct myself. This isn't nature. These dust devils happen because of us humans. They are chimeral; they are monsters.

The Greek tragedian Sophocles was the first to describe dust devils in writing, in his play *Antigone*, in around 450 BCE:

'And so the hours dragged by until the sun stood dead above our heads, a huge white ball in the noon sky, beating, blazing down, and then it happened—suddenly, a whirlwind! Twisting a great duststorm up from the earth, a black plague of the heavens filling the plain, ripping the leaves off every tree in sight, choking the air and sky. We squinted hard and took our whipping from the gods.'[17]

How long does Moynaq have to keep getting beat?

'The biggest source of damage is the dust,' Otabek tells me. 'It lays down in all neighbouring countries: in western Kazakhstan, in Qaraqalpaqstan, in Turkmenistan; it also goes to the Caspian countries.

'What the government is saying now is, "Okay guys, we have this problem, let's not talk about why it happened. Let's talk about what we should do in order to at least mitigate or minimise the damage that it is causing now." That's why the United Nations has, by the initiative of our president,' – ever the political operator, of course Otabek makes sure to give credit to the president – 'created a special fund called Human Security Fund for the Aral Sea, and the fund is spending lots of money to mitigate that risk. And one of the best ways – and probably the only way to do that – is afforestation.'

Scientists agree: plants are the only thing that can stop the dust.[18] The roots of one saxaul tree can fix 10.6 cubic metres of soil, reducing wind erosion, stabilising sand dunes and, thereby, improving air quality.[19] These, and other desert plant species like saltwort, another halophyte able to grow in these harsh conditions, offer the best chances of greening the desert.

My Aralkum guide Sarsenbay told me that the Soviets were the first to start foresting the Aral, dropping seeds from the sky by aeroplane to regreen the infant desert. I can find nothing about this in the historical and scientific literature I consult, and it seems difficult to square with the state's decades of disavowal. But perhaps there was an opening in the late 1980s, in the few short years when *glasnost* thawed this official denial. Airplane afforestation would help explain how the seabed we drove over in the southwest, from Ustyurt Plateau to Moynaq, was green all the way, sometimes sparsely, with metres or more between bushes, but sometimes too dense to walk through, as around the gasworks that had arrived. Yet I also knew that airplane planting is usually ineffective, indiscriminate as it is as to where the seeds are laid and giving them no help to survive in such barren ground as this. Nonetheless, whatever the origins, it was good to see life returning to the sand.

Now, organisations such as the International Fund for Saving the Aral Sea, with support of global bodies such as the Food and Agriculture Organization of the United Nations and the World Bank, have set up a process to expand this afforestation process at scale.

Afforestation today is a part-manual and part-mechanised process – the jobs it creates in an impoverished region a significant part of its benefit. Women gather seeds from existing saxual plants, which are grown in furrowed rows in dedicated nurseries until they're a year old and tough enough to be planted on. Out on the bare salt marsh and sand, tractors pull special ploughs fitted with two low seats beside each furrow, and a man sits on each, placing the young shrubs into the groove the plough has cut. On *barchan* sand dunes, reeds are cut from reedbeds on reservoirs and the Amu delta, bundled up, and lain in grids to stabilise the wandering sands. The sand piles up against the saxaul trees too, forming drifts a quarter of their height.[20] In this way, as the shrubs come to maturity over about a decade, windblown dust transport is, gradually, reduced.

By 2012, a tenth of the dry Aral seabed had been afforested – 400,000 hectares – absorbing an estimated 4.6 million tons of carbon dioxide a year. Uzbek agricultural scientists report that six-year-old plantations reduce wind speed by 90 per cent, and salt and dust transfer is stopped almost entirely.[21] And because the plants used are so local, at least one in two is managing to stay alive.

'It's quite stunning that in a matter of 50 years, what used to be a marine ecosystem is going to turn into a forest ecosystem,' says Benoît Bosquet, director of the World Bank's Environment and Natural Resources Global Practice.[22]

He sounds a warning note: change this fast needs careful planning and scrutiny. The Uzbek government and international funders need to consider how this environment can meet the needs of the local population: how can this landscape give them the jobs and income they lost when fishing went? Are there opportunities for wind turbines and green energy schemes, not just gas? And can this saxaul landscape become something more diverse, extending to food, forage and useful forest products once the first wave of salt-loving halophytes have started to improve the soil?

'Monoculture is rarely, if ever, a longterm solution,' he reminds us. Still, he believes it's justified: 'There are very few alternatives to stabilizing the soil of the Aral Sea and prevent those destructive storms.'

Otabek, of course, had a plan: 'We are bringing 500 seeds of the Joshua tree from the Mojave Desert,' he told me in 2019. It seems I am not the only person to see the parallels between what has happened here and California's own saline, evaporating Salton Sea, out on the far hydrological periphery of Los Angeles. The Joshua trees live just next door. Why Joshua trees? 'That was my idea. I was mesmerised when I saw it for the first time on U2's album, *Joshua Tree*. And then last year we had a road trip and we went to California, Nevada, Arizona, so I saw it by my own eyes and I was really impressed.' He waxes lyrical on the poeticism of these 'bizarre' trees emerging from the total desert environment.

'Most interestingly, climate conditions in Mojave Desert and Aralkum are almost identical,' he told me. 'The soil quality, the air temperature, the humidity: it has almost identical conditions and we have all prerequisites to grow the trees there.'

He wanted to test as to whether the Joshua tree will actually survive in this part of the world, saying he'd already contacted USAID and the US embassy to set plans in motion, and hoped to be able to report back next year.

'Will this really happen, this strange, psychic link between one dying sea and another?', I wondered at the time. Otabek may be a visionary but he's not necessarily a horticulturalist: Joshua trees are slow growers, averaging just three inches a year, and the salt of the Aralkum is unlikely to do them good. I perhaps lacked his faith in techno's ability to change the world as well, for all I love to dance to it. But Otabek himself, on the other hand, I couldn't rule out the possibility he might change that landscape nonetheless. The man had ability as a catalyst and a cultural entrepreneur: for all the dreamy talk, if he could pull off the political and logistical feat that was Stihia, he was evidently fairly capable.

After it was all over, we said our goodbyes on the steps of the Nukus Museum of Art, home to a world-renowned collection of dissident Russian art. Here, in the 1950s and 1960s, a man named Igor Savitsky was able to collect and display avant-garde paintings that had been banned in the Soviet Union because the location was so remote, no one really knew what he was up to.

I spoke for a few minutes with another of the Stihia organisers I'd not had a chance to talk to before. A lean, angular man, aged around thirty, with short-cropped hair, Aleksandr wore a kind of bitter, moral anger that contrasted with Otabek's relentless entre-preneurial optimism.

Aleksandr was getting treatment at a hospital in Tashkent where the kidney ward was full of people from Qaraqalpaqstan because they were still being poisoned by the salt. He showed me photographs he'd taken in Moynaq, of houses where people were still getting their water from wells, straight up from the ground, kids still drinking from toxic groundwater saturated by a hundred years of pesticides and fertilisers. All this talk of development and modernisation, all the pomp of the Presidential visit – and yet what was really changing?

'My interest in Stihia is not music or techno, or doing a festi-val,' he told me. 'From the landscape I've seen, and knowing the story of the Aral Sea – as well as knowing the current state of things and the first 27 years of independence: government is abusing this catastrophe to beg for money from different funds as much as they can. Money is disappearing in Tashkent with high-rank officials.

'Even though it's not gonna help bring water back, to me it was worth doing Stihia for people who live there. They've been abandoned and forgotten for many decades, striving to live in these conditions they have to endure. And to me, it's not just to bring music and do a party there, at the city on the last point of the road, at the dead end. It's at least to give them the feeling that they are not left alone in this salty pesticide sand pit.'

What future is there for Moynaq and the humanmade desert where a sea used to be?

In summer 2022, Uzbek President Shavkat Mirziyoyev announced constitutional amendments that would strip Qaraqalpaqstan of its status as a sovereign republic and remove its right to secede from Uzbekistan via referendum. Tens of thousands of people came out onto the streets, in Nukus and across the region, in largely peaceful protests to which the state responded with grenades, gunfire and violent suppression. Hundreds of people were injured with many seriously maimed, and at least 21 people were killed – while over 300 people were arrested and 91 were reported 'missing', their fates still unknown. Activist Polat Shamshetov was one of 22 people imprisoned for their role in the protests. In February 2023, he died in custody. His friends suspect he had been tortured.[23]

The same government that silences Qaraqalpaq dissidence with one hand emphasises its commitment towards the socio-economic development of the region with the other. Since 2017, the state has invested several hundred million US dollars in developing jobs, industrial plants and transport infrastructure in the region. A Saudi energy firm has won a contract to build a wind farm with 25 turbines, with two more facilities planned to launch in the next five years. The state also proclaims how 'Active work is underway to protect and restore the ecological system of the Aral Sea region,' with more work 'greening the bottom of the dried-up Aral Sea' following the 1.7 million hectares that have already been planted. They talk of introducing 'electronic cartography' and creating a database of endangered flora and fauna, plus a genetic bank of plant seeds. There's talk of an 'Academy of Young Scientists', a 'Youth Technopark' – even an 'International Innovation Center of the Aral Sea Region'.[24] It sounds quite good if taken at face value, which one almost certainly should not.

But there will not be a Stihia 2023 in the little town by the vanished sea.

In February 2023, Otabek Suleimanov announced that the festival that year would be relocating to Lake Tudakul near Bukhara. Holding a 'festive event' like a rave would clearly not be appropriate or respectful at this time, he told Eurasianet news. 'The festival's mission remains the same – boost sustainable development in Moynaq. We were, are and will be committed to this mission,' Suleimanov said. He promises to return.[25]

When the water is gone, the hot sand glimmers like a mirage and it's hard to see what's really going on. Which is to say I write through a haze of uncertainty as to what any of this might come to mean. Much is of course my ignorance as an outsider, a mere brief visitor without the language and relationships and time to report this out in the way I wish I might. But wish as I may, the state would never have allowed such reporting. In 2022, the one international journalist in the country, Agnieszka Pikulicka-Wilczewska, was thrown out of Tashkent by the state authorities in retribution for her politically inconvenient reporting. (I'd started following her after Stihia: she'd also been in attendance.) Similarly, investigative reporting on environmental justice matters would almost certainly meet a similar response. So is the reforestation real? What's the dust really doing to people's health? These remain unanswered questions.

Elsewhere in this book, dust proves a method to see into concepts and processes that seem too vast to comprehend. Here in Qaraqalpaqstan, it might delineate instead a space of lasting uncertainty.

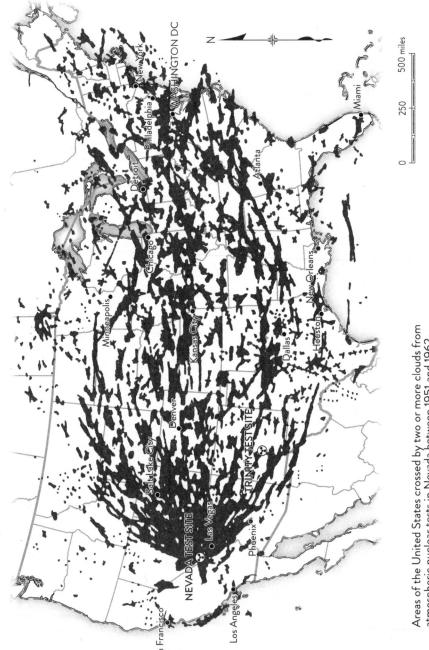

Areas of the United States crossed by two or more clouds from atmospheric nuclear tests in Nevada between 1951 and 1962. Source: Richard Miller, *Under the Cloud: The Decades of Nuclear Testing* (Two-Sixty Press, 1999)

Chapter 6

Fallout

At 45 seconds after 05:29 a.m. on 16 July 1945, a new sun ruptures up through the New Mexico atmosphere, birthing a new geological era – and a new geopolitical order in its wake. An electrical circuit clicks shut and thirty-two detonators fire exactly at once. A shell of chemical explosives ignites and explodes with shattering force – smashing into a curved layer of a second, slower explosive, Baratol, which detonates in turn. The geometry of the operation is precisely honed: the two blasts interact, unite and turn perversely inward, merging into a sphere of accelerating intensity, an explosion collapsing in on itself, plummeting towards the heart of the bomb. The shockwave passes through a dense natural uranium tamper plug, squeezing, compressing, heating, liquefying; it passes through the yet-heavier plutonium core, a sphere just under ten centimetres in diameter, and then it ignites the small, golf ball-sized initiator at the very centre of the gadget.

In the initiator, radioactive polonium-210 spits out super-energised alpha particles that punch neutrons free from neighbouring beryllium. Thus liberated, these neutrons stream out into the core and crash into giant, unstable plutonium-239 nuclei. The centre cannot hold. The plutonium fissions, each atom splitting into two or three lighter elements (zirconium and xenon; strontium and caesium); a blast of gamma radiation – and two or three more neutrons, which career onwards in a chain reaction of phenomenal pace. Eighty generations flash by in a few millionths of a second. The plasma ball reaches tens of millions of degrees, a temperature never before seen on the face of this planet.[1]

In his Pulitzer-winning 1986 book, *The Making of the Atomic Bomb*, Richard Rhodes describes the explosion, in its first milliseconds, as an eye. The moment of implosion sees 'the small sphere shrinking, collapsing into itself, becoming an eyeball', one that 'briefly resembled the state of the universe moments after its first primordial explosion'. It's an oddly bodily metaphor for something so entirely beyond human perception – as the explosion continues, the nuclear weapon releases a 'double flash of light, the flashes too closely spaced to distinguish it with the eye'. But that proves to be his point.

'Further cooling renders the front transparent;' he continues, and 'the world if it still has eyes to see looks through the shock wave into the hotter interior of the fireball.' But even the machines could see it all: the brightness of the ball of fire went unmeasured; the temperature of the radiating surface in the first few millionths of a second unrecordable. The director's report on the test can only say that it was approximately a sun, a humanmade sun conjured on Earth, a device not wholly fathomable in its grandeur or its power.[2]

The scientists in their observation bunkers 10,000 yards from the blast, and the VIPs at Compania Hill, twenty miles northwest, had been instructed to look away, yet no one complied, the physicist Edward Teller reported. 'We were determined to look the beast in the eye.' He put on a pair of welding goggles, 'then looked straight at the aim point.'[3]

'Suddenly, there was an enormous flash of light, the brightest light I have ever seen or that I think anyone has ever seen,' the physicist Isidor Rabi recalled.[4] 'It blasted; it pounced; it bored its way right through you. It was a vision which was seen with more than the eye.' The first 0.1 second of the explosion could be seen clearly only by something inhuman and unmoved: the cameras monitoring the blast show a perfectly symmetrical dome expanding outward and up, a ruffle of dust and flame rising at its base. A moment lasting barely two seconds, but which those watching described as an eternity.

Once its temperature fell below 5,000°C, the fireball could cool no more where it stood. It dimmed and became 'something more like a huge oil fire', the physicist Otto Frisch recounted, 'slowly rising into the sky from the ground, with which it remained connected by a lengthening grey stem of swirling dust'.[5] It was surrounded by a spectral blue glow of ionised air, and Frisch was reminded of his friend and colleague Harry Daghlian, who had dropped a piece of uranium in the lab during testing and who, as his assembly went critical, would have seen this glow, one of the first people ever to do so. He died of acute radiation sickness mere days later.

The fireball rolled, flashing yellow and scarlet and green. The Oscura mountains to the east were lit 'golden, purple, violet, gray and blue'.[6] As the cloud rose, it formed the now-familiar mushroom shape, stretching up, up, into the atmosphere to a height of perhaps 12,000 metres.

'It was an awesome spectacle;' Frisch wrote in his 1979 memoir, and 'anybody who has ever seen an atomic explosion will never forget it. And all in complete silence; the bang came minutes later, quite loud though I had plugged my ears, and followed by a long rumble like heavy traffic very far away. I can still hear it.'

'Naturally, we were very jubilant over the outcome of the experiment,' Rabi continued. Bainbridge went round the southern observation bunker congratulating everyone for the success of the implosion bomb design, which the scientists hadn't been wholly certain would work. Oppenheimer arrived back at Base Camp, a lean figure in a wide-brimmed western hat.[7] 'His walk was like *High Noon* – I think it's the best I could describe it – this kind of strut,' Rabi said. 'He'd done it.'[8]

'Our first feeling was one of elation, then we realised we were tired, and then we were worried,' the Austrian–American physicist Victor 'Viki' Weisskopf recalled.[9]

Another, delayed shockwave came over the scientists, and they grew silent in the dawning apprehension of what they had achieved. Rabi reflected that: 'A new thing had just been born; a new control; a new understanding of man, which man had acquired over nature.' It made him pause. 'I thought of my wooden house in Cambridge, and my laboratory in New York, and of the millions of people living around there, and this power of nature which we had first understood it to be – well, there it was.' He ran out of words.

Oppenheimer remembered a line from the *Bhagavad Gita*, which he had studied at Yale: 'Now I am become Death, the destroyer of worlds.' He thought of Prometheus, too, he says, 'of that deep sense of guilt in man's new powers, that reflects his recognition of evil'.[10]

The horror might be rationalised. The nation was at war and they were a part of the war effort, and the line of scripture Oppenheimer quoted was from a moment where Vishnu has sought to persuade the Prince that he must only do his duty. Or perhaps the fission reaction they had for the first time set in motion was not so much something *made* as something *found*, a terrible potential lying imminent within the heart of the atom. The discovery might have been new, Oppenheimer thought, 'but even more we know that novelty itself was a very old thing in human life, that all our ways are rooted in it'.[11] The bomb was perhaps inevitable to modernity: it could not but be discovered. If they had not found it, then the Russians would have. But the fact remains, it is they who have done it. The responsibility was crushing.

'Now we are all sons of bitches,' Bainbridge told Oppenheimer.[12]

***

It was codenamed the Trinity Test, though the reason why seems to have been lost.[13] The name was the brainchild of J. Robert

Oppenheimer, the director of the Manhattan Project, in reference to the writings of metaphysical poet John Donne – but as to which poem Oppenheimer had in mind, sources grow unclear. It may have been 'A Litany' ('O blessed glorious Trinity / Bones to philosophy, but milk to faith'); it may have been 'Holy Sonnet no. 14' ('Batter my heart, three-person'd God'). But perhaps it doesn't matter – both are poems of obliteration, spoken by a despairing subject 'now grown ruinous', a man of naught but 'condensèd dust' who begs for divine violence as redemption: 'Your force to break, blow, burn, and make me new'. We know Oppenheimer was familiar with Donne's writing via his lover Jean Tatlock, who had killed herself the year before. When pressed, though, he claimed not to recall where the name came from at all. Perhaps more than any other figure in the bomb's development, Oppenheimer knew what he had unleashed on the world. Perhaps he wanted to forget.

The Trinity Test in July 1945 was the first humanmade nuclear explosion. It would be followed not even three weeks later by the acts of war for which it was the prototype: the US bombing of Hiroshima and Nagasaki in Japan, which killed approximately 200,000 people, most civilians. The bombing ended the Second World War and ushered in the Cold War arms race, as the Soviet Union raced to match America's newly infernal destructive power.

I have chosen not to write at length about the Hiroshima and Nagasaki bombings: my diffuse and tiny lens on the world can offer no insight. Hiroshima was simply annihilated, its buildings razed flat, those at the hypocenter vaporised. A history professor remembered looking down on the city from Hikiyama Hill: 'Hiroshima didn't exist—that was mainly what I saw—Hiroshima just didn't exist.'[14]

People died from flash burns caused by the blast, and from the firestorm that engulfed each city; they died crushed by falling masonry and lacerated by broken glass. Many thousands would have received a lethal dose of gamma radiation, but most succumbed to other injuries before radiation sickness could claim them. Both bombs had been designed to detonate in the air, 500 to 600 metres above the cities, a decision which maximised the physical devastation while limiting residual radioactivity.[15] A black rain fell on the land, but the long-term contamination was less catastrophic than might perhaps be expected. Southern Japan was not left an uninhabitable wasteland. Cancer rates rose, although only by 9 per cent.[16] So dust – that is, radioactive particles – is not the main story there. I write instead of the nuclear tests that followed, where it is.

There have been a thousand times more nuclear blasts during putative peacetime than have ever occurred in war. Since 1945, a total of 2,056 nuclear weapons have been detonated, their fallout embedding a bright line of radioactive isotopes into the planet's geological strata, one marker among many of the beginnings of the modern age.

Eight countries have conducted tests: primarily the United States (1,054) and the Soviet Union (715), but also France (210), the United Kingdom (45), China (45) and India, Pakistan and North Korea (6 each). South Africa and Israel have built nuclear weapons capabilities but have never officially tested them. An additional suspected nuclear test – the 'Vela Incident', observed by a satellite over the Indian Ocean on 22 September 1979 – occurred without proper attribution to a country or nation-state.

Most of the first detonations were atmospheric tests, taking place above ground. Bombs were detonated from atop 200-metre towers, lifted with balloons, floated on barges and dropped from planes, in airburst explosions designed to maximise wide-range destruction by reflecting the shockwave off the ground, thereby doubling it. A handful of tests between 1958 and 1962 took place at much higher altitudes, even in space, in order to generate a powerful electromagnetic pulse that could wipe out electronics for hundreds of kilometres. Underwater tests measured the potential of nuclear bombs for naval defence. In October 1961, the Soviet Union tested the *Tsar Bomba*, a hydrogen bomb with a yield of 50 megatons (that is, an incomprehensible 3,850 Hiroshimas). Its fireball alone covered an area of over sixteen square kilometres.

In 1963, the Partial Test Ban Treaty was introduced, banning atmospheric testing due to rising concerns about radioactive fallout. So testing went underground. America's Project Plowshare (1961 to 1977) explored the use of nuclear explosions for civil engineering purposes: nuclear explosions might widen the Panama Canal, excavate water storage and transport infrastructure or cut a pass through mountains in the Mojave Desert to accommodate a new highway and railway line.

In Russia, Nuclear Explosions for the National Economy operated along similar lines, seeking to use the shockwaves from nuclear explosions to seismically identify oil and gas fields, stimulate their production and extinguish gas blowouts that had gotten out of control. As the public appetite for nuclear modernity started to sour – a generation of schoolchildren having been taught to duck and cover and rehearse their imminent destruction – the Atomic Energy Commission sought to 'highlight the peaceful applications of nuclear explosive devices'[17] and the merits of the 'friendly atom'.

Atmospheric nuclear testing will kill perhaps 2.4 million people, International Physicians for the Prevention of Nuclear War estimates.[18] This is ten times the immediate toll of Hiroshima and Nagasaki, yet it's a death toll that has gone substantially unnoticed because radioactive dust is such a diffuse and delayed killer. Half or more of these deaths haven't even happened yet, forty years after above-ground testing was banned – but the particles produced by these explosions haunt our world still, and slowly and randomly exert their consequences.

Barring a relatively small number of people directly exposed to massive radiation doses by their close proximity to the blasts, each individual killed by nuclear testing would never know they were a victim: whether their thyroid cancer or leukaemia was sparked by fallout or by the many other sources of radiation we're exposed throughout our lives, from medical X-rays to natural radon gas and cosmic rays. In this way, the deadliness of radioactive dust is easy to deny – as authorities have done for decades.

***

America's approach to nuclear testing was riven with contradiction, simultaneously both a massive public spectacle and an exercise in the deepest denial. It's as if the US government wanted the mushroom cloud without the fallout: all of the bomb's colossal geopolitical power and none of the actual, material consequences. It wasn't even a cover-up: if you never systematically measure nuclear radiation across the nation, you don't have to hide any records of it.

A cultural icon like the atom bomb does not just arrive fully formed in the world, however. It's actively and deliberately made through choices of language, image and public communications, as government and media teach the public to interpret what exactly it is they are looking at. Cultural historian Peter B. Hales writes of how magazines like *Time*, *Life* and *Newsweek* gave the nation accounts of the Japanese bombings and subsequent nuclear tests in terms that were 'profoundly aesthetic, rather than ethical, moral or religious in tone'.

The military, and later the Atomic Energy Commission, strictly limited access to the test sites, and the information and images that journalists were able to access. Few of the photographs released from Hiroshima and Nagasaki show the ground, the city, the target or the destruction: the human consequences and the lives lost are not rendered visible. Instead, the atom bomb is typically shown as a cloud of smoke, seen from a great distance, looking 'considerably more like a giant storm than a human event'.[19] It is reduced to a kind of 'abstract visuality', Hales writes, and depicted as more a 'marvel of nature' than a matter of calculated death and annihilation. Through this sleight of hand, he argues, 'the question of responsibility for the effects of the explosion remained slippery'. In this way, the Allies could believe that nuking two cities of civilians was not a war crime.

By thinking with dust in this chapter, then, I want to see if we might see what goes unseen: how nuclear tests were simultaneously both a public spectacle and a cover up. The mushroom cloud was celebrated, even turned into a perverse tourist attraction, while the radioactive fallout that blew northeast over following days, dispersed and invisible, was ignored and denied.

Rancher William Wrye lived twenty miles northeast of the Trinity Test site. 'For four or five days after that, a white substance like flour settled on everything,' he told local journalist Fritz Thompson. 'It got on the posts of the corral and you couldn't see it real well in the daylight, but at night it would glow.'[20]

The contradiction was there right from the start. All the accounts of the Trinity Test emphasise its visual awe and, as, one viewer put it, the spectacle of 'an enormous flash of light, the brightest light I have ever seen or that I think anyone has ever seen' – yet at the same time the critical instant was fundamentally unseeable, operating at electromagnetic wavelengths and energy levels far beyond human capabilities to endure.

Richard Rhodes nods at this in his account with a phrase that has haunted me ever since: 'the world if it still has eyes to see'. Because it doesn't.

Both metaphorically and literally, something in the apocalyptic spectacle of the nuclear blast takes people's eyes. The Trinity Test witnesses screwed their eyes tight shut else they would be blinded. The Sheahan family at nearby Groom Mine reported seeing wild horses in the aftermath, their eyes burnt clear out their skulls.[21]

We see the nuclear testing era as if through a glass darkly, then – a glass fogged with Cold War myth, propaganda and denial. We are dazzled by the image of nuclear war, in order to obscure the realities of such a thing. To better see the cloud cast by the nuclear testing era, then, we might try seeing *with* the cloud – that is, using dust as a lens to bring into view the people, places and consequences that are otherwise overlooked. The metaphor's a bit of a paradox, but perhaps that might help us think through these contradictions more clearly. Bring what's hidden to light.

Dust helps us see beyond the point where the plasma cools and the mushroom cloud diffuses into the atmosphere, letting us follow the material traces of the explosion as far as the winds carry them: thousands of miles north and east, right across America, into the bones of a nation and the sediments of our planet. In tracing its origins and consequences, we are prompted to think through the full lifecycle of nuclear materials, from uranium mining and refining to the indefinite challenges of storing nuclear waste. Dust widens our time horizons as we think through nuclear decay, half-life after half-life, risk tending to zero but never reaching it. It helps us understand radiation as both a deterministic and a random, stochastic risk. Most importantly, thinking with dust is as always political. It carries us beyond national borders to see the commonalities between nuclear tests worldwide, in their carelessness, the populations they deem expendable and the colonial logics used to decide who that is. It turns our attention towards the question of who has to deal what's discarded: that is, who has to clean up nuclear test sites, and who must live with nuclear dust in the long term.

It might help us see the full picture, not just the glossy *Life* magazine photographs.

Focus on the bomb and you miss the actual disaster.[22]

***

The Trinity Test took place in New Mexico, about two hours south of Albuquerque. The site was close enough to the Los Alamos nuclear weapons laboratory where the bomb had been developed to be convenient – and otherwise as empty and remote as physicist Kenneth T. Bainbridge, in charge of location-scouting, could find. The Jornada del Muerto Desert basin is a flat, dry landscape of creosote and yucca stretching between the Sierra Oscura to the east and the narrow green line of the Rio

Grande to the west. It was originally Mescalero Apache country which was then settled by ranchers – who were later evicted themselves, in 1942, when the United States Army took over the land to use as a bombing range. (Three years later, the main bedroom of the McDonald's ranch house would be used as the assembly site for the nuclear device, its windows taped up to keep dust out and make an impromptu 'clean room'.) The site was remote enough that the authorities could cover up the fact it was a nuclear bomb test at all, the press told only that there had been an explosion at a weapons depot. Plans were made to evacuate nearby towns if the wind turned, on grounds that 'chemical tanks' had been hit.

In practice, nobody was any the wiser. The Trinity bomb had been detonated from only thirty metres above the ground, which limited the blast field but maximised radiation because the explosion uplifted huge amounts of newly radioactive desert sand. Relatively little fallout dropped on the test site itself: instead it rose in the dense white mushroom cloud, five miles high, which dispersed in the wind in about an hour, dropping radioactive fission products all the while. Chief medical officer Stafford Warren reported that 'while no house area investigated received a dangerous amount, the dust outfall from the various portions of the cloud was potentially a very dangerous hazard over a band almost 30 miles wide extending almost 90 miles northeast of the site'.[23] The Japanese bombings just three weeks later would be airdropped, producing the opposite effect: more blast, less radiation. This is how Nagasaki and Hiroshima are habitable today.

After the Second World War, the United States conducted five nuclear tests in the Marshall Islands in the central Pacific in 1946 and 1948 – then realised they would need a different site. The Atomic Energy Commission's concern was not for the people of the Marshall Islands – who would be subject to several more tests and a lifetime of consequences – but only logistics: these remote atolls were difficult for the Americans to reach. They were often threatened by storms and typhoons which might postpone the tests, or cloudy and overcast, preventing photography. Besides, out in the middle of the Pacific, it was difficult to control who else might be watching. They needed another site, and in 1950 the nuclear scientists came together again at Los Alamos to choose another sacrificial landscape.

They decided upon the Nellis Air Force Gunnery and Bombing Range in Nevada. The clear, dry, desert landscape offered similar practical advantages to the New Mexico site, but with one catch: the site was only 65 miles northwest of Las Vegas. For the next forty years, tests would only be conducted when the wind was blowing to the north and east, so that the city – and the tens of millions of people living in California, to the west – would be safe from the inevitable dusting of nuclear fallout. The people to the east and northeast, on farms and ranches and in scattered rural communities, would not be so lucky. 'Those were just people in little Utah ... where's that?' Downwinder campaigner Mary Dickson later commented to an oral history interviewer, her voice weighted with sarcasm. 'They were Mormons and cowboys and Indians, who cares?'[24]

The Nevada Test Site lies on Western Shoshone land; to the east, where the wind blows, is Southern Paiute territory. The Shoshones' 1863 peace treaty never ceded ownership of their land to the United States, only rights of access and use – for railway and the telegraph, mining and ranching, and the establishment of military posts and travellers' stations. White American prospectors arrived within the year, searching for silver and, in 1889, Patrick and Avis Sheahan settled at remote Groom Mine, three days' travel from Indian Springs, and set up a modest family-run operation high up on the mountainside, the views wide open for twenty miles or more to the south.

In early 1951, a man from the Atomic Energy Commission turned up saying there's going to be some nuclear testing down on Frenchman Flat, about 35 miles to the south. He didn't say much more, and so the Sheahans and their workers went about their business until 2 February, when they were shaken from their beds before dawn by a thundering blast. Windows broke, the front door burst open. The eight kiloton Baker 2 bomb had just been detonated, and as the sun rose into the sky, the fallout swept in, a rain of dirt and dust as heavy as a thunder shower. Falling metal rang on the corrugated roofs of the mine buildings.[25] The Sheahans must have been terrified – but they had been told the tests were a 'temporary affair', and considered it their 'patriotic duty' to cooperate.[26]

This first test series was seen by the authorities as a success: the scientists hadn't measured any radiological problems and they'd received just one public complaint, from a woman in Las Vegas whose home had been damaged by the shock. So the Los Alamos scientists used what they had learnt about blast heights, weather patterns and dust transport not to improve the safety of future tests, but to scale them up: in the words of environmental historian Leisl Carr Childers, to 'operate closer to their self-imposed environmental limits'.[27]

But there is no true safe limit for radiation exposure: that's the nature of its random, stochastic risk. It just takes one atom to decay and zap the wrong bit of DNA to create a cancer mutation. Instead, as Carr Childers writes, 'acceptable exposure is a flexible and, more importantly, a social determination—one deeply informed by the entwined imperatives of war and industry'. And America was indeed at war, with no time for the precautionary principle. Even the National Committee on Radiation Protection didn't object to above-ground nuclear weapons testing and demanded no safety measures. Its chair, Lauriston Taylor, said, 'I see no alternative but to assume that [an] operation is safe until it is proven to be unsafe. It is recognized that in order to demonstrate an unsafe condition you may have to sacrifice someone. This does not seem fair on one hand, and yet I see no alternative.'[28]

The next test series, Buster-Jangle, in autumn 1951, was bigger: a 21-kiloton bomb, then a 31-kiloton one. The AEC turned up again at Groom Mine and told the Sheahans that women and children should leave the property when tests were taking place. They taught Dan Sheahan and his son, Bob, how to use a Geiger counter and gave them equipment to monitor the radiation levels in fallout: vacuum devices to sample particles in the air; trays with sticky flypaper surfaces to catch the dust as it fell. The Sheahans felt uneasy: they no longer saw wild rabbits running among the sagebrush and galleta grass, and their neighbours' horses started getting these strange white, raw spots on their backs.[29]

Carried by winds in the troposphere, the fallout from these tests landed primarily on the East Coast. The Eastman-Kodak plant in Rochester, NY, complained that the radioactivity had damaged thousands of dollars of photographic film. The AEC put out a press release insisting that it was fine.[30]

Each test series was hotter and dirtier than the last. Tumbler-Snapper in 1952 saw the scientists test four tower-mounted bombs, which detonated much lower than the airdropped weapons, at just ninety metres above the ground. So far, the fallout from the bombs had come from three sources: radioactive fission products of the nuclear reaction, un-fissioned uranium or plutonium that hadn't reacted and weapon residues vaporised by the fireball. Tower-mounted bombs, though, also irradiated and then pulverised the metal tower structures – and blew out a crater in the ground below, blasting the radioactive dirt up into the sky. The tests were taking place on dried-up lakebeds and salt flats, too, where, as we have seen, the dust was especially fine. The resulting radioactive particles were not just microscopic but nano-scale, ranging from about ten nanometers to twenty micrometres in size. Your average bacterium is about one micrometre (a thousandth of a millimeter) in diameter: as dust goes, radioactive fallout is some of the very tiniest. Alpha and beta radiation are easily stopped by skin, but the dust gets inside people's bodies, in the air they breathe and the food they eat. This is how it harms.

Some combination of bad luck and carelessness with the weather patterns that spring meant that Tumbler-Snapper produced fully fifteen times the civilian radiation exposure than previous test series. In the forty-year history of nuclear testing in the continental United States, these eight bombs resulted in 29 per cent of the total exposure to radioactive iodine-131, a product of the fission reactions.[31] Carried eastward by the wind, the fallout drifted down on to pastures and grasslands thousands of miles away, where it was eaten by cows and goats and passed into their milk – that all-American beverage, a glass of fresh cold milk, at that time served in schools across the country and on family dinner tables every evening. Those who were just children at the time of the tests would be hardest hit by the cancers that followed.

Iodine-131 has a relatively short half-life: just eight days. Had the milk in affected areas been dumped for a week or two – as it sometimes was, when the AEC chose to warn a county of particular fall-out risk – then those people would not have died. But thyroid cancer takes decades to develop, and in the early 1950s the AEC's priorities were rather shorter term: assuaging public fears and building legitimacy for nuclear science. The fifth shot in the Tumbler-Snapper series on 7 May 1952, codenamed 'Easy', had produced a fine rain of radioactive dust across the Salt Lake City area, sufficient that people who owned Geiger counters had heard them click. But the quantities were 'so small as to be of no consequence,' the AEC announced in a press release. 'At no time has any radiation from AEC test operations been harmful to any human, animal or crop.'[32] This wasn't true, of course, but, when the risk is dusty – diffuse, tiny and delayed in its effects – it is all too easy to deny.

The brothers Kern and McRae Bulloch were ranchers and sheepherders, as were their father and grandfather before them, a family of Mormons who had settled in Cedar City, Utah. They grazed their flock each winter out in Eastern Nevada, in valleys and passes about 40 miles northeast of the test site. It was sometimes a lonesome life, Kern would later say, but one he wouldn't trade for the world.[33]

Early one morning in March 1953, the Bullochs were taking the sheep over Coyote Pass and into Tikaboo Valley. Kern was sitting on his horse, feet out of the stirrups and leg cocked over his saddle horn, when: 'All at once, that son of a bitchin' bomb went off.'

The fireball is blinding, terrifying everyone. Kern's horse rears and nearly throws him, the sheep scatter. They watch as 'a doggone cloud comes up, just like a mushroom, and spreads out. I didn't know what was in it'. He watches it come up the valley towards them and 'over the top of us, where we were'.[34]

This keeps happening. One day, as they collect their sheep, a military jeep drives up and army personnel jump out, protective dust covers over their boots. An officer addresses them: 'You're in a hell of a hot spot! Get these sheep and yourselves the hell out of here as fast as you can.' But 2,000 head of sheep don't move quick, especially when the ewes are heavy with lamb. As they journey back west into Utah, at just 6 miles a day, the sheep's faces start to scab with lesions and white spots develop on their backs. They collapse on the trail. The ewes start to abort.[35]

Fifty years later, you can still hear the upset in Kern Bulloch's voice as he tells his story for the radio. 'God, they just started dying like flies. And we couldn't figure out what in the hell was the matter with them sheep. There'd be a hundred head of sheep dead, and we'd draw 'em on a tractor and haul 'em off, and we'd come back in to feed them the next morning and there'd be another hundred dead, laying there dead around those mangers. We couldn't figure out what was happening. Those Atomic Energy vets come into our yard, they autopsy them and say, "Oh this is a hot one. Jeez, this is a hot one!"'

They were all 'hot' ones. The nuclear programme's veterinarians would run a Geiger counter over the piles of bodies and the needle would swing clear off the scale.

'They covered it up, no question about it. This lawyer got me in a room one day, and he says to me, "We admit we killed your sheep." He says, "We admit it. But you're going to lose this case in court, because every pregnant woman, every sick person, will sue the government – and not only that, they'll stop these tests. And these tests have got to go on."'

The risk was right in front of their eyes, but too inconvenient to see. And so the tests went on.

Over 500 nuclear bombs would be tested worldwide in the decade from 1963, mostly by the United States and Russia, which had been testing in Kazakhstan since 1949. The UK began testing in 1952, in the southwest Australian desert, the Montebello Islands off the north coast of Australia and Kiribati, in the Pacific; France began in 1960, in the Algerian Sahara and again in the Pacific. The geography is nakedly colonial. 'They test where they think there are populations that don't matter,' Mary Dickson, a campaign for nuclear affected people known as 'Downwinders', commented to a Utah oral history project in 2017. 'Whether it's indigenous people or small populations or uneducated people, they think they can do it in places where the people don't matter, that they deem who is to be sacrificed.'[36]

As the bombs got bigger, the murderous carelessness first inflicted on sheep and sheepherders was repeated at increasingly larger scale. The microscopic dust of the fallout cloud drifts down on farms and small towns in southern Utah, and it shrouded the city of Semipalatinsk in Kazakhstan. Its very dustiness is what made it so deadly.

Nuclear fallout does not kill quickly and decisively, like a bullet, but slowly and stochastically, a random, statistical build-up of risk over months and years that cannot be fully measured for decades, even centuries, until the full toll of cancers it causes can be counted. It's a fuzzy risk – and therefore a deniable one. And deny the authorities did, in every country. From America to Algeria, Australia to Kazakhstan – the remote, often desert environments where bombs were tested were describe as 'uninhabited', when their occupants were instead nomadic, Indigenous or too humble to count. The risks that swept downwind with the fallout cloud, meanwhile, went under measured and unacknowledged.

***

Thinking with dust also brings into focus another, seemingly contradictory, aspect of the nuclear era: its paradoxical visibility and its function as spectacle. The harm may be slow and invisible, but the detonation is, as we have seen, anything but: dazzlingly bright and polychromatic, concluding in the apocalyptic icon of the mushroom cloud. Far from being metaphorically brushed under the carpet, government television programmes beam this dusty image into 100 million homes across America. Why? What was this spectacular visual image being used to communicate – and what did it conceal?

'From 1953 to 1961, the yearly centerpiece of the civil defense program was a simulated nuclear attack on the United States directed by federal authorities,' writes anthropologist Joseph Masco.[37] By this point, the purpose of nuclear testing was not just to test nuclear weapon materials and design, but to develop nuclear strategy. What would a nuclear attack mean for families and homes in the suburban American heartland? How might smaller nuclear weapons be used tactically in combat on the ground, alongside other military forces? The US Army designed two programmes, Desert Rock (1951–7) and Operation Cue (1955) to find out. That was not, however, all they achieved.

Operation Cue saw the Army construct a handful of suburban homes in the middle of the desert, complete with furniture, household goods and shelves filled with packaged food. The purpose of the operation was, in part, to test questions of civil defence: what radioactivity doses might people experience? Would household atomic bomb shelters protect them? (*Barely.*) Might tinned foodstuffs survive the nuclear apocalypse? (*Yes.*) And so families of fully dressed mannequins posed, stiff but smiling, in typical domestic activities such as sitting at the kitchen table or watching TV, as the Apple-2 test shot detonated close by and a pressure wave and heat and debris smashed into the houses like they were toys. The Federal Civil Defense Administration captured it all on camera. The resultant film was watched by one hundred million people.

Seeing the so-called 'Survival Town' ravaged by nuclear shock not only communicated the power of the bomb to Americans and their enemies alike but instructed them in the language of its imagery. The film's narrator tells the viewer that 'the towering cloud of the atomic age is a symbol of strength, of defense, of security for freedom-loving people everywhere', and that nuclear testing is essential to America's national survival.[38] And so, as anthropologist Joseph Masco details, 'Each year Americans acted out their own incineration in this manner, with public officials cheerfully evacuating cities and evaluating emergency planning, while nuclear detonations in Nevada and the South Pacific provided new images of fireballs and mushroom clouds to reinforce the concept of imminent nuclear threat.' The ragged and burnt clothing of the mannequins was displayed in the JCPenney department stores that had supplied the garments, with a sign declaring rather ominously: 'This could be you!'[39]

This spectacle of dust and destruction – what Masco calls this 'national contemplation of ruins', a concept we will come back to in the next chapter – aimed to transform Americans into obedient, well-prepared nuclear age citizens, comforted by futile little personal safety rituals and a garage stocked with canned foods. Through this public vision of national annihilation, the government sought to engineer popular consent to ongoing testing and nuclear armament.

Yet 'this effort to think through the disaster colonised everyday life as well as the future, while fundamentally missing the actual disaster,' Masco writes. It misses the fact that America was literally nuking tens of thousands of its own people.

US Army commanders believed that the average soldier possessed a supposedly 'mystical fear of radiation' of which he needed to be cured were he to be any use on the nuclear battlefield. Their solution: to ship troops into Nevada to observe nuclear tests, in the hope that watching their almighty power might make them learn to love the bomb. In 1957, Russell Fjelsted, a lieutenant in the Air Force Strategic Air Command, was among this group ordered to watch the Diablo test. He described to the Utah Downwinders Project how he and two or three hundred other men and women were instructed to stand in a twelve-foot-deep trench, bend over and cover their faces with their hands – as a 17-kiloton nuclear bomb was detonated seemingly only 5,000 feet away.[40]

He was told to hide his eyes, but the explosion insisted upon being seen. Fjelsted remembers, '. . . I saw through my eyelids and I saw through my fingers, where these spaces are, I saw the guy's shoes in front of me . . . And that's how brilliant this light is.'[41]

They wore no protective clothing, only their regular uniforms, and the promised film badge dosimeters to measure individual exposure levels were never issued. When looking back, like many people involved in the nuclear testing programme, Fjelsted struggles to square his patriotism and trust in the institution he served with the evidence of his own experience. 'I don't know whether we were guinea pigs or whatever, but it was kind of like that,' he says. 'I wasn't physically affected by it, I don't think. I did have prostate cancer when I was about 70 years old,' he adds. That could have been from other causes: it's impossible to tell for sure.

Witnessing the bomb blast – specifically *seeing* its devastation – was central to both operations. 'I had to see Operation Plumbbob through many eyes,' the narrator of the FCDA film announces, 'not only my own; but as a reporter through the eyes of the average American man and woman'.[42] Western culture is ocular-centric, privileging the visual sense as a source of knowledge: we say things like 'seeing is believing' and 'I see', when we understand. Yet there's something more here, a deep and paradoxical desire that both the nation should see the unseeable – and that the consequences of the nuclear programme should be kept quite out of sight. An explosion at first so hot it is beyond the visible spectrum, an explosion so deadly it would burn out the eyes of those who get too close. An explosion on every TV network and newspaper cover, a spectacle of unfathomable power and destructiveness that is presented as a way to keep the nation safe.

In 1973, a fire at the National Personnel Records Center in St Louis destroyed the records of many of the veterans who participated in these tests. And so their risk remains unknown.

***

Shere Finicum-H was a young girl in southwestern Utah. She described to the Downwinders of Utah oral history project how she was out driving with her father one day, heading to the family ranch. 'As we came up over Cedar Ridge, I could see this immense white cloud on the horizon, something I had never seen before. And I was just like, "What is that, Dad? Daddy, what is that?" He started trying to explain to me. . . . He was talking about things like war and things I didn't understand. But the one thing that really stood out in my mind to me is he said, "I hope you never see this for real in your lifetime."'[43]

She *was* seeing a real nuclear explosion – but that was the power of the US testing programme: the power to render the bomb as somehow only a symbol, as if unreal. A destructive power imagined to only exist as strength and military potential – and not at all as public risk.

'That made quite an impression on me,' Shere Finicum-H continued. 'I'll never forget what it looked like . . . I hold that in my memory of that, of that sight, and what an immense – in a way, a magnificent sight to see.'

US historian Peter B. Hales writes of how this public imaginary was achieved through the military and Atomic Energy Commission's tight control of access and information. The only photographs anyone saw of nuclear explosions were ones the government had chosen to release. The 1945 Hiroshima and Nagasaki bombings had been carefully media-managed, and the longform reports that followed in *Life* magazine, *Time* and *Newsweek* focused on visual description over ethics or human cost. 'Most notable was the emphasis on natural imagery,' Hales writes. 'By choosing such analogies, the writers ... bridged a previous gap between what was human and what was natural – the atom bomb became a man-made marvel of nature, and thereby the question of responsibility for the effects of the explosion remained slippery.'[44] In this way, the atomic bomb becomes sublime – and the mushroom cloud specifically, this cloud of dust, its defining symbol.

It also became a pop culture icon. In 1946, designer Louis Réard unveiled a tiny two-piece swimsuit at the Piscine Molitor, a popular swimming pool in Paris. He named it the 'bikini', after the Marshall Islands atoll where, four days earlier, the United States had initiated its first peacetime nuclear weapons test. Réard reportedly hoped his swimsuit's revealing style would create a similarly 'explosive' commercial and cultural reaction.[45]

The Nevada tests were even marketed as a tourist attraction. In 1952, Las Vegas dancer Candyce King was dubbed 'Miss Atomic Blast' as she entertained Marines fresh off atomic manoeuvres at Yucca Flats, newspapers describing her as 'radiating loveliness instead of deadly atomic particles'.[46] Casinos marketed their north-facing views, the Las Vegas Chamber of Commerce issued a calendar of scheduled testing dates, and tourists flocked to the city to stand on hotel rooftops, sip 'atomic cocktails' and watch mushroom clouds swell in the early dawn. For a moment, Cold War dread might be imagined as something like a thrill.

The atomic spectacle lasted for eighteen years. The 1963 Partial Test Ban Treaty eliminated above-ground and atmospheric testing, requiring all subsequent nuclear blasts to take place deep underground. This should have prevented the release of radioactive material; it didn't entirely. The Storax Seden test on 6 July 1962 was a dusty one, creating the largest humanmade crater America had ever seen while dispatching 880,000 curies (33 PBq) of radioactive iodine-131 into the atmosphere. The fallout settled over Iowa and South Dakota, contaminating more Americans than any other nuclear test. The Baneberry test in December 1970 suffered from geological miscalculations and another dust cloud rose, dropping fallout first locally, then as radioactive snow over northeastern California. Two men exposed to high levels of radiation died of acute myeloid leukaemia less than four years later. The courts found the government had acted negligently but was not responsible for causing their deaths.

Tiny, diffuse and delayed, dust is so very deniable.

***

There's a deep ambiguity around the death toll of the bomb – both individually and overall.

As mentioned earlier, atmospheric nuclear testing will kill perhaps as many as 2.4 million people. It's a devastating number: ten times the immediate toll of Hiroshima and Nagasaki combined. It is frightening because it's largely unknown, strange and unsettling for the fact that many of these deaths haven't occurred yet. The figure comes from a 1991 modelling exercise by International Physicians for the Prevention of Nuclear War (IPPNW), a federation of medical groups in sixty-three countries around the world. They took UN estimates of global radiation exposures and applied a public health model that converted population dose levels to cancer deaths. As such, they were able to estimate the number of human cancer fatalities caused as a result of above ground nuclear testing between 1945 and 1980.[47]

What's striking about the IPPNW claim is its timescale: it's calculated outwards to infinity. They estimated less than a fifth of these deaths (430,000) would occur by the year 2000. Instead, these are potential deaths in years, even centuries to come: a slowly ticking timebomb over the heads of two million people. Carbon-14 is the primary culprit. It's an activation product of nuclear explosions, as stray neutrons stream out and knock into ordinary carbon-12 atoms, rendering them unstable and radioactive. Its significance lies in its half-life, the time it takes for its radioactivity to decay: at 5,730 years, it is on any practical human timeline eternal. The consequences of the Cold War will stay with us forever.

Other analyses of the human impact of nuclear testing have concentrated on other isotopes – and reached very different conclusions. The US National Cancer Institute's 1999 study analyses population exposure to iodine-131 from the Nevada tests, estimating that this would produce an additional 49,000 cases of thyroid cancer (mostly among people who were children during the above ground tests in the 1950s), plus 11,000 other cancers, such as leukaemia.[48] The US had 400,000 cases of thyroid cancer a year, they pointed out, making this just a 12 per cent increase – a number small enough to make such sacrifice seem almost reasonable.

As previously mentioned, iodine-131 has a half-life of just eight days, making it a relatively short-term hazard: in less than three months, its harm decreases by over a thousandfold. As such, INPPW estimate that it makes up just 2.02 per cent of people's collective radiation dose by the year 2000. Radioactive isotopes with longer half-lives, such as caesium-137 and zirconium (both around the thirty-year mark) produce the majority of exposure in this time frame as they remain a hazard for so much longer. Looking out into the deep depths of the future, carbon-14 becomes the primary troublemaker. And so, decades later, a person might ingest a radioactivity via their ambient house dust, or a stray particle could zap them with a burst of gamma radiation as it finally, randomly, decays. The ionising radiation breaks molecular bonds and creates free radicals in the body, causing cellular damage and breaking DNA strands. Cells acquire genetic mutations, which, through generations of cellular reproduction, proliferate into cancer. It takes ten or twenty years, or more. By the time a person is sick, it's impossible to pinpoint the origin.

The IPPNW's 2.4 million estimate is no certainty. Its gaze out to a far future time horizon means that only small statistical corrections to the underlying model – perhaps the underlying dose-exposure model gets updated, or their estimates of world population may prove over-generous – could have big impacts on the overall number. Perhaps medicine will take great strides forward in detecting and curing cancer, we can but hope.

Nonetheless, looking across the various estimates made, however variable the numbers and wide the error bars, it seems likely to me that at least a million people have died as the result of nuclear testing. A million lives sacrificed for what? The possibility of perhaps deterring a worse war? America, the Soviet Union and China perhaps each sacrificed hundreds of thousands of their own citizens for this purpose;[49] Britain and France, meanwhile, offloaded the risk on to the countries and people they had colonised, alongside tens of thousands of military recruits. No country fully measured either the radiation doses experienced by people near to the test sites, or the fallout that drifted with the winds to drop its lethal payload sometimes thousands of miles away. Secrecy was presented as a military necessity. Politically, it was easier not to know. Yet even if radiation exposure had been measured, uncertainties would linger; they're inherent in the very nature of nuclear risk.

The relationship between radiation exposure and health effects isn't neatly linear. The body metabolises some radioactive isotopes more readily than others, and medium matters: are you inhaling the isotope, ingesting it in water or consuming it via irradiated food? The decade-long time delays further blur any confident attribution of cause. Over a lifetime, people encounter so much other radiation – much of it intended to heal us, in the form of medical diagnostics such as CT scans, radiography and radiotracing, and most of the rest of it natural, in the form of radon gas and high-energy cosmic rays from space. Even the maths can mislead, if we are not careful. One recent epidemiology study warns: 'The interpretation of results for specific cancer types requires some care since, simply by chance, one would expect about one finding that is statistically significant at the 5% level for every 20 cancer types studied.'[50] Consequently the true fallout of

211

the bomb, its human cost, will never be known by most of those who paid for it with their lives. It *cannot* be known individually.

This strange, dusty violence.

For many British military veterans in particular, the uncertainty makes it crueller: they are certain that their cancers and their family's illnesses are consequences of tests they witnessed as young men in shorts and shirtsleeves on some Pacific atoll, and the fact that the UK government denies them acknowledgement and compensation is a further lifelong wound.

In February 2022, I attended a seminar in Manchester organised by LABRATS, the campaign group for people affected by nuclear testing globally. A panel of veterans – old men now, well into their eighties, dressed smartly in navy blue blazers adorned with their service medals – described witnessing atomic tests in the Pacific in the 1950s. It was an unexpectedly emotional event. It had been sixty-five years since the events they described occurred, but what they had seen was burnt into the mind's eye, and the feeling of betrayal still fresh, just as if it were the previous morning.

Archie Hart served on the HMS *Diana*, a British naval destroyer ordered to enter the blast zone of an atomic bomb test on the Montebello Islands, off the northwest coast of Australia, just three hours after the explosion, expressly in order to pick up as much contamination as possible. He describes the dust fall 'as like a gentle rain from heaven. But it wasn't gentle: it was ionising radiation.' They were given no protective equipment worth the name, just a cotton 'anti-flash suit' made out of old 'Second World War material that the moths had been at'. They were treated as guinea pigs.

In 2008, crew members mounted a legal challenge for compensation, citing the health issues so many had faced: cancers and rare blood disorders. Infertility and stillbirths. Serious birth defects and health problems among their children, suggestive of genetic damage.[51] In the years since, Hart found his body growing a hundred or more lipomas (non-cancerous soft tissue tumours), some the size of tennis balls – and he was treated for bowel cancer in 2002.[52] In 2012, his compensation claim was finally rejected by the Supreme Court, on the grounds of lack of evidence. 'Other countries have given compensation to veterans, but the parsimonious British Government can't do it,' Hart told his local newspaper, the *Warrington Guardian*. 'It's shameful.'

Despite the horrifying individual accounts of the LABRAT veterans, systematic evidence of harm from Britain's bomb tests has been harder to prove. Epidemiological research comparing nuclear exposed servicemen to a control group found for many years either no real difference in cancer rates and mortality causes between the two groups, or only small impacts that couldn't be distinguished from statistical quirks. The fourth wave of this research, in 2022, for example, found mortality rates among the nuclear vets was 2 per cent higher, but the fact this was occurring from higher rates of cerebrovascular disease (e.g. stroke) was a problem: there's not an established biological mechanism for how radiation increase stroke risk. Only one cancer, chronic myeloid leukaemia, shows up at clearly higher rates in the nuclear veteran group overall – while other previously identified patterns have failed to hold up over time, and are now thought to be chance. It's not easy data to make social sense of. The stochastic risk of nuclear dust is fundamentally difficult to weave into the narrative of a human life, its effects random, nanoscale in size, slow and often decades delayed.

Nonetheless, deep in the discussion section of the research paper, the LABRATS' experiences are validated. The study notes how certain groups of service members – those men 'most likely to have been exposed to internal radionuclides' have higher rates of other cancers too, such as prostate cancer, and tumours of the brain and central nervous system. These are the men aboard the HMS *Diana* who sailed through the fallout plume or those partic-ipated in tests at Maralinga in South Australia, which took place on a dusty desert plain rather than an ocean atoll. Men who ingested fallout dust in the air they breathed and the water they drank. Alpha radiation is weak and easily stopped by just the skin, but that doesn't matter if it's already inside you.[53]

Ultimately the injustice doesn't lie in the dust itself, of course, but the authorities that exposed these men to so much of it. The monstrous carelessness of young men used indeed as 'lab rats' for an experiment they were never properly informed about, their dose monitoring badges not distributed and their blood and urine test results often 'missing'. Geographer Becky Alexis-Martin writes: 'Uncertainties about the long-term effects of the nuclear tests make psychological stress and fear an important and continuing legacy of nuclear weapons testing.'[54] It's not just the bomb but the lack of care that harms, because that – the human side – is the betrayal.

In November 2022 – the 'Plutonium Jubilee', seventy years after Britain's first nuclear detonation – Britain's nuclear veterans finally won the acknowledgement they were seeking: a Nuclear Test Medal, to honour the unique trauma of their service. Their fight for compensation is ongoing.

Not everyone wants to focus on fighting injustice, however. For others affected by nuclear testing, uncertainty may perhaps be a mercy. Better sometimes not to know that family members, loved ones, or oneself are victims of the Cold War. Some downwinders reel off family cancer cases and tragic early deaths, then say things like, 'we have had no evidence of demise within the family community'.[55] It may feel easier to think of your cancer as 'just bad luck' rather than facing up to the ways the government saw you as expendable.

There may be lessons in this uncertainty for the future. Measuring the human cost of a warming planet is a similar problem to assessing the toll of nuclear testing: often a rather bloodless, statistical matter of population modelling and 'excess deaths', each with a plurality of potential causes. A death from what's termed 'heat stress' isn't just the discrete condition of heatstroke, but also the elevated potential for heart attacks, strokes and breathing difficulties. What gets measured and attributed – and what doesn't? What gets overlooked, evaded, denied? Deaths at the scale of the nuclear testing toll ought to be a cataclysm on a scale equalled only by war – but instead they prove too diffuse to be seen. It happens again and again: gun violence, overdoses, smoking. In just two years the US has seen over a million Covid deaths 'without the social reckoning that such a tragedy should provoke,' as science journalist Ed Yong has extensively reported.[56] If this much mass death can be rendered essentially permissible (through attribution to old age, endemicity, comorbidities and poor individual decision-making), then of course America and every other nuclear testing country can repeatedly nuke people for several decades and get away with it. TV news cameras don't turn up when someone dies fifty years after a bomb goes off.

But – as a substantial body of human rights scholarship argues – memory is necessary. It's 'necessary for coming to terms with and righting the wrongs of the past and thus preventing future violence', sociologist Amy Sodaro writes: 'there is an ethical duty to remember.'[57]

Bearing witness to the losses of the nuclear era can help us fight better for those facing other dusty deaths from air pollution today. We pay our respects to the nuclear dead by paying attention.

***

It's easy to be dazzled by the spectacle of the mushroom cloud, but those living under the shadow of its fallout were not the only people affected by the nuclear age. From beginning to end – from extraction to disposal – nuclear modernity was dirty, dusty work.

When Britain and France declared war on Germany in 1939, government officials in Europe and America quickly began to recognise the strategic importance of uranium: who controlled this metal would control the ultimate forces of both nuclear weapons and nuclear power. The Manhattan Project began clandestine programmes to purchase uranium from mines at Shinkolobwe, in what's now the Democratic Republic of the Congo (DRC), and

Bear Lake, in Canada, while contracting with Union Carbide to identify new sources of the mineral in the hope of controlling 90 per cent of world supply.

Postwar, the Manhattan Project's activities were transferred to a new Atomic Energy Commission, which offered a guaranteed price and market for uranium ore. This set off a 'uranium rush' in the western United States, particularly in an area running northwest from near Albuquerque in New Mexico to the Arizona–Utah border. Predominantly Indigenous land, it encompasses tribal reservations belonging to the Laguna Pueblo (who know themselves as K'awaika), Acoma Pueblo (Áak'u) and the Navajo Nation (Diné). The New Mexico mines alone produced nearly half the uranium in America.[58]

In April 2022, I visited Laguna in the company of Gregory Jojola, environmental director for Laguna Pueblo. It's a place right at the heart of the nuclear archipelago: the Trinity Test site sits 100 miles to the southeast, while Los Alamos, where the bomb was developed, lies a similar distance to the northeast. Here, in the midst of the foothills and valleys of Mount Taylor – or Tsibiina, a sacred place for many tribes in the region – the Anaconda Minerals Company operated what was for a time the largest open-pit uranium mine in the world: the Jackpile-Paguate mine.

Meeting at a diner on old Route 66, Gregory and I drove a twenty mile loop up around the mine site so I could learn more about the profound impact it's had on the landscape and community – and see the work they're doing today to help mitigate the ongoing risk of the nuclear age, first-hand.

Uranium dust wasn't the only airborne hazard to be aware of, however. The Covid pandemic has hit Indigenous communities in the United States hard, with under-resourced healthcare systems, poor infrastructure and pre-existing health inequalities making Indigenous people 3.5 times more likely to be hospitalised than the average American. Gregory and I drove in separate cars, didn't go indoors and stayed masked-up, even outside. I was grateful to be able to visit at all – some reservations remained closed to non-residents entirely.

We drove through a semi-arid landscape of buff and sand and tan punctuated by glossy, deep-green juniper bushes, then pulled up by the side of the road. Across a dry riverbed stood a human-made mesa, an utterly flat-topped tableland made out of low-grade uranium ore.

'They call it protore,' Gregory explained. Some ores aren't rich enough in metal content to be profitable at the time, but may be in future, 'so the better-grade stuff they piled in these easy-to-access areas because they thought they're gonna come back when the price of uranium went up', he said.

In three decades of mining from 1953 to 1982, 400 million tons of rock were moved within the mine area. Twenty-five million tons of ore were shipped out on the railroad, the rest piled up in 32 waste dumps and 23 protore stockpiles, turning four square miles of land into a brand-new topology of plateaus and scarps, surrounding three big, open scars on the landscape – the three uranium mining pits.[59]

A few more minutes in the car and there they were before us – and I suddenly understood how the Jackpile-Paguate mine got its name. The village of Paguate is right on top of it. Houses sit on a rise overlooking the north and south pits, less than a quarter mile from where diggers and trucks would have hacked away at the earth in a cacophony of noise and dust. At times, the mining operation ran for 24 hours a day, seven days a week. It ruptured not only the earth, but the entire Laguna social world.

Dorothy Ann Purley grew up in Paguate, playing in the fields and orchards of what was then a pretty self-sufficient farming community. In a testimony she delivered at a 1994 New Mexico environmental conference, Purley remembered the explosive blasts – two a day, at noon and 4 p.m., which 'would shake the homes so severely that dishes rattled and glassware would fall off the cabinets. You could feel the land under your feet move. If the blast was close enough, we could hear the rocks that had been blasted fall back to the earth. What was even worse was the smell of sulfur. We were told that it was only blasting powder, nothing to be afraid of. The smell would linger in our homes for hours. A fine mist of dust would settle on our tables as we walked back into our homes to finish our lunch or dinner.'[60] The explosions shook people's houses to pieces: walls buckled, roofs sagged, the floor of the Purley's home gave way. 'If we had only known the dangers in the beginning,' Purley said.

Laguna Pueblo had welcomed the mine – they were told that 'It was all for the war efforts,' Gregory Jojola said. 'There's a lot of veterans from World War Two out here; a lot of the tribal leadership are veterans. So when the government comes knocking and they say you got to do this for your country', it was a done deal.

Besides, 'there was plenty of money to be made,' Dorothy Ann Purley noted. The mine offered employment opportunities – and not just casual labour but steady, year-round work. The Pueblo of Laguna was held in high esteem for being self-sufficient,' she wrote in her testimony, and therefore jumped at this chance to better their prospects.

'With the mine in full operation, life seemed great. People were living comfortably with their fat paychecks.' (Indigenous miners were often paid only two-thirds the standard wage – but to people who'd previously been living substantially outside the wage economy, this was still a big lift.)[61] 'Almost everyone had a new automobile. We tried not to notice the land being destroyed,' she added.

The land, people's homes, their way of life – their language, their culture, even their perception of time. The Laguna and Diné activist, attorney and judge June Lorenzo – who grew up in Paguate – has written about how men returned home each week with money in their pockets, tipping the power dynamics with their wives and families. Clocking in for an eight-hour shift meant people couldn't participate in traditional ceremonial life as they once had. Culturally and spiritually significant landmarks were 'eliminated' so that Anaconda could build an access road or get to the ore underneath.[62]

And, of course, the mines destroyed people's health. People started dying from cancer during the 1970s. Purley reported: 'It seemed like an epidemic.'

Until regulation in 1969, uranium mines operated with next to no health and safety measures: no face masks or breathing equipment, no dosimeter badges to monitor exposure – barely any training in how to handle dynamite. Mines lacked sufficient ventilation for either the dust or the build-up of radioactive radon gas which accompanies uranium ore. Even the silica in the sandstone rock cut people's lungs like glass, giving rise to emphysema and silicosis. Men in their thirties and forties were dying in a community where, a generation before, cancer had been so rare that medical researchers had wondered if they might be immune to it.

It wasn't just men who were exposed. Miners went home in their dirty work overalls, which their wives washed with the family laundry – giving them rashes.[63] Mining dust landed on meat and fruit laid out to dry and cure in the sun which people then ate.

'Our younger generations were afflicted by leukemia and tumors,' Purley continued in her testimony. 'Babies were born with birth defects.'

People didn't realise they were being exposed to radiation – they did not even know that radiation was a threat. The Diné or Pueblo languages had no word for 'radioactivity'. Mining companies told the mine supervisors about the dangers but instructed them not to tell their workers – or issued information in English, which many people couldn't read. During the 1950s, the US Public Health Service had been monitoring the uranium miners to understand the health effects of their work, but they'd gained access via the mining companies, who demanded their silence. The USPHS never informed its test subjects of the health hazards they faced or told them that their illnesses might be radiation related.[64]

***

In February 2021, strong winds once again carried Saharan dust north, over southern and central Europe. The sky in the Jura turned a deep rusty orange; Alpine ski resorts found their slopes covered in a fine layer of sand. Scientists took samples – and found the dust was radioactive. It contained caesium-137 isotopes, a legacy of French nuclear tests in Algeria, sixty years before.[65] The dose was very small, an estimated 80,000 becquerels per square kilometre, the Association for Control of Radioactivity in the West reported – lower than natural radon levels and posing no meaningful health risk to Europeans. ('What about Saharan people?' Maïa Tellit Hawad, a French scholar of Tuareg origins, wondered.[66]) Still, the irony was sharp. 'This radioactive contamination, which comes from far away, sixty years after the nuclear explosions, reminds us of the perennial radioactive contamination in the Sahara, for which France is responsible,' ACRO reported. France's colonial exploits had come back to haunt it.

Let's follow the dust back to its origins in southern Algeria where, in the 1960s, France carried out the first seventeen of its 210 nuclear weapons tests.[67] We've seen the mess made at the start of the uranium mining process, and the dirty fallout in the hours and days immediately after the bomb detonates. I want to look now at the longer-term aftermath – and the criminal carelessness of not cleaning up a nuclear bomb site, leaving irradiated sand to simply blow in the wind.

France was the fourth country in the world to start nuclear testing. Their nuclear programme took a tactical approach, simulating wartime environments in order to understand the effects of the bomb as a weapon of war. In the first test, *Gerboise bleue* (Blue Jerboa, named after the jumping desert rodent), the army placed vehicles, tanks and artillery around the test site to see how a nuclear attack would affect military equipment – irradiating tons of machinery, and turning it instantly into dangerous radioactive waste. Planes were flown through the fallout cloud in order to take radiation readings, contaminating the aircraft and their pilots. Three of the four tests in the first series also took place from tall, hundred-metre steel towers, which were completely pulverised in the explosion, scattering highly irradiated metal fragments widely across the vicinity. Sometimes the contamination was even the point. A further forty tests, the Augias and Pollen series (which did not entail nuclear chain reactions) sought to simulate accidents involving plutonium at a small scale, to model the consequences of a larger disaster. How much land would be poisoned? How far would radiation spread?

Nuclear testing accidents were another major source of environmental contamination. France's second test series took place underground, in tunnels bored deep into a granite mountain in the Hoggar highlands in southern Algeria. (The country had gained independence from its coloniser just a few weeks before, but part of the deal had been to allow France continued access for nuclear testing.) During the Beryl Test on 1 May 1962, the concrete plug meant to contain the explosion failed catastrophically. A huge 'blowtorch' flame shot out of the tunnel, followed by a cloud of ochre dust that soon turned an ominous black, alongside fragments of flying lava and vitrified slag. The particles and radioactive dust rose up in a cloud 2.6 kilometres tall, producing radioactive fallout detectable for several hundred kilometres downwind. A crowd of officials and soldiers had been invited to spectate: everyone ran for the showers to attempt to decontaminate themselves. Minister of Scientific Research Gaston Palewski would die of leukaemia twenty-two years later. Cause and effect can't be proven for certain – but he was pretty sure he knew what had killed him.

The impact on rank and file soldiers and Algerians in the vicinity was never publicly released. 'The showers cleaned our bodies and clothes,' said electrician Jean-Claude Hervieux to a *Deutsche Welle* reporter in 2013, 'but not what we breathed in or swallowed.' When he asked the government for his radiation tests, the information he got back was 'bizarre', seemingly falsified, misrecorded or reported destroyed.[68] The documentary filmmakers Larbi Benchiha and Elisabeth Leuvrey have reported that following the Beryl accident, radioactive dust and fallout travelled sixty kilometers to the village of Mertoutek in the Tamanrasset Province.[69] Seventeen people died suddenly in the aftermath and residents are still reportedly suffering ill effects.

Health records weren't the only data to go missing or never be collected in the first place. Researchers discovered through classified documents that the French Ministry of Defence seems to have no records on the radiological condition of the launch bases when they were returned to Algeria in 1967.[70] Both main test sites were never properly cleaned up and decontaminated. Plutonium waste was just buried in the sand, sealed inside concrete-filled metal barrels. Large amounts of scrap metal and unwanted machinery were just left out in the open, visible to this day. Two concrete control bunkers sit, half-buried by dunes. Much of the most heavily contaminated areas at the epicentres of each blast were simply left open to the sun and the sky, letting the wind blow and carry radioactive sand particles where they may.[71] (The proper remediation process is essentially to strip the topsoil, bury it in pits and tarmac over the lot, which is of course highly environmentally damaging but does at least keep radioactive dust at bay.) Not in French-occupied Algeria. Uninformed about the danger, local people saw the scrap metal as a rich bounty and gathered it for reuse in house building, fencing and making utensils, even jewellery. Copper wire sold for a good price as scrap.

Algeria's former Minister of Veterans Affairs, Tayeb Zitouni, criticised France in 2021 for 'refusing to hand over the topographic maps which make it possible to determine the burial sites of polluting, radioactive or chemical waste not discovered to date'. ('France has provided the Algerian authorities with the maps it has,' the French responded. Perhaps no maps were ever made, all the better for keeping such a toxic legacy hidden.) No clean-up and no compensation. 'The French party has not technically led an initiative to clean up the sites, and France has not done any humanitarian act to compensate the victims,' Zitouni decried.[72]

'The desert was seen as an "ocean",' wrote Jean Marie Collin and Patrice Bouveret in a 2020 report on French nuclear waste, a place so vast and empty that you could just casually throw things away and they would sink into inconsequentiality, out of sight, out of mind.[73]

Deserts get chosen as nuclear test sites because they are empty: this seems obvious – except that it's wrong. It's not that deserts are empty – though they are often sparsely populated, I'll allow. More precisely, it's that the people who live and travel through the world's deserts are rendered invisible by colonial powers who decide they do not count as fully and equally human.

Back in the seventeenth and eighteenth centuries, colonial land claims were justified using a rationale of natural law: the principle 'that which has no owner can be taken by the first taker'. The perception wasn't precisely that the American, Australian and African interiors were empty land, with nobody there – but rather that this land was *res nullius*, nobody's thing and nobody's property.[74]

Under the colonists' European legal norms, land rights were acquired through use – that is, through settlement, construction and farming. But the early anthropology of the nineteenth century cast Indigenous peoples as 'primitive', only just advanced beyond a naive 'state of nature'. Their nomadic hunting and gathering – a way of life necessary for survival in the arid interiors of these continents – was not legally recognised as a pattern of use that bestowed ownership. (Meanwhile, the concept of land as property to be bought and sold was not a concept that existed in Indigenous worldviews either.) And so, in this way, the first colonial occupiers designated Indigenous land as free for the taking, producing a mindset that held firm right to the end of empire and, in the case of French and British nuclear test programmes, well beyond.

Maïa Tellit Hawad writes of the ' "invention" of the Sahara as a periphery to be conquered, exploited and disciplined'. The marginalisation of desert places in twentieth-century modernity and onwards isn't an inherent physical property of these landscapes so much as something actively made by colonial dispossession. But the dust might help us to see otherwise. Hawad writes how the 'red wind' returning to Europe 'reminds us of what centuries of extractive capitalism have tried to make us forget: our worlds are ineluctably intertwined on an environmental scale'.[75]

***

In May 2022, I spoke with the New Mexico Environmental Law Center's executive director Dr Virginia Necochea and staff attorney Eric Jantz, to think further about nuclear dust as a matter of environmental injustice. They didn't hold back.

'When we talk about dust, although it impacts everyone, these environmental consequences do not impact communities equitably,' Necochea emphasised. 'It's not random. It's an intentional siting of certain industries in specific communities – and New Mexico is a prime example of how it continues to operate.'

The fact uranium deposits are found so substantially on Indigenous reservations might legitimately be a coincidence of geography – but the fact these mines aren't being adequately cleaned up isn't accidental at all.

Eric Jantz described how 'the communities where these operations are located are by and large BIPOC communities and very under-resourced. So they don't have the political power to insist and demand, and get *results* from those demands, for expedited clean-up. If you look at Moab, Utah, at the Atlas tailing site and the Durango mill, those are in largely white and affluent communities – and they're getting the gold-plated remediation. But in places like Shiprock on Navajo Nation, Church Rock on Navajo Nation, Tuba City: all those places get the nickel-plated version.'

Indigenous communities have also been fighting for decades not to end up with a disproportionate share of America's nuclear waste back on their reservations, too. Tribal sovereignty makes their lands exempt from state law and many environmental regulations – but they're also very poor. And so comes the federal government or private companies offering millions of dollars to communities where a third of people live in poverty, an inequality making it seem less like compensation than, perhaps, coercion. It's difficult to say no – but increasingly tribes are doing so. In 2007, the Skull Valley Goshutes in Utah refused a government bid to store spent nuclear fuel on their land, and the Western Shoshone in Nevada have fought against the long-term deep geological waste repository Yucca Mountain for decades. Resisting 'nuclear colonialism' might come at an economic cost – but, tribes are saying, it's cheaper than rendering their land a sacrifice zone for thousands of years into the future. Enough has been sacrificed already. Now it's time to restore.

***

The Jackpile–Paguate mine may now be long-closed, but in Laguna Pueblo people still worry about the dust.

Forty years on, the site is now listed by the Environmental Protection Agency (EPA) as a 'Superfund' site on the National Priority List – meaning it's among the 1,300 most hazardous areas of environmental contamination in the entire United States. Initial remediation work in the 1980s had terraformed the landscape to try to protect it – but the dirt was sloped too steeply. The sides of the waste piles are crinkled with rivulet channels as rain erodes low-grade uranium ore down into the rivers below, and onward through the watershed. We stood overlooking the Rio Paguate, while Gregory Jojola described how the EPA has done aerial surveys with gamma detection equipment to monitor radiation levels today. 'This area right here is pretty hot,' he said, pointing. 'That hill right there: hot! That slope right there: hot.' A cattle association runs cows in this area – 'They're not keeling over or anything!' he qualifies. 'They seem fine. Cows are pretty hardy.' But is the land truly safe for livestock? That's what people need to know.

'People are just real concerned about the health hazards that still remain,' Gregory added. Part of the anxiety stems from just how little real information they have. Environmental monitoring was lacking while the mine was in operation or for decades after, and so 'people are concerned about what they're breathing'. They're worried about their water, both for human use and for agriculture and livestock. And they're worried about the health impacts that remain today: 'There's never been any real, extensive health studies done. A lot of the mine workers are getting older, some are getting cancer, some are getting kidney problems, lung problems, but it's hard to pinpoint it as the mine.' Again, the specific violence of nuclear waste: it denies people the dignity of ever really knowing what killed them.

'Some of the elders might rather not know at all,' Gregory says. '"We've been here all our lives," they say. "We grew up with it and we're gonna die here, so don't tell us!" But I really want to educate the younger ones, because they're going to take it over. And they deserve to have their questions answered,' he says.

The Laguna Pueblo environmental department are working with a team at the University of New Mexico to answer questions like this and many more. The UNM METALS Superfund Research Program studies the effects of uranium exposure on tribal communities in the Southwest. I've met many brilliant scientists while reporting this book, but in their deep commitment to make change for the people and places they work with, the UNM METALS team were some of the most inspiring. In an ordinary-looking bungalow in Albuquerque, Johnnye Lewis – a badass in her seventies with bright red bobbed hair – has built a one-stop solutions factory to identify how nuclear-affected communities can live more safely on their ancestral lands.

The risk is profound. In 2015, the team's Navajo Birth Cohort Study found that dust in 85 per cent of Navajo homes contained uranium, alongside other toxic metals such as arsenic, manganese and lead.[76] As dust does, it was getting into people's bodies: all the babies studied had uranium in their system at birth, and levels of the metal rose as they started to crawl and toddle and explore the world. Urine tests on 700 mothers and 200 babies found that one in five had levels of uranium above the 95th percentile nationally. But in the lack of any substantial previous health studies, there wasn't the science to prove what risk this actually posed. What types of environmental exposure were highest risk: was it air, was it water or food, was it in the home or outdoors? It was beyond doubt that uranium mining waste was a health hazard – as demonstrated by the community's very high rates of cancers, emphysema, kidney disease and also rare childhood developmental disorders such as 'Navajo neuropathy'. But what was uranium actually doing in the body to cause this? That causal pathway was unknown.

In the years since, an interdisciplinary team of researchers from across health science, chemistry, engineering, geography and anthropology have begun to build a comprehensive, beginning-to-end picture of the impact of uranium mining in this region – and how to address the ongoing risks it poses for people and the environment today.

Yan Lin uses geographic information science to map the exposure risk from abandoned uranium mines, as wind, rivers and groundwater transport particles away from their original sources.[77] Then chemist Melissa Gonzales examines the size and mineralogy of transported particles, to translate this data into intelligence about community exposure and toxicity risk.[78] Dozens of students – a significant number from Pueblo and Navajo communities themselves – go out measuring uranium levels in water wells across Navajo Nation, as part of their practical field research methods courses. To understand how this uranium impacts the body, Katherine Zychowski doses mice with representative dust samples to identify the precise physiological pathways involved. It turns out that the primary harm of uranium isn't that it's radioactive, but rather because, as a heavy metal, it displaces zinc and this disrupts how proteins function. In response, Debra MacKenzie is running a clinical trial of zinc supplements as a protective measure against uranium exposure.[79] Meanwhile, engineer José Cerrato is exploring how locally abundant minerals such as limestone can immobilise heavy metals, providing cost-effective remediation methods to prevent them contaminating local water sources – and Eliane El Hayek is exploring bioremediation solutions, looking for plants that are able to grow on a toxic waste dump to hold the uranium safely locked in the ground.

This is true decolonised science: not just a land acknowledgement at the start of your conference presentation (though they certainly do those too), but work that is driven by the needs and priorities of the people who live every day with the issues investigated. Johnnye explained how the METALS centre got started when 'twenty communities in the eastern part of Navajo came to us because they saw one of the first environmental justice grants come out, and asked if we would work with them to answer questions about exposure to uranium. That kicked everything off.' Today the centre employs community engagement director Chris Shuey and artist-in-residence Mallery Quetawki to make sure that this research answers community questions and is shared back into direct, tangible community benefits.

Chris explained: 'Amongst the work we've done in Laguna was concern over wells, if we're growing all these crops near the village of Paguate, are they uptaking contaminants? Is it unsafe to grow? The results of the plant and soil tests that we did early on was that they didn't show much of a problem, as the agricultural lands are upstream of the mine.'

When coronavirus arrived on the reservations, Mallery turned her hand to illustrating how RNA vaccines worked, using familiar Indigenous symbols and metaphors to help people understand this new technology and so feel safe getting the Covid vaccine. Well-water monitoring data is reported back to the people who live nearby, and there are plans to set up an air quality warning system, so people can mask up and stay indoors on days when the windblown dust risk is bad.

That said, it is all still mitigation and not solution: the scientists can't make uranium safe, and they can't make it go away. But by thinking laterally and identifying what they *can* change, 'those are ways we'll have more impact than moving the waste out or capping the whole state,' Debra MacKenzie said.

I confess I had at times wondered if the problem of nuclear contamination across the US, let alone the rest of the world – the whole end-to-end picture of it from mines to mills, and weapons testing sites to long-term storage – might not be too big to solve. The EPA has identified over 15,000 uranium-associated locations across the United States. Many small sites can be made safer relatively cheaply: a few tens of thousands of dollars to close mine entrances, move still radioactive ore stockpiles and nine waste below ground or cover them over with rock and topsoil, and then revegetate the land to help prevent long-term erosion.[80] Sites the size of Jackpile, meanwhile, are orders of magnitude more costly. Collectively, it requires billions.

Is this possible, I asked the New Mexico Environmental Law Center team?

They weren't having any of my doubts. 'The sense of being overwhelmed is one of the strategies that governments and corporations use to maintain the status quo,' Eric Jantz rejoined. 'This issue of financial resources is a lot of crap, frankly. For example, in the situation with Red Water Pond Road, we're talking about GE, who has for all intents and purposes an infinite amount of money at its disposal.' (General Electric's 2021 revenue was 74 billion dollars.) 'But let's say for the sake of argument that the private entities who caused the pollution were not involved. Half a percent of our defense budget [$801 billion in 2021] could deal with this issue in a very meaningful way.' His case adds up.

Chapter 4 touched upon the associations of dust, cleanliness and whiteness – here, we have come to see it more deeply as a medium of colonial power and violence. Nuclear testing fractures the world in two: into those who have to suffer the dusty consequences of the atom bomb at every stage, from mining to explosion to the aftermath. And those who could render these problems invisible, by exporting them half a world away.[81]

It makes me think of the words of a friend, the philosopher and editor Robin James: '"Who cleans up after other people?" is a more important question than "Who governs?"'[82] We could clean up so much of the radioactive mess we've made – if we wanted to. America in the Indigenous southwest and the Marshall Islands; the UK in outback Australia, on land once deemed empty because its Aboriginal inhabitants weren't counted as people; Russia in nomadic Kazakhstan; France in its former colony of Algeria; China in its western provinces populated by the Uighur minority. But all of this would require recalibration – seeing deserts not as desolate and these remote places not as unpeopled. It would require a shift in economic and political thinking from air, water and soil as costless 'externalities' to recognising them as something irreplaceable and infinitely valuable. It would require all of us to heed Indigenous lessons about land, not as a possession to be exploited, but as something vital to our existence.

It will never be possible to return these nuclear landscapes to the way they once were – the defining act of modernity that is the atom bomb cannot be undone. Yet we can take far better care with its wreckage.

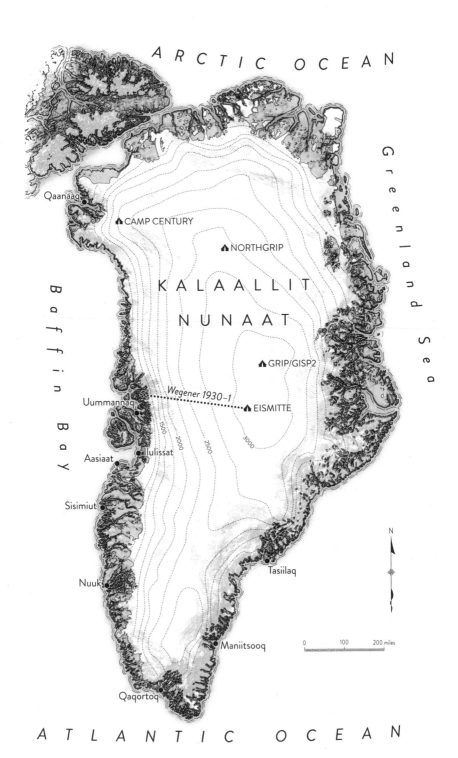

ARCTIC OCEAN

Greenland Sea

Baffin Bay

Qaanaaq

▲ CAMP CENTURY

▲ NORTHGRIP

KALAALLIT

NUNAAT

▲ GRIP/GISP2

Wegener 1930–1

Uummannaq

▲ EISMITTE

1500

2000

2500

3000

Aasiaat

Ilulissat

Sisimiut

Tasiilaq

Nuuk

Maniitsooq

N

0   100   200 miles

Qaqortoq

ATLANTIC OCEAN

Chapter 7

The Ice Record

Greenland – known by its inhabitants as Kalaallit Nunaat – might seem a country on the edge of a void. It is said that the Inughuit who live in the far northwest once believed themselves the only people in the world.[1] The largest island on the planet, it supports a population of just 56,000 people clustered in a handful of towns and sixty-odd small settlements on the narrow fringe of habitable land around its jagged, fjord-sliced coastline. Ever since it was first settled 4,500 years ago, the country has looked out to sea. The sea is Greenland's roadway, its larder, its mythos, its economy. The Atlantic waterways brought first Vikings in the medieval warm period around 984 CE, then whaling boats, then Danish colonists. Here, as in the most arid of deserts, water is life – it is the world.

Yet at the heart of the Greenland landmass lies not liquid water, but ice. The ice sheet in the centre, a 1.7 million square-kilometre expanse second only in scale to Antarctica, supports no plants or animals, and no human life either. Inuit legend holds that it is the home of *erqigdlit*, monstrous beings with dogs' heads but human bodies who were banished inland and viciously attack visitors to their terrain – a tale likely used to keep children from wandering too far from home.[2] The only visitors to the ice sheet are ephemeral: the occasional wandering polar bear or a kite-skiing expedition, plus half a dozen scientific research camps so isolated they might as well be space stations.

It's the ice sheet where we will be spending time in this chapter. I want to persuade you that all human history and more is contained within. Dust in the ice is how we come to understand how climate has changed in the past, and how it may change again – but dust is not just a yardstick, it is also an agent. Scientific research is revealing how dust on the ice will shape our planet in the centuries to come, as it accelerates the deluge as ice sheets melt.

Through its tiniest emissary, the modern world gets everywhere.

***

Flying from London to San Francisco or Seattle, your route will cross over Greenland, and if the clouds collaborate (they often don't), you may be lucky enough to glimpse it through the window. The first sign of land ahead is fragments of coastal sea ice dispersing into navy blue sea, the patterns fractal and fascinating. Next, you fly over the coastal mountains, glaciers carving impossibly slow, fluid curves into the land as they flow into fjords and the sea. Then, you will pass over the ice sheet itself.

It may look dazzling white – a pristine and sterile landscape, alien in its inhospitability. It's not. Look closely and you'll see other colours. First, a deep cobalt blue, perhaps from glacial lakes melting the top layer of snow, or perhaps from the *piteraq*, a wind that roars at over 100 miles per hour and strips the snow from the surface, revealing the older, compressed ice beneath. (*Piteraq* means 'that which attacks you' in Greenlandic.)

The snow may also be pink. In 1818, a British expedition seeking the Northwest Passage to the Pacific, led by Captain John Ross, reported 'crimson-coloured snow' near Cape York, in Baffin Bay, streaking the ice in tints of varying degree, or pooling in meltholes in 'so dark a red as to resemble red port wine'.[3] They returned to England with samples which were first identified as oxidised iron from meteorite debris, before subsequent investigations by microscope revealed the actual culprit was algae, *Chlamydomonas nivalis*, a species that carries an intense red, carotenoid pigment to protect itself from sunburn. It's sometimes called 'watermelon snow', as it's supposed to smell slightly sweet – though it's not as tasty as that sounds: the algae is in fact a laxative.

The ice sheet may also be brown or black from dust. That's actually how the algae gets there too: in dust on the wind, as the next chapter shall explore more fully. But up on the roof of the world, Greenland too is part of this global mineral circulation, collecting dust from around the planet – from windblown sand from the Taklamakan and Gobi deserts, to fertile sedimentary loess soils from Central and Eastern Europe. A tiny fraction is tephra, the rock fragments and glass shards emitted through volcanic activity in the seismic shock zones of the Pacific Rim.

Some of the dust that falls on the ice is known as cryoconite, from the Greek '*kryos*' and '*konis*': 'cold dust'. The first Westerner to observe and name it was Arctic explorer Nils A. E. Nordenskiöld, in 1870, who described 'cylindrical ice pipes' about one to two feet deep and a few inches in diameter, containing 'remarkable powder' at the bottom.[4] The dark colour of these granules absorbed more heat from the sun, thereby melting the particles into the ice – producing the holes Nordenskiöld encountered. He wasn't a fan: it hampered his attempted crossing of the Greenland ice sheet by making the ground too uneven to walk on. Nordenskiöld – and later in the nineteenth century another Scandinavian explorer and scientist, Fritjof Nansen – studied the composition of cryoconite granules and the 'dirty, grayish, or even brownish' dust on the ice surface between the holes, to see what it contained. They found both mineral components (quartz, feldspar and augite) and biological elements (bacteria and algae).[5]

A century on, scientists in Greenland made an astonishing discovery: each kilogram of cryoconite wasn't just earthly sludge but contained about 200 interplanetary dust particles and 800 'cosmic spherules', tiny micrometeoroids that had melted as they plunge from the chill of space into the friction of Earth's atmosphere. In this way, they wrote, the 'massive deposits of cryoconite found on the Greenland ice sheet constitute the richest and best-preserved mines of cosmic dust grains found on the Earth so far.'[6] Most of this cosmic dust will have originated in collisions between comets and asteroids in our solar system: you might see it in the 'zodiacal light' of false dawn as sunlight scatters off its silicate minerals. But a fraction will be interstellar, from far, far out between the stars. What a thing to find in the ice sheet's crevices and gutters – yet it's not even the only cosmic lesson the cryoconite contains. Japanese scientists have also defrosted tardigrades – the internet's favourite microscopic 'water bears' – from Antarctic ice, where they had hibernated in suspended slumber for thirty years. The melted cosmic dust grains carry in their very chemistry a record of the atmosphere they once plummeted through; meanwhile, these extremophiles hint at the kinds of life that might survive in quite other atmospheres altogether, on planets far different to our own.[7] What began as empty nothingness is, you start to realise, a microcosm. The ice is alive – and dirty. Visibly dirty.

The western edge of the Greenland ice sheet is known as the 'Dark Zone'. In 2013 and 2014, glaciologists Johnny Ryan and Jason Box shared grubby and disturbing aerial photos of this area as part of their crowdfunded scientific expedition, the Dark Snow Project.[8] The ice looked like a ploughed field, brown and rutted in every direction extending for mile upon mile out to the horizon and beyond. In some photos the ice is cracked into plates, like dried-up clay on a lakebed, riven by paler crevasses where meltwater drains down deep beneath the ice. The splayed elephant's foot of a glacier protrudes towards the sea, striped with dirt. It is troubling to look at. You feel instinctively that something must be wrong: ice isn't supposed to be anywhere near this dark.

'I was just stunned, really,' Jason Box told *Slate* magazine in 2014.[9] The ice was the darkest it had ever been. His analysis of NASA MODIS satellite data found that high-latitude forest fires had been more powerful that year than ever before, as boreal forests burnt at unprecedented rates. He believed the two things were connected, as the wind carried ash and soot from fires in Siberia and Canada's Northwestern Territories over to Greenland. The Dark Snow Project sought to understand what proportion of the ice sheet dirt was from these fires, and what from other sources, such as windblown mineral dust or factory emissions.

The problem with dirty snow is it messes with a major climate feedback loop: albedo. A measure of reflectivity, albedo is judged on a scale of 0 per cent (all light is absorbed) to 100 per cent (total reflection). Fresh snow has a very high albedo of up to 90 per cent, which is why it looks bright white: light bounces right back off it. As climate scientist Marco Tedesco says in *Ice: Tales from a Disappearing Continent,* his memoir of working on the Greenland ice sheet, 'looking straight at the sun or staring at fresh snow on a sunny day are more or less the same thing'.[10] (Ultra-dark, wraparound sunglasses – 'glacier goggles' – are essential kit for ice sheet research.)

Because the snow is so reflective, radiant energy from the sun bounces off, keeping it cool – and allowing more snow to settle. It's a 'virtuous circle': a positive, self-reinforcing feedback loop. Fresh snow falls, which increases how much sunlight is reflected, leading to cooler temperatures that allow more snow to settle and not melt, which keeps things cool, and so on. Problem is, the opposite feedback loop also holds true. Darker bare ice or dirty snow absorbs more of the sun's energy and gets warmer, so snowfall melts instead of settling – leaving the surface still bare, still dark – and therefore warmer.

Albedo is in its essence a simple principle – and yet in practice it accumulates, feedback loops piling on top of other feedback loops into an extraordinary interlocking complexity. The problem is that each mechanism compounds, like interest, and the warming and melting accelerate, bringing ice-sheets – and all of the Earth's systems that connect to them – closer to risky ecological tipping points. Closer to the possibility of collapse with no return.

Let us count the ways the ice is warming. Dust keeps getting itself involved.

We start at the tiniest of scales. Each year, as the Arctic summer progresses and the snow gets older and warmer, its ice crystals get bigger. They get more jagged and visually rougher, and so they're a little bit less reflective than fresh new snow – that is, they're a little darker. This darker snow absorbs more sunlight, thereby warming it: feedback loop number one.

As the winter snow accumulation melts away, the darker, solid multi-year ice beneath it is exposed – which has a dramatically lower reflectivity of between 30 and 60 per cent, meaning that it too absorbs more energy from the sun, and warms.[11] Feedback loop number two.

Meanwhile glaciers have ground down on the rock below for eons, producing large amounts of fine sediment that scientists call 'glacial flour'.[12] Strong winds pick up these particles, carrying them locally and regionally – where they join sand from the deserts of Africa and Asia that has surfed the planet's jetstreams to form a layer of dust atop the ice. Icelandic volcanoes add ash and tephra, and then there's soot – pure black carbon – from diesel engines, wood-burning stoves and forest fires, which absorbs the most heat and light of all. This windblown dirt reduces albedo even further, down as low as 20 per cent, until the ice is absorbing four times as much light as it reflects.[13]

This dust has its own feedback loops too. 'Now that seasonal snow cover in the Arctic is retreating earlier than before, bare soil is available earlier in the spring for dust transport,' Marie Dumont of the French National Centre for Meteorological Research has explained.[14] Dust blows onto the ice sheet earlier in the year and starts the melting process sooner. Additionally, the changing climate means the land is warmer, the exposed rocks drier, and winds are stronger – meaning more dust is transported onto the ice as well.[15] Using satellite remote sensing data, Dumont and team created a simulation model of the entire Greenland icecap surface. They found that a decrease in the albedo of fresh snow by just 1 per cent would lead to the loss of an extra 27 cubic kilometres of ice per year – doubling the pace of its melt and starting to expose more bare ground as glaciers withdraw inland.[16] So that's feedback loop number four.

Next, rising temperatures mean more water is liquid, and for longer, offering a cosier environment for algae to grow – while the greater quantities of dust gives them more nutrients to nibble on, too. 'Consequently, as dust content increases, microorganisms are likely to increase, too,' Dumont told Christa Marshall. Recent research finds that they thrive on the phosphorus released from the local rocks, which produces yet another positive feedback loop: dust from exposed rocks spurs algal growth which darkens the ice sheet, leading to yet more melting and more rock being exposed. Loop five. They may be microscopically tiny, but in large number they are mighty: these miniature plants on the glacier cause up to 13 per cent of the surface melting in this region.[17] The ice sheet that at first seemed like a great white nothingness is instead alive with hundreds of millions of cells per square metre – a living, breathing, biologically dynamic environment. In one of the world's harshest environments, life is finding a way – a life sustained by what once might have seemed the dullest and most inert form of matter, but which turns out to be in its own way vital, too.

It would be something quite beautiful were it not also so destructive in its consequences. Perhaps it still is, even so – the ice painted in these sweeps of deep red, and orange, and green. Another name for *Chlamydomonas nivalis* is 'blood snow'. But let us return to the ice sheet's spiralling array of vicious circles. They continue.

From 10 to 11 July 2012, the Greenland ice sheet experienced an unprecedented (and as yet unequalled) event: 97 per cent of the surface was melting.[18] The ice was melting at 80 degrees north, only a few hundred miles from the North Pole; it melted even at Summit, the highest central point on the ice sheet, a mountainous 3,200 metres above sea level. The fundamental characteristic of an ice sheet is that it should be frozen – yet for these two days, it wasn't. A ridge of high pressure had lodged itself in the upper atmosphere, bringing warmer Atlantic air flooding over the island while soot from a bad fire season in Siberia and North America that summer exacerbated the heating effect by darkening and thus further warming the ice as well.

As we see each year now on America's West Coast, rising temperatures lead to more and more severe forest fires. Yousuke Sato and colleagues at Japan's Nagoya University warn that current atmospheric models 'underestimate the dirtiness of Arctic air' more than fourfold – producing 'profound impacts' on global heating.[19] Other scientists agree that soot is a cause for concern: Kaitlin M. Keegan has found that black carbon not only caused the 2012 melt event, but Greenland's previous widespread melt in 1889, too.[20] Her team also found yet another vicious feedback loop in which the meltwater refreezes as a darker ice layer in the snowpack, 'causing lower albedo and leaving the ice sheet surface even more susceptible to future melting'.

That's six interlocking feedback loops, by my count – there are more, perhaps countless intersections between mineral dust, carbon, microbes and snow surface darkening, but we haven't got forever to tally them. Melt ponds on the ice sheet grow, day by day, a brilliant, blinding, terrifying blue, until they drain in a gulp and water rushes in not a stream but a river, down through the moulins, the great well shafts that sink vertically into the ice, melting luge track curves into its frozen immensity. This vein of liquid water descends treacherously to the bottom of the ice sheet, to melt it from beneath. Thus lubricated, the great glaciers themselves accelerate, and flow faster to the sea.

***

How much ice will be lost from Greenland and the Antarctic is one of the great open questions of the century. Dust is a crucial variable shaping that answer.

The Greenland ice sheet is on the verge of a tipping point, a recent study by Earth system modellers Niklas Boers and Martin Rypdal warns. As the ice melts, its decreasing height exposes the surface to warmer air found at lower altitudes, causing further, accelerating thawing. But this does not mean inevitable oblivion, the scientists emphasise: ice equivalent to one to two metres of sea level rise is probably doomed, but this will take centuries to melt (and the Greenland ice sheet as a whole would take a millennium). By 2100, we may expect between 30 centimetres to 1.1 metres of sea level rise – of which ice sheets are likely to contribute up to 42 centimetres.[21] The chief contributor is not meltwater but thermal expansion: as water gets warmer, its atoms have more kinetic energy, and thereby take up more space. Beyond that the future grows unforecastable, and there is everything yet to play for. The fate of the ice sheets is not yet sealed, and the faster businesses, states and global high-consumers can switch away from fossil fuels and climate-wrecking carbon consumption, the more ice can be preserved. There may yet be a stable state for a smaller ice sheet, Boers and Rypdal tell us – though none of us will live to see it.

Yet even in the most benign of futures with just a metre or two of sea level rise, Greenland's melt will nonetheless rewrite the world's coastlines and, with it, the contours of our civilisation. Our cities will have to retreat from the waterways that birthed them, as lower Manhattan is deluged by storm surges and Miami a gonner. Pacific Island nations such as Kiribati will either have to relocate en masse, perhaps to New Zealand, or be dispersed into a global flow of hundreds of millions of climate refugees. Meanwhile, the vast flows of fresh water unleashed from the melting ice may potentially disrupt the circulation of warm ocean currents such as the Gulf Stream.[22] Seawater at high latitudes is meant to be salty, and therefore dense, meaning it sinks: this powers the 'thermohaline circulation' that makes warmer surface currents flow from the tropics towards Europe's Atlantic coasts. Yet into this system comes 268 billion metric tons of melting ice per year – or twice that in 2019, a dismal record.[23]

Oceanographers call it the 'Greenland freshwater flux anomaly' and measure it in cubic kilometres. The cutting edge of research is to understand where this freshwater travels and whether it will produce a phase shift in the Atlantic circulation – potentially ending ten thousand years of post-Ice Age climate stability, and heralding a strange and unstable world. The Gulf Stream keeps the British Isles temperate, letting palm trees grow in Cornwall and fuschias bloom in West Cork. These may become things of the past.

This transformation of the planet's polar regions will not be a smooth process, but one punctuated by sudden phase shifts. Each year it feels as though the reports from the Arctic summer field-work season come back worse: glaciers moving faster, more ice lost, yet more predicted. As I was writing this chapter in December 2021, Antarctica was making headlines. Glaciologist Erin Pettit was warning that the Thwaites Eastern Ice Shelf might finally collapse within as little as five years. The ice shelf – the part of the Thwaites Glacier that extends out to float over the sea – acts like a dam slowing the flow of ice off the land. Yet rifts and fractures are spreading across it like cracks in the windshield of a car. At some point it will fracture, and the Thwaites – which already contributes to 4 per cent of global sea level rise – will accelerate.

As we'll also see later on in this chapter, dust down deeper in the ice sheet tells the story of the Younger Dryas period when average temperatures in the Greenlandic Arctic shifted by as much as 5°C in just thirteen years. Feedback loops feedback on themselves and multiply. In this way, Greenland is not a remote island but a powerful actor in the Earth's environmental systems. This makes our easily overlooked subject, dust, a powerful actor too. It starts to have enormous consequences – like Lorenz's butterfly that flaps its wings in Brazil and gives rise to a tornado in Texas.[24] Complex systems and feedback loops give rise to non-linear behaviour. They make the world more dangerous and unpredictable.

Greenland's ice is still there, and it is still vast. But it's haunted by its future – which makes it something deeply, terrifyingly uncanny to witness.

***

I'd wanted to go to Greenland since I was eighteen, for the same reason I used to dream of subarctic Siberia: the vastness. I was fascinated by the question of how people might live in a climate that seemed, to my temperate European eyes, to be on the edge of human possibility, quite indifferent to their wellbeing. Finally, in 2016, I was able to make it happen, travelling once again with the always-adventurous Wayne Chambliss.

As I planned the trip, Greenland's impossibility grew ever more present tense. Month by month, 2016 set record lows for Arctic sea ice coverage. Come spring, the surface of the ice sheet started melting two months early due to freakishly warm temperatures: 16.6°C in Nuuk, the capital, on 11 April – twenty degrees above average. It made me feel as though we were going to see this place while we still could.

After three days of travel in increasingly diminutive aircraft, my friend, Wayne, and I arrived in Ilulissat, population 4,500 people and 4,000 sled dogs, and as such the second largest town in Greenland. It was late May, the air temperature a couple of degrees above zero, and the days never ending: we were at 69°N, and this was the Arctic summer. The next day, we hiked out to the Ilulissat icefjord.

The fjord was about a mile away from town. It's a UNESCO World Heritage site, which at the time mostly manifested in a boardwalk over a large bog that you took down towards the shore, before coloured dots painted onto the local granite marked a hiking trail of sorts. (There's now an elegantly twisting minimalist pavilion, designed by the Danish architectural studio Dorte Mandrup which specialises in 'complex and challenging sites'.) We passed the ancient settlement of Sermermiut – and then just stood on the headland and looked.

The mouth of the fjord is about seven kilometres wide, and it was choked with giant icebergs. Tall, pointed Himalaya-esque peaks, towering cliff faces and tempting snow-smoothed ski-slopes. Bergs so big they had their own ponds, caves and arches. Countless 'bergy bits' – the marvellous and entirely legitimate technical name for icebergs under five metres tall – and growlers, the littlest ones under one metre. A litter of snow and ice fragments covering the sea surface like rubble from the explosions at the glacier's calving front. Barely a metre of clear water to be seen.

Thirty-eight gigatons of ice flow down this fjord per year, three times the water usage of the entire United Kingdom.[25] Sermeq Kujalleq drains 6.5 per cent of the Greenland ice sheet and is the fastest moving glacier in the world, at up to 40 metres per day in the summer. (It is accelerating.) From its calving front, the bergs take about fifteen months to travel 70 kilometres down the fjord to the open sea. The largest bergs, nearing a kilometre in size (most of it, of course, underwater), get caught on the submarine ridge at the end of the fjord, producing a spectacular traffic jam – until they break up enough to drift off into Baffin Bay and the North Atlantic. In early autumn 1911, this glacier likely calved the iceberg that sank the *Titanic*. Catastrophe is integral to the geography of this place.

As we took it all in, a deep blue vertical pressure seam on a 70-metre-tall berg exploded in front of us. Wayne went down to the shoreline and fished out a chunk of clear, hard ice. I bit into a corner: it tasted *old*. Ice at the bottom of the Greenland ice sheet dates back up to 130,000 years, since before the last ice age. The eruption up to the surface of a deep gob of geologic time.

This is not and has never been a static landscape. Ice is dynamic. It always flows downhill and spills out into glaciers. It always melts a little in summertime and replenishes from winter snows. The icefjord has existed as long as the ice sheet; it is not itself a symptom of sickness. But the mind shies away from trying to comprehend the ice sheet in full systems perspective – hydrological cycles, global warming, human responsibility – and absolutely blanks at the question of what it all might mean.

It was easier to contemplate these bergs than the dazzling white oblivion that was the ice sheet itself. It was all too immense, too much. The ice was too big to think about – both demanding our attention, demanding we stop and sit and look, and simultaneously, stilling the language we had to think about it. It is hard to focus at such times: you find your mind twisting away from confronting the existential mass, the gravity of what is right in front of you. It is easier to think smaller thoughts.

It's a romantic problem to have. As Ruskin said: 'Sublimity is, therefore, only another word for the effect of greatness upon the feelings; – greatness, whether of matter, space, power, virtue, or beauty.'[26] Greenland was all those things at once.

Wayne mentioned eco-theorist Timothy Morton and his concept of 'hyperobjects'.[27] It was fitting for this landscape: I spent much of my time in Greenland trying to, in some way, comprehend this vastness, this silence, these glaciers and how they enmeshed in yet vaster global hydrological systems. The consequences. I'm not sure I got anywhere. I still don't really know what that comprehension would even feel like. Morton would say that the point of a hyperobject like an ice sheet is that we can't properly perceive them at all. Dust may be a method for bringing immensities down to human scale, but I can't promise you'll always succeed.

***

The first scientific research camp on the Greenland ice sheet was named Eismitte, 'Ice-Middle'. It was set up in July 1930 by the German scientist Alfred Wegener and a team of thirteen Greenlanders and three other Europeans as part of an expedition to seismically measure the thickness of the ice cap and take the first winter climate measurements.

Wegener is an interesting, transitional figure. His 1930–1 expedition was 'a watershed event,' writes biographer Mott T. Greene, 'bridging the old (sailing ships, ponies, dogs, exploration of uncharted territory) and the new (motorboats, motorized sleds, radio communication).'[28] His experiences of suffering and starvation frame his journeys as 'the last pulse of the great age of heroic exploration', even while his motivations were somewhat different: not the military Romanticism of the polar explorers but a markedly more modern drive towards scientific research and knowledge. (Eight years prior to this journey, Wegener had originated the theory of continental drift.) Like Knud Rasmussen a generation before, his journeys were slightly more of a partnership with local Greenlanders and certainly quite explicitly reliant on using Greenlandic knowledge, clothing and technology in order to survive. In contrast to Scott of the Antarctic's wool and canvas, Wegener wore head to toe fur, Inuit style: the best technology for the conditions.

Despite being an old Greenland hand, Wegener found his 1930 Eismitte expedition – his fourth – not going quite to plan.[29] A late spring thaw trapped the team at their first camp, 30-odd miles north of Uummannaq and halfway up Greenland's west coast, for nearly six weeks. With two dozen Icelandic ponies hauling a hundred tons of supplies 400 kilometres inland, they reached the prospective Eismitte site at 70.5°N on 30 September, much later in the year than planned. There isn't really much of a shoulder season in Greenland. Contemporary field camps pack up and fly out in mid-August, after just a six-week summer drilling season. Late September isn't autumn as much as winter.

Lacking sufficient stores to last out the months of Arctic night, the men needed to make a return journey to the coast and back out again in order to resupply – a further 800 kilometres round trip. Wegener, meteorologist Fritz Loewe and the fourteen Greenlanders appear to have made it back to West Camp without serious incident. On the return journey east, however, heading deep into the centre of the ice sheet in what was by now winter, all but one of the Greenlanders chose to return to base camp. Wegener, Loewe and 23-year-old Inuit Rasmus Villumsen continued, arriving at Eismitte on 30 October 1930. Temperatures during the journey had fallen to a bone-chilling -60°C and Loewe's frostbitten toes had to be amputated by penknife. He stayed put at Eismitte for the winter, alongside Ernst Sorge and Johannes Georgi. There wasn't enough food for five, though. So Wegener and Villumsen had to return 400 kilometres to the coast for a second time.

They had to travel fast and light – taking just two sleds and no food for the sled dogs, only the minimum for themselves. The dogs they killed, one by one, to feed to the others. Wegener survived his second Greenland expedition in 1913 in this way, going down to the very last dog before they were rescued.

This time, though, the ice won out over will. The next spring, another expedition heading out to Eismitte found a pair of skis stuck upright in the snow, a broken ski pole lying between them. Digging down a way, they found Wegener's body. Villumsen had buried him laid atop a reindeer skin and stitched into two sleeping bag covers. The men who found him said he looked 'relaxed, peaceful, almost smiling'.[30] Villumsen had taken just Wegener's tobacco and his diary, the official expedition record, in presumed hope of making it onward and back to the western camp. His body was never found.

Both men remain, then, on the ice sheet – or rather in it, blanketed under hundreds of metres of snow, their quiet mausoleum shifting ever-so-slowly westward as the North American plate drifts, and the ice sheet flows on down from Summit to the sea.

That winter in 1930, Ernst Sorge hand dug a 15-metre-deep pit next to the ice cave he and Johannes Georgi were living in. He spent seven months in that frozen hole, two of these months in complete polar darkness, analysing the strata formed each year as snow compresses and crystallises first into neve, then firn, then ice. His research demonstrated the feasibility of measuring the annual snow accumulation cycles – which glaciologists have been doing in increasingly elaborate manner in Greenland and Antarctica for the last seventy years.

Ice sheets are dynamic, plastic environments. To either side of the central divide, the ice slowly sheers and deforms as it flows into the sea, until at the glacier's edge it is a ragged, crevasse-ruptured terrain too rough to walk across. In the centre, though, the layers of the annual snowfall cycle are preserved, one atop the other, like the rings of a tree. To drill straight down is to drill back into time.

People had been drilling holes into glaciers since the 1840s to take ice samples and measure temperature, but it wasn't until 1950 that they started to try to extract ice cores: cylindrical tubes of ice, typically about 10 centimetres in diameter, capturing a continuous vertical cross-section of an ice sheet.[31] Drilling was difficult and tedious work: it might take an hour just to lower the drill string into the hole, screwing on each section of drill pipe as it descended, before the scientists could drill for just a few minutes. They'd then spend another hour or more pulling the drill back out of the hole again, unscrewing everything as they went, in order to extract the next metre of ice core they had obtained. After all that, the drill might be opened to find they'd captured little more than ice chips – and so these first cores they extracted were not the hoped for continuous record, but a mess of gaps and breakages. Progress was quick, however, and by the late 1950s ice core teams in both Greenland and Antarctica were getting down to over 300 metres and extracting cores that were about half complete, a considerable improvement.

The United States' presence in high northwestern Greenland at this time was not solely a matter of innocent scientific enquiry. Following the Second World War, the US Air Force had constructed Thule Air Base 1,200 kilometres above the Arctic Circle as a major location for reconnaissance flights over Russia, missile early warning systems, and housing nuclear bomber planes. Ten thousand Americans duly arrived on a base that's still there today. (President Trump's seemingly bizarre 2019 statement about 'buying Greenland' was in partial reference to this quasi-colonial presence.) One hundred and thirty Inughuit people were forcibly relocated from their homes and hunting way of life in a village they called Pituffik. Their descendants are still seeking reparations today.

A decade later, in 1959, 240 kilometres east up on the ice sheet at Camp Century, the top secret American Project Iceworm sought to establish a nuclear missile site under the Greenland ice, putting medium range ballistic weapons within striking distance of the Soviet Union without either Greenland or Denmark knowing anything about it. Two miles of trenches were excavated under the ice and an entire town was built, with a ten-bed hospital, shop, movie theatre and chapel, all powered by a mobile nuclear reactor. Yet geologists found that the ice at this location was much more mobile than anyone had expected, and within just three or four years the trenches had shrunk and deformed inward, making the site unusable. Project Iceworm was scrapped, and Camp Century would be crushed by the moving glacier in less than a decade. Its legacy was twofold: 200,000 litres of diesel, polychlorinated biphenyls and radioactive waste abandoned on the site – and the world's first truly deep ice core, which hit bedrock in summer 1966 at an unprecedented 1,387 metres.[32] The ice at the bottom was dated at over 100,000 years old, and provided an unprecedented insight into the climate throughout the whole of the last glacial period in North America.

Each decade, the holes got deeper and the ice older. In 1985, after six drilling seasons, the Russian *Vostok* 3 hole in Antarctica reached 2,202 metres, collecting ice dating back 150,000 years from the Ice Age before last. In Greenland, the 3,000 metre mark was reached by two European and US teams working side by side in the early 1990s, not competing but cross-referencing their findings as they drilled back into a never before seen world. In 2004, the EPICA project at Dome C in Antarctica pulled up a core that covered a mammoth 800,000 years and eight glacial cycles.[33] The ice core was not that much longer than its Greenlandic cousins, at 3,270 metres – but, as Dr Liz Thomas, a paleoclimatologist and head of the BAS Ice Cores group, told me, 'the centre of Antarctica is actually classified as a desert: it receives less than a millimetre of snowfall in a year'. (In Greenland, it's more like 70 centimetres in the centre.)[34] As such, drilling gets you 'a lot more years for the same depth,' she added.

In 2016, ice cores retrieved from the Allan Hills in Antarctica were found to contain ice an astonishing 2.7 million years old – older than all humanity. It was a clever, seemingly counterintuitive choice of site: instead of aiming vertically down through the stable, less-flowing centre of the ice sheet, the Princeton team drilled horizontally into a shallower region where an underlying rocky ridge tipped the ice flow onto its side and jumbled up the stratigraphy. Strong winds at this site also stripped it bare of fresh snow and even younger ice: as such, there was a chance that very old ice might sometimes be nearer the surface. They struck lucky, extracting a cylinder of frozen water that told the story of the very start of Earth's cycle of ice ages and interglacials.

'We're just scratching the surface,' geochemist Ed Brook told *Science* magazine: they reckon a future attempt might be able to find ice that's fully five million years old. This would date back to a time when carbon dioxide levels were the same as today's, at over 400 parts per million.[35] During this period, in an epoch known as the Pliocene, temperatures were 2–3°C higher than now – and global sea level 25 metres higher. Greenland was forested and the Arctic sea ice ephemeral.

This sounds all too terrifyingly familiar – which is why geoscientists go to such great lengths to drill into ice sheets. To measure Earth's ancient climates is to draw a possible map of the future. How does a hotter, higher carbon dioxide planet work? How much rain falls? Where do its winds blow? And what do they carry? Desert sand or savannah pollen? Geologist Richard B. Alley describes it as akin to decoding the 'Operator's Manual' to Planet Earth. 'Land and water, air and ice, soils and plants – if we can figure out how they work, how they are wired together and depend on each other, maybe we can then make wise decisions about global warming, ozone depletion, and other globe-girdling questions.'[36] Land and water, air and ice, soils and plants – and dust. Far from inert, dust is an active a part of the Earth's system.

But for now, let's keep our gaze focused on the ice cores – these fragile, frozen chronicles that carry such terrifying insight into the past, and therefore our future (and therefore our responsibilities now, urgently, in the present). They are beautiful in their clear, crystalline grace, hard-won from some of the most extreme places on this planet, and tremendously rare. At the time of writing, just 22 ice cores have captured a full surface-to-bedrock transect.[37] But they are also unimaginably valuable: climate modelling depends on the high-resolution lens they provide, revealing the Earth systems of the past. Without this precise, year by year data, it would be impossible to forecast what's next.

***

Extracting a 10-centimetre ice core from a 3-kilometre hole can be a challenge. Eric Wolff of the British Antarctic Survey has led its ice core programme for the past thirty-plus years and is a veteran of six Antarctic and two Greenlandic research seasons. He described how most ice cores have a section called the 'brittle zone' which 'tries to explode' as its air bubbles expand in the dramatic change of pressure.[38] But even with other sections of the core, 'I get really nervous that you're going to drop it,' he added. Wolff knows this is irrational: under thousands of years of pressure, snow crystals sinter together so well that 'you could actually stand on either end of a 3-metre piece of ice and it would be fine.' But the ice is so precious it seems fragile, nonetheless.

Once out of the drill hole, each section of ice is taken to a field laboratory built right into the ice sheet itself. Field logistics crew use snowblowers to dig trenches and snow caves up to six metres deep to use as processing facilities, which stay at a helpfully chilly -20°C and keep the ice core out of the surface sun. Working in shifts around the clock, a processing line of a dozen scientists inspect, photograph, cut and measure each chunk of ice as it emerges: melting a piece to measure its chemical composition or subjecting a length to electrical conductivity analysis to analyse impurities. The already narrow core gets sawn lengthways into perhaps five to twelve sections to enable different types of analysis, and also so it can be distributed around the world (ice core research is all about international cooperation, but this means everyone wants a slice). But at least one section may be left in storage at the Antarctic base too, so, as Eric Wolff explains, 'if everything went horribly wrong on the ship on the way back – if the freezers broke down and the core melted – there would still be something left'.

Back in the lab – whether it's the British Antarctic Survey's facility in Cambridge, the US National Ice Core Laboratory in Denver or the Niels Bohr Institute in Copenhagen – the first step in ice core analysis is to build a timeline. In the upper, younger parts of an ice core, dating can be as simple as counting the annual layers, just as you would the annual rings on a tree stump. On the Greenland ice sheet, it snows all year round, but in the summer the snow gets heated by the sun and sublimes (that is, it turns straight from ice to water vapour), leaving large, airy hoarfrost crystals behind.[39]

Melt events on the central ice divide where drilling takes place are rare – they used to occur only every few centuries – but three have occurred in the last decade, including unprecedented rainfall on 14 August 2021, which froze down into a distinctive hard, icy layer. To make a visual record of this stratigraphy, the ice core section was polished smooth and then photographed with a line scanner to capture its stripes. Paler, cloudy bands revealed the presence of mineral dust transported onto the Greenland ice sheet at particular times in the last glaciation period. Hard, dense ice appears darker. Deeper down, as the snow got compressed into glacial ice and the ice sheet flowed and spread out, its layers became less reliably visible to the naked eye. Analysis continued. Scientists tested the electrical conductivity of the melted ice core to measure impurities and find winter dust layers, or looked for hydrogen peroxide, which was only produced in summer sunlight-driven reactions. Meanwhile, mathematical flow models helped estimate the relationship between ice age and depth.

Dust is crucial to understanding ice core chronology. Occasional volcanic eruptions might leave a yellowish stain in the ice, or irregular blotchy debris from bigger pieces of tephra rock fragments. The WAIS ice core from Antarctica has a remarkable dark, brownish-black line almost a centimetre wide from a massive eruption 21,000 years ago, which must have completely covered the ice in a deep layer of ash.

'The volcanic eruptions are fantastic in terms of our dating,' BAS's Dr Liz Thomas told me, 'because you can link it to other well-dated ice cores,' and thereby see where a section of ice fits into the environmental record. 'In the more recent times, you can actually link it historically as well,' Thomas continued. 'You know what date Tambora erupted [1815], or Krakatoa [1883]. And you have the historical records of how big the eruption was and how long it lasted, so you can link back to what you see in the ice core.' Volcanic debris is also particularly helpful for the Antarctic, which gets a lot less snowfall than Greenland, meaning its narrower annual ice layers can be less distinct.

But a big volcanic eruption will blast cubic kilometres of pulverised rock up into the stratosphere, where it circles the globe on high-level air currents and lands on every ice sheet on the planet – letting scientists match ice cores from opposite ends of the world. Two specific layers of dust – the Saksunarvatn tephra and Vedde Ash, both remnants of Icelandic volcanic explosions about ten and twelve thousand years ago – enable common referencing between the Greenland ice cores and sedimentary rock records across the North Atlantic. Since 1945, atomic bomb blasts have done much the same, the radioactive isotopes they spray out proving a highly distinctive signature. For geologists, then, dust literally defines shared 'horizons'; it creates knowability across vast distances in time as well as space. We might say it's a sort of paleo-communication: it allows geological archives to talk to each other.

Once the ice core has been turned into a neatly labelled timeline, it's time to start exploring what it can tell us about the climate and environmental systems at each point in time. Thomas explained to me how this works: 'The easiest way of thinking about it is that whatever there is in the atmosphere has the potential to be deposited on an ice sheet.' Gases, aerosols and organic contaminants are chemically 'scavenged' out of the atmosphere and become part of the snow as it falls. Meanwhile, 'dust and larger things that have been lifted off the ocean may be transported by the winds, and then just dropped and deposited on the ice. The following day, or whenever more snow falls, it traps this material – and so you end up with a time capsule of all this information about what was in the atmosphere at the time.'

Glaciologist and climate scientist Richard B. Alley calls ice cores a 'two mile time machine'.[40] They're thrillingly direct: real air from hundreds of thousands of years ago is preserved unaltered, making the ice sheets the only source of direct, non-proxy paleoclimate information. In a sterile cold lab at a chilly -20°C, segments of ice core may be melted very carefully on a chemically inert, gold-coated plate. Potentially contaminated water from the outside of the core is rejected until only the inner part, which has never been in contact with the contemporary environment, is exposed. This is melted, 'debubbled' to extract the ancient air trapped within and then subjected to dozens of types of analysis. How much carbon dioxide was in the atmosphere at that precise moment in time? How much oxygen? How much methane? Comparing the frequencies of the rare, heavy versions of oxygen (oxygen-18) and hydrogen (hydrogen-2, or deuterium) against 'normal' water provides a proxy for past temperatures. Heavy isotopes require more energy to stay airborne and so evaporate less and precipitate out soonest as rain or snow, meaning that when conditions are cold, such as during a glacial period, fewer of them make it as far as the central ice sheet. Cosmic radiation produces radioactive beryllium-10, which provides a proxy for past levels of solar activity, and how much of the sun's warmth reached Earth.

Some of the melted ice core will be sent for mineral dust analysis. An Abakus laser sensor machine counts the number and size of particles in the ice and a mass spectrometer helps identify what elements they are. So after it's done the initial dating job, dust then also provides evidence on ancient 'paleoclimates' at each point in time.

It turns out that the Ice Age world was an exceptionally dusty place, with between two and twenty-five times as much dust in the atmosphere as there is today.[41] During the last glacial period the climate was harsher: colder, of course, but also more arid, with strong winds and frequent storms. Huge amounts of water were locked up in continent-spanning ice sheets, so sea levels were 125 metres lower than today – and whole swathes of continental shelf became land, opening up more land area for dust transport. When the wind blew, then, it had the potential to lift huge amounts of sand and soil into the atmosphere – and with less rain, more dust stayed aloft, journeying thousands of miles to leave its signals in the ice of Antarctica and Greenland.

Analysing the size of the dust could offer clues as to the wind speeds, Eric Wolff explained, 'when it's windy, it's easier to transport large particles', and different source locations can be 'fingerprinted' by their chemical composition, as the geology of each place has a characteristic ratio of strontium and neodymium isotopes. In this way, scientists can identify how wind currents crisscrossed the planet to carry the dust to the ice sheet.

Then just as you can taste the salt in the air when you're near the seaside, the salt levels in the ice reveal how far a place was from the sea when each stratum of snow was laid down – and therefore how sea ice changed over time. Invisible shards of glass show how active volcanoes were at each point in time; levels of black carbon soot reveal forest fires. Pollen grains reveal something of the plant life in neighbouring, more temperate areas. Sometimes the analysis can be quite ambiguous: did dust levels increase because the source regions had become more desertified? Or is it because the winds are stronger, or there's less rain on the way, so the dust isn't deposited out? 'It's still an argument in our community as to which of those is most important,' Eric Wolff told me – as climate modellers race to update their computer simulations with new findings from the ice record.

But sometimes, the evidence in the dust can be astonishingly specific. In 2018, an international team of hydrologists, ice core scientists, classicists and archaeologists announced that lead pollution levels found in the ice core of the North Greenland Ice Core Project (NorthGRIP) mapped exceptionally closely to historical events, 'including imperial expansion, wars, and major plagues'. Why? The economies of ancient Rome, Carthage and Phoenecia ran on silver-based currencies, which were smelted from ores that contained lead. As such, lead pollution levels provided an index of how much money was being minted at any one point in time and could be taken as a proxy of economic activity. Lead emissions dropped at the outbreak of the first Punic War, in 264 BCE, as miners were drafted into the military, but then rose as Carthage was forced to mint additional currency to pay mercenaries.[42] Julius Caesar's efforts to restore order and stability to the Roman Empire produced a rise in lead emissions, as the economy flourished – which was then halted or reversed by his civil wars.

'You could look at pollutants and organic compounds and be able to trace that down to a particular power station or biological source,' Liz Thomas explained. One of her current projects looks at tiny single-celled algae called marine diatoms, as well as sea salt levels, providing information about wind strength and speed that lets climate scientists model past atmospheric circulation. Some signalling impurities are present in the ice at the magnitude of just parts per billion – as if the whole ice sheet was a clean room. It's hundreds of kilometres from any soil, and thousands of years prior to any measurable human action.

'Detection methods have to be very sensitive,' the Niels Bohr Institute's Centre for Ice and Climate says, 'able to detect concentrations similar to a pinch of salt in a 50 metre swimming pool filled of pure water.' [43]

It's detective work, really – cold, arduous, multi-year detective work in some of the remotest places on Earth – seeking out these microscopic 'proxies' and time series that cumulatively let climate scientists deduce what the planet must have been like long before the first *Homo sapiens* ever roamed its grasslands.

Dr Peter Neff is a glaciologist at the University of Minnesota with a PhD on dust and how it is transported to the Antarctic ice sheet – and also a popular TikTok creator, as @icy_pete. I reached out to him to understand a little bit more about what it actually felt like to work in such an astonishing environment.

'We work so hard, typically for a decade before any fieldwork takes place for a large drilling project to bedrock,' he explained. 'So the feeling when you succeed, hit the bottom and see green ice and mud 760 metres deep or deeper,' as he did in 2012–13 season on Roosevelt Island in Antarctica. 'It's a relief that the tedium of drilling and producing core is done, excitement at seeing that unique material. And *man* is there a release of stress, especially for the scientific leader that night at the dinner table . . . folks can finally let their hair down and we all share a nice glass of whiskey that was stored away for just such a time.'[44]

It has to be strange to encounter a material that seems so familiar and is yet so precious – to drill a hole and extract these pristine rods of pure, fossilised time. I asked Liz Thomas what this was like.

'Back in 2004,' she told me, 'we drilled into a place called Berkner Island, and the ice at the bottom of that, we believe, probably goes beyond the last interglacial, down to maybe 140,000 years. I was involved in the project team and was actually the person who drilled the very last core. We pulled the core up and we started getting inclusions and sand and little pebbles, so we knew that we were pretty much on bedrock. That was 948 metres of ice – and so that's probably the oldest ice that I've actually, physically myself drilled and handled.

'It's kind of surreal, I have to say. The whole experience is quite surreal: you're in the middle of nowhere, you're incredibly remote; just a tiny field camp, a little blip on this massive, massive continent – and then you're pulling up ice that's been buried underneath there for 140,000 years.' She hesitates. 'It's quite a strange feeling.'

Strange, like the back of the mind awareness that you are dealing with an object that is somehow much much vaster in its implications than it might first appear.

In 2012, the Vostok drill in Antarctica pierced the roof of the subglacial Lake Vostok, which has been isolated from the rest of the world for fifteen or more million years. Scientists hypothesised that the never-before-seen lifeforms and geochemical processes it might contain would give insight into the icy oceans of the moons of Jupiter. Ice cores are wormholes that short-circuit space-time.

***

Ice cores also help scientists understand the scale, speed and threat of climatic tipping points – by providing year by year, granular data about previous rapid climatic shifts in the past.

The first deep bedrock core at Camp Century in northwestern Greenland, which hit dirt in 1966, provided an unprecedented complete record of the last ice age from beginning to end – a thousand centuries of climatic shifts. In the 1950s, a Danish geophysicist named Willi Dansgaard had been studying the different isotopes of oxygen in water molecules – and would go on to invent the ice core dating method mentioned in the previous section. But in order to do this, he first needed an ice core – and after a few years' wait enforced by military secrecy, he finally persuaded the Americans to give him some samples. By plotting the change in oxygen-18 ratios versus normal oxygen-16, Dansgaard and his team were able to show how temperatures in Greenland had fluctuated over the past 100,000 years.[45] They found a world of wild instability and oscillation at multiple, over-lapping scales.

Looking at the last millennium, Dansgaard and colleagues observed how oxygen-18 ratios shifted on a cycle of around 120 years, which they attributed to fluctuating radiation energy from the sun producing periods of warmer and cooler climate. Scaling up to look at the period since the end of the last ice age, 11,700 years ago, they could see a further eleven distinct shifts in oxygen isotope levels, operating on a 940-year period – again attributable to the sun. Scaling up once more, now looking across the 100,000-year span of the whole ice age, they found a third cycle averaging 13,000 years – attributed this time to a 'wobble' in the Earth's axis of rotation, which again affects climate by making different places on Earth slightly nearer or further away from the sun.

Subsequent work has demonstrated that further types of variation in Earth's orbit produce even longer types of climate periodicity, known as Milankovich cycles. In particular, the earth's orbit actually changes shape a little, from more circular to more elliptical, over a 100,000-year period. In 1976, scientists declared that these changes in the earth's orbital geometry – influenced by the gravitational pull of Jupiter and Saturn – were the fundamental cause of ice ages.[46]

Just pause to take that in for a moment. What an illustration of the ecological maxim that 'everything is connected to everything else' – on enormous, interplanetary scale. It's a scale that dwarfs us and renders us irrelevant, in a way that can, in the right light, be somewhat frightening.

Yet it's important to emphasise that these oscillations are different to our present global warming: they are natural phenomena. One caused the 'Little Ice Age' in Europe between 1300 and 1850 (not actually an ice age, just chilly), which let Londoners hold frost fairs on the frozen-over River Thames and forced the Vikings out of Greenland. Multi-decade megadroughts in the American southwest during the preceding Medieval Warm Period led to the collapse of the Anasazi civilisation. Environmental historians argue that climate oscillations contributed to the fall of the Khmer Empire in Cambodia, too, as multidecadal El Niño oscillations weakened the monsoon cycle.[47]

But not only does our planet wobble its way through time and space, and the sun wax and wane a little in the amount of radiant energy it produces. The second insight found within the ice was that when the climate changes, it can do so astonishingly rapidly.

'As the world slid into and out of the last ice age,' the geoscientist Richard B. Alley tells us, 'the general cooling and warming trends were punctuated by abrupt changes. Climate shifts up to half as large as the entire difference between ice age and present conditions occurred over hemispheric or broader regions in mere years to decades.'[48]

As the Earth started to warm up after the last glacial period, there was a 1,200-year blip in which it flipped back into much colder, glacial conditions: a time known as the Younger Dryas. When this ended, around 11,700 years before present, the climate in Greenland changed shockingly rapidly.[49] The overall transition into our present-day warmer Holocene took around forty years, scientists believe – occurring not as a gradual, steady ramping-up of temperatures but a series of sudden steps. In Greenland, temperatures increased by approximately 15°C over the entire transition period – but about half of this change occurred in less than fifteen years. Two deep ice cores drilled in the centre of the Greenland ice sheet tell the story: the Greenland Ice Core Project (GRIP) and Greenland Ice Sheet Project 2 (GISP2). Alley describes how snowfall in Greenland doubles, methane levels increase by half as wetlands expand – and dust drops several-fold as the gales of the glacial period die down.

Around this point but at a lower resolution – decades – 'there is an order of magnitude drop in dust concentration, reflecting a reduction in dust flux to the ice sheet from arid or poorly vegetated regions.' The record is so precise that the intervals between events can be dated to within just two years. A sharp shift in the deuterium ratio shows that temperatures increased 11,645 years before present. A decrease in the size of dust particles arriving on the ice (as well as the amount of sea salt) shows that windspeeds suddenly declined that year, too. The climate got profoundly weird. Alley describes how 'several records show enhanced variability near this and other transitions, including "flickering" behavior in which climate variables bounced between their "cold" level and their "warm" level before settling in one of them'. Overall, scientists say, it's evidence that 'climate in the North Atlantic region is able to reorganize itself rapidly, perhaps even within a few decades', and 'recent climate stability may be the exception rather than the rule'.[50]

These are lessons for the future.

Now, it's important to emphasise that the Younger Dryas wasn't a *global* cooling (or rewarming). It's the last in a sequence of what are known as Dansgaard-Oeschger events, after the Danish climatologist of a few pages earlier, which have a seesaw relationship with Antarctic temperatures – as the northern hemisphere warms, the southern hemisphere cools, kept in a kind of (slightly delayed) equilibrium by Atlantic Ocean currents. As such, these warming episodes are different to our present age of rapid heating, which is of course global – and all the more concerning for it.

'It's a warning to us that the ocean circulation system is not to be taken for granted,' Eric Wolff told me. Some of the current ice core research is trying to establish what exactly happened to this 'thermohaline circulation' of cold, dense, salty ocean water currents during the last interglacial 130,000 years ago. Might it then weaken or turn off not just during cold periods, but during warming ones as well? I asked. 'It might have a threshold beyond which something bad happens,' Wolff said. 'And because we don't really know what that threshold is, it's probably a good idea not to get near it.'

In 2021, a new analysis of the bedrock at the bottom of the Camp Century core revealed a shocking finding: the massive

Greenland ice sheet had melted completely away in this location within the past million years. The ice core had been kept in a freezer for fifty years, geologists Andrew Christ and Paul Bierman explained: 'until our Danish colleagues found the soil samples in a box of glass cookie jars with faded labels: "Camp Century Sub-Ice" '.[51]

Analysing the dirt within using the latest advanced techniques – from counting radioactive atoms in a particle accelerator to measuring the luminescence of grains of quartz sand to learn when they last saw the sun – revealed the ice sheet's vulnerability. At some point, probably during a warm interglacial period 400,000 years ago, it melted – drastically. The twigs, leaves and moss fragments they could see under the microscope were a direct glimpse of an ice-free northern Greenland. In this way, they wrote that an 'Arctic military base built in response to the existential threat of nuclear war inadvertently led us to discover another threat', equally existential: 'the threat of sea level rise from human-caused climate change'.

In 2001, the UN Intergovernmental Panel on Climate Change (IPCC)'s third report introduced the possibility of 'large-scale changes in Earth systems that could have severe consequences at regional or global scales'. Three potential tipping points were identified: the shutdown of North Atlantic and Antarctic deep cold water formation, which powers ocean circulation; big changes in carbon dynamics, such as whether tundra and peatland operated as sinks or emitters of carbon dioxide – and the disintegration of the West Antarctic and Greenland Ice Sheets.[52] The probabilities of these events were highly uncertain, the IPCC acknowledged – but they were thought only to be likely in the most severe warming scenarios of +5°C above preindustrial levels.

That story has changed – and evidence from ice core research has been crucial to how. Tipping points are no longer far off hypotheticals, but now 'too risky to bet against', a group of climate scientists wrote in *Nature* in 2019. 'Evidence is mounting that these events could be more likely than was thought, have high impacts and are interconnected across different biophysical systems, potentially committing the world to long-term irreversible changes.'[53]

As we have seen earlier in this chapter, dust is deeply implicated in the melt of the Greenland ice sheet through the mechanism of albedo. And through this, it connects to everything else on the planet, too. Niklas Boers and Martin Rypdal's modelling of the Greenland ice sheet's tipping point describes how, if the Greenland ice sheet melts completely, it would cause a global sea level rise of over seven metres. This would increase the amount of fresh, non-salty water in the North Atlantic, which could potentially collapse the whole system of Atlantic water circulation. This might, in turn, trigger a cascading series of effects that could impact additional ecological systems on the brink of tipping points, such as the Amazon rainforest and the tropical monsoon rains.[54]

Disaster begets disaster.

It would be a crime to let the ice sheet melt – a crime against humanity for an estimated 410 million people to be forced from their homes, farms and livelihoods by rising waters, their memories lost to salt.[55] The first in what will be a tsunami of human rights court claims and asylum cases have already begun.[56] Yet it strikes me as something else, too. The ice sheets hold all human history in their dusty depths, and a million-year ecological legacy in their air bubbles and clathrates. The prospect of its loss carries echoes of a moral crime too, akin to burning the library of Alexandria.

***

Tuesday lunchtime, Wayne came back from buying some beers with a suggestion – let's fly to the edge of the ice sheet.

I'd come out to Greenland with a dream: to walk to the ice cap. The distance was forty or forty five kilometres each way, as the crow flies, hiking cross country on tricky ground: fording rivers, traversing bogs. But there was (barely) a dotted line drawn on a map, and a huge poetic appeal: enough to spend serious money on flights and gather the kit Wayne and I would need: subzero sleeping bags, emergency GPS tracker. On arrival, though, we slowed down. We recognised more deeply that this was unfamiliar country, and we needed to take some real time to learn about it, its weather and its risks. Even short hikes took hours as we kept stopping simply to absorb the landscape, to try to take it all in. We decided not to rush. A hard march would be missing the point.

Yet if we weren't going to hike to reveal the icecap slowly and arduously, then here was another way to encounter that immensity. We had come so far to see it, after all – and yet arriving from Iceland we had glimpsed the ice sheet amid snowy mountains for only a minute before white cloud concealed the scene. Cloud enfolded our landings first in the capital Nuuk, then at Ilulissat too. This would be a first, perfect encounter. I took no persuading.

We walked to the airport in glorious sunshine under a bright blue canopy of sky. 5 p.m. at the terminal, we looked around: who were we supposed to be meeting? An Air Greenland flight landed and shipped out 30-odd people in ten minutes flat, all chattering happily. We had the airport almost to ourselves. Then Ricky arrived. Wayne commented that a guy this tall, blond and matinee-idol handsome couldn't be anyone else but the pilot.

Having qualified last year, Ricky (a Dane) had come to Ilulissat to fly eight-seater prop jets as preparation for a career, he hoped, flying even smaller planes in even further-flung places. He greeted us cheerily, then opened a door in the corner of the airport and led us straight out airside to a plane no taller than he was. We buckled into our seats. Headsets on. Then Ricky outlined our flight plan to air traffic control as he lined us up straight down the *800-metre* runway for takeoff. The plane rose and turned 210 degrees to the left, out over Disko Bay, the sea golden, to fly to the glacier front over the hills inland from Ilulissat. Twenty minutes later, a series of sweeping turns over the main Sermeq

Kujalleq glacier face where the icebergs calve. Ricky took us down low, to 200 metres, to get a better view.

The ice was compressed into countless sharp crevasses and ridges by the force of a thousand kilometres of ice behind. Where it met the water, it would alternately just fall apart into snowy pulp and calve into vast bergs, several square kilometres at a time. The pressure cracks radiated back in serried rows as 110,000 square kilometres of ice bore down. Within the fissures lay occasional meltwater pools – a cobalt blue so exquisite it hardly seemed natural.

On the Arctic Sea Ice web forum, glaciologists and geoscientists had been noting with concern that the melt ponds were visible a month ahead of last year. They reckoned 2016 would see record glacial retreat, the calving front receding back deeper into the ice sheet itself and melting it away. The ice something so vast and yet not quite infinite enough.

The feeling of awe turned out to be a lot like bewilderment. My photographs looking out into the centre of the ice sheet didn't work; they were just formless bright. To have walked that ninety kilometre circuit to the ice sheet and back would have been a pilgrimage.

Chapter 8

Dust Is Part of the Earth's Metabolism

S ometimes the heavens bleed.

Blood rain is a phenomenon known since antiquity. *The Iliad* recounts how Zeus wept blood over the battlefield in portent of the death of his son, Sarpedon, or to throw the Greeks into confusion. Pliny writes of a 'rain of flesh' over Italy in 461 BCE and Plutarch describes a rain of blood coming to the city of Fidenae as a bad omen, accompanied by plague and barrenness.[1] Little changes. BLOOD RAIN pours down in Russian freak weather event prompting "biblical plague" panic', blared the British tabloid the *Daily Express* in July 2018, reporting on a rufous rainstorm in Norilsk, northern Russia.[2] Roman statesman and philosopher Cicero, a more practical man, was perhaps the first to question such a framing of events. He suspected the rain may be red from 'the admixture of water with certain kinds of soil'.[3] He was, of course, correct.

Often, it's dust.

While beloved by UK newspaper editors, truly blood-red rain is actually quite rare. The Norilsk rain was genuinely gory-looking: Russian metals processor Nornickel had been carrying out what they claimed was a 'clean-up operation', removing tons of iron oxide from the roof and floors of their processing plant, which was picked up by the wind to produce a deep rust-coloured hue.

Other really red rains throughout history – and there haven't been that many, one historical study finds just 168 events – have been attributed to algae such as *Chlamydomonas nivalis,* which produces the crimson snow we encountered in the previous chapter.[4] The phenomenon we see in the UK entails sand from the Sahara being picked up by the wind and carried northward, until it precipitates out as at best yellowy-orange, leaving a thin film of dirt on glass skylights and parked cars. (The sunsets are good, however.)

In 2020, one such dust storm gained worldwide press attention – and earned the nickname 'Godzilla' – due to its gigantic size and the enormous amount of dust it carried. Over the course of four days in June of that year, strong winds picked up 24 million tons of dust and carried it high into the atmosphere, up to five or six kilometres in height. Here it met the great easterly jet streams that encircle the planet at the mid-latitudes, which blasted the dust cloud over 8,000 kilometres across the Atlantic Ocean to the Caribbean, Central America and parts of the southern United States. It was the biggest dust episode in a generation.[5] The air filled with a pinkish haze from Puerto Rico to Trinidad. Air quality monitors reported hazardous conditions. People were warned to stay indoors and use air filters and face masks if they had them.

Yet, curiously, the media did not write up the Godzilla plume as a monster, but rather something of a marvel. The *Daily Mail* described how 'the dust can also enhance sunsets, suppress tropical storm development and plays an important role in our ecosystem', and Umair Irfan at Vox went deep on how dust fertilises the Amazon.[6] For a few days, dust caught people's imagination as a symbol of *life*, not just desertification and environmental depletion. Back in early 2016, it's how dust caught my imagination too. The first sentence I wrote in the email newsletter that became this book was, 'Dust is part of the earth's metabolism' – an insight with depths I have kept excavating ever since.

So let us finally explore dust as an agent in Earth systems, the integrated – and deeply interacting – energy flows and physical, chemical, and biological processes that make our planet able to sustain complex life. Dust, I will argue, is as integral to our planet's functioning as water, or soil, or ice. By following its paths round the globe – and beyond – we might come to see how we, too, are inescapably part of Earth's systems as well.

***

The idea of planet Earth as a holistic, intricately intercoordinated system is an idea as old as humankind. In *Timaeus*, written in around 360 BCE, Plato describes the universe as necessarily alive and intelligent, like an organism, the microcosm and macrocosm fundamentally connected. Indigenous knowledge systems around the world assert much the same idea. In modern Western science, the notion might be dated to 1807 when explorer and scientist Alexander von Humbolt began to describe his vision of a comprehensive *physique générale du monde*, a 'general physics of the world'. In *Essay on the Geography of Plants*, he sought to explore the interrelationships between Earth's natural forces: vegetation and solar radiation; elevation and weather; vulcanism and earthquakes; atmosphere, temperature and deep geological history. He made the case that the 'general equilibrium obtaining in the midst of these disturbances and apparent disorder is the result of an infinite number of mechanical forces and chemical attractions which balance each other' – that is, planet Earth is an integrated, balanced system.[7]

But it was the Cold War that gave birth to Earth System Science as it exists today, as military funding for global environmental surveying and monitoring expanded dramatically, supporting the development of satellite-based technologies such as remote sensing and geo-spatial positioning.[8] Intellectually, the strategic demands of conflict produced an emphasis on modelling and systems thinking. This became unified in the field of cybernetics, which offered an inter-disciplinary framework to discuss the operation (and regulation) of dynamic systems, from biology and ecology to computing, social systems, even the mind. Yet even as this techno-scientific systems thinking valorised a high-modern way of thinking about the world as a canvas amenable to human control, it produced a kind of counter-systems thinking about its costs, too.

In 1962, the biologist Rachel Carson published *Silent Spring*, demonstrating how cavalier usage of pesticides was ricocheting through the food chain and devastating bird and insect populations. A team at MIT developed a computer model, World3, to simulate the interactions between population, economic growth, food production and ecological limits – which they published as the report *The Limits To Growth*, in 1972. It became the most-read environmental book in history. James Lovelock published his 'Gaia hypothesis' that same year: he argued that the planet's ability to self-regulate showed that living and non-living things together formed a complex system operating in mutually beneficial relationship – in effect, as a single organism. Meanwhile, astronauts on America's *Apollo* space missions took era-defining photographs: *Earthrise*, in 1968, and *Blue Marble*, in 1972, showing our planet as beautifully, indivisibly whole. To some it was a message of fragility: the Earth, a precious jewel, protected only by the delicate swirls of its atmosphere. To others, it was not a break from instrumentalist thinking but its vindication: the planet was knowable, and we might grasp it all at once.

'Science has become part of our way of life and how we approach nature,' environmental scholars Cristina Inoue and Paula Franco Moreira write. Sometimes that approach is in the spirit of high technocratic modernism, the whole world framed in terms of command and control and laid open for human intervention. Yet at the same time, Western ecological consciousness is also 'something that resulted from modernity's historical planetary dynamics, a heir of its scientific revolutions,' they say.[9] Ecology is the Earth's systems thinking too. The NASA scientists I spoke to while researching this chapter want us to understand the exact same thing as the ecologists and Indigenous activists we'll meet in Chapter 9. 'How interconnected it all is,' they all say.

***

What types of dust are we talking about when we think at Earth system scale? This book takes a catholic line on what qualifies as 'dust' – as you have seen – but our focus in this chapter is on mineral dust, by mass the most abundant aerosol in the Earth's atmosphere.

Around five billion tonnes of mineral dust is emitted from deserts and other areas of bare soil and rock each year.[10] Most of it comes from the 'dust belt', which extends around the planet's middle, from the Sahara, in North Africa, to the Middle East, Central Asia, to India and China. While most of this book dwells on dust as humanmade, it's important to recall that, globally, three-quarters of mineral dust is entirely natural: just the wind and dry bare soil, and planetary processes doing what they do.[11] As we've seen earlier, prime dust sources are very often former lakebeds or flat, low-lying areas of land which have repeatedly flooded. Sediments carried in the water form deep deposits of fine clays and silt soils, which provide the tiny, weakly bonded particles that, when dried out, are most ready and eager to become dust. But this dust isn't *all* made up of silt. Tiny creatures known as diatoms provide an object lesson in dust as part of the Earth system. Both fed by dust – and then becoming dust themselves – they reveal the extraordinary ecological interlinkages that exist in the most mundane of places.

In 1833, Charles Darwin was voyaging on the HMS *Beagle* through the Cape Verde islands in the mid-Atlantic. Though they were 400 miles from land, he wrote, 'The atmosphere was so hazy that the visible horizon was only one mile distant' – and 'a very fine dust was almost constantly failing, so that the astronomical instruments were roughened and a little injured'.[12]

The wind direction made clear that the dust must have come from the coast of Africa. He collected a sample and sent it off to a Professor Ehrenberg in Berlin to examine. 'The little packet of dust collected by myself would not have filled a quarter of a teaspoon, yet it contains seventeen forms,' Darwin reported back with wonder: seventeen different species of *Infusoria* diatoms.

During the African Humid Period between 14,500 and 5,500 years ago, Milankovich solar cycles and albedo feedback loops produced a much wetter climate in North Africa, just as we saw in Greenland in the previous chapter. Ancient rock paintings from this time show pastoral scenes of abundant grazing animals: antelope and aurochs, giraffes and elephants – even semi-aquatic hippopotami. The German explorer Heinrich Barth, crossing the desert from Tripoli to Timbuktu in the mid-nineteenth century by caravan, was struck by these incongruous scenes, writing that this artwork 'bears testimony to a state of life very different from that which we are accustomed to see now in these regions'.[13] In the south, in present-day Chad, Niger and Nigeria, a vast Lake Mega Chad covered some 361,000 square kilometres, an area greater than the UK and Ireland combined. And living in the fresh water, drifting gently in the currents, were countless trillion diatoms.

Diatoms – a type of single-celled algae, or phytoplankton – are extremely tiny (between 2 and 200 micrometres in size), and yet have an outsize impact on the functioning of our planet. They achieve this through sheer number: diatoms live pretty much everywhere that has both light and water; not just lakes and oceans, but bogs and rock faces, on moss and even the feathers of waterbirds. By photosynthesising like plants, they produce 20 per cent of the oxygen on earth.[14] Unique among living things, diatoms' cell walls are not made of carbon but silica, like glass – and when they die, they're deposited out of the water just like grains of sand on a beach. As such, the Sahara Desert is unexpectedly rich in tiny fossilised diatom shells, so numerous that they make up a white, powdery mineral known as diatomaceous earth, or diatomite. Under the microscope, diatoms' tiny, fossilised shells make opalescent patterns like tiny jewels in a kaleidoscope.

North Africa is the largest source of dust into the Earth system, producing some 200 million tonnes per year – 60 per cent of the mineral dust aloft at any point.[15] It starts in the Sahara, where strong harmattan winds jostle the sand grains and pick up flakes and fragments up into the air. The dust rises until it touches the trade winds and is whisked westbound, in great trailing ochre streamers that show up on satellite images like smears of ochre out into the ocean.

But mineral dust is relatively large as airborne particles go, and the wind cannot hold it up forever. Approximately 70 per cent – 140 million tonnes a year – falls out across the tropical Atlantic Ocean, where the dust provides nutrients such as iron and phosphorus to marine ecosystems. Algae – the diatoms, and their cousins the dinoflagellates and coccolithophores (what names!) – scavenge these minerals as food, and multiply, over and over, until the microscopic is made visible from space in great green swirls and fractal fingers. Their colour comes from chlorophyll, like plants – and by photosynthesising like plants, they produce fully half the world's oxygen.[16] The air you're breathing now isn't made solely in scenic forests, but much of it far out to sea. And as these algae die and sink to the ocean floor, their hard shells sequester significant quantities of carbon. They are compressed over millions of years into minerals such as limestone – and also the oil and gas that power the planet today.

Algae terraformed the planet. They first enabled human life at all, and then the modern lives we live today, giving first Britain, then the West, then the world the power of hundreds of millions of years of fossilised sun. And, as both cause and consequence of these algal abundance, dust is an integral part of this vital bio-geological process. Blown out from deep in the Sahara, the bodies of algal ancestors feed the blooms that swirl through the ocean today.

There is a slo-mo morbidity to this process: dust can bring life and death inside the same, single particle; it is quite indifferent. In the weeks after major Saharan dust events, red tides wash up on American shores, as iron-hungry *Trichodesmium* algae gorge themselves and multiply a hundred-fold beyond the regular confines of their ecological niche. This alters nitrogen levels in the water, enabling another nutrient-limited plankton, *Gymnodinium breve*, to multiply as well.[17] It is toxic: killing fish and poisoning shellfish along the Florida coastline. As the wind whips the waves, these plankton cells get blown into the air, causing health problems for people and other animals that breathe it in.

Meanwhile, high up in the troposphere, above the Earth's weather systems, bacteria and viruses surf its global currents on aerosol particles and fragments of mineral dust, before falling from the skies on continents new. Scientists believe this is one dusty transport that has positive consequences, increasing the diversity of downwind ecosystems and their abilities to adapt to environmental changes.[18] Yet dust also brings disease. In the Caribbean, the Saharan winds carry spores of the fungus *Aspergillus*, making corals and sea fans sicken and die.[19] 'Our hypothesis is that much of the coral reef decline in the Caribbean is a result of pathogens transported in dust from North Africa,' US coastal geologist Gene Shinn told NASA's Earth Observatory blog in 2001, continuing: 'You have to live in the Virgin Islands to fully comprehend the amount of red dust people clean from boat sails, decks, and window screens there.'[20]

On a few occasions, whole swarms of African desert locusts have been caught up in the dust and have made it all the way across the Atlantic alive. In October 1988, aided by the winds of a tropical cyclone, they landed in Barbados, in the eastern Caribbean, causing, doubtless, some consternation.[21]

***

Dust interacts with the Earth's biological systems in the form of algae and the Amazon. But it's critically important for reasons of physics, too: it modulates the planet's fundamental energy flows in two ways, directly and by influencing the formation of clouds.

The first one's pretty simple. Energy comes to planet Earth from the sun and is reflected back into space – and dust gets in the way. Scientifically speaking, dust impacts a factor called 'radiative forcing': the number of watts of the sun's energy the Earth gains (or loses) per square metre. Ever-ambiguous, dust does this in two contradictory ways. First off, dust in the atmosphere is cooling – particles block the sun's rays and reflect them back into space, reducing the amount of heat the Earth absorbs. (White-coloured sea salt's actually the best aerosol at doing this, with an albedo of 97 per cent.) Yet particulates can also be warming – carbon soot, being black, doesn't reflect at all and just absorbs all the solar power it can get. (Think of being a Goth on a sunny day.) One of the big challenges in dust research right now is to figure out the overall planetwide balance of scattering versus absorption that takes place, as it's obviously crucial for accurate climate modelling. The 2007 IPCC study reported that the impact of human-produced dust on the Earth's energy balance was between -0.3 to +0.1 watts per square metre: that is, it might be negative (cooling) or positive (warming), they couldn't say precisely.[22] More on the NASA study that will solve this problem, shortly.

The second way dust influences the planet's energy flows is by 'seeding' clouds. Solid aerosol particles act as nuclei encouraging gaseous water vapour to condense into droplets – or ice crystals if the air is cold enough. At any one time, about 60 per cent of the Earth's surface is covered in clouds, which both cool the planet by shading it from the sun's radiation, and warm it by preventing trapped heat escaping. Like dust, whether an individual cloud does one thing or another depends on its size, colour, height in the atmosphere, and a variety of other atmospheric conditions. The planet stays at a stable temperature because the heating and cooling properties of clouds are in balance: this is the law of conservation of energy. But anthropogenic changes to the atmosphere, in the form of black carbon and mineral dust, as well as carbon dioxide, methane and other pollutants, warm the planet by disturbing this fundamental equilibrium between Earth and space.

Different types of aerosols also make different sorts of clouds. In the lower atmosphere, 'hygroscopic' or water-attracting aerosols, such as sea salt, sulphate and even wind-lofted phyto-plankton, produce low-lying cumulus clouds which primarily reflect sunlight. More aerosols make brighter, whiter clouds, making them extra-cooling. The larger, insoluble particles of mineral dust, by contrast, encourage ice crystals to form, producing wispy, high-altitude cirrus clouds which trap the Earth's warmth and radiate it back down to the surface.[23] In the northern hemisphere outside the tropics, between 75 and 93 per cent of cirrus clouds have these dusty origins. Most of this dust comes from Central Asian deserts: though they emit only 13 per cent of the world's dust, the Asian summer monsoon provides the necessary lifting power to get this dust into the upper troposphere where cirrus clouds fly.

Both directly (via reflection) and indirectly (through cloud formation), aerosols are thought to be net coolants, counteracting as much as a third to half of the warming produced by human-produced greenhouse gases since the 1880s. In 2010, NASA suggested that just a 5 per cent increase in cloud reflectivity could compensate for the entire humanmade rise in greenhouse gases since the industrial era.[24] (You start to see why interest grew in solar radiation management – that is, geoengineering.) Dust is an integral part of the fundamental flows and equilibrium of our planet, an element as important as oxygen, or water or ice. Culturally, dust may be a metaphor for time, decay and forgottenness. But out here in the world, it's a vigorous and active thing.

What makes dust so interesting, University of Reading meteorologist Claire Ryder told me, is that it doesn't just affect the Earth system one-way: it's got a feedback loop relationship. 'Dust is generated by the weather, by winds and surface moisture, so it's dependent on the climate, but it also has an impact on the climate: you've got this complete circle of links and feedbacks.' So as the climate changes, the world's dust will too.

Human factors add yet another layer of feedback: as the climate changes and new regions aridify, people change how they use that land – with potential consequences for its dust generation. The 1930s Dust Bowl was one such catastrophic example, where lower rainfall levels and economic depression combined to push farmers to put more marginal land into production, thereby increasing dust emissions and turning a drought into a decade of disaster.

Another example can be found in 1990s' Iraq. Saddam Hussein ordered large areas of the Mesopotamian Marshes to be drained in order to punish the Marsh Arabs who lived there for their anti-government rebellions. Some 9,000 square kilometres of marsh had shrunk to just 760 square kilometres by 2002, and sandstorms rose tenfold.[25] A study found that the peak daily dust load blown up by the summer Shamal winds, some 2.5 million tons, was 'in the same order of magnitude as those derived from simulations downstream of the Bodélé depression in Chad, known to be the world's largest dust source'.[26] So much dust is the result of human decisions: it's easy to write it off as a mere natural disaster but no, a key message of this book is that dust is intensely political. Whether or not Iraq's marshes are restored will be equally a political decision.

As awareness grows of dust's contribution to climate change and its impact on human health, the appetite to address the world's dusty hot spots and engage in environmental restoration and remediation activities needs increase. As I have mentioned, 75 per cent of the world's mineral dust is of natural origin, or winds and deserts doing their thing as they have since time immemorial. But that remaining 25 per cent is somewhere in the vicinity of 250 million and one billion tonnes per year. To this is added around eight million tonnes of black carbon from fossil fuel combustion, biofuels and biomass burning for land clearance – a particulate that may be lighter in mass but not in consequence, its supercharged warming ability making it the second most dangerous human-produced emission after carbon dioxide.[27] We are already geo-engineering our planet. It is our obligation to recognise this responsibility, and to go about it with considerably greater care.

***

Imagining the Earth as a holistic system creates the possibility that the dynamics of the entire planet might be contained within a computational model.

In 1956, a meterologist named Norman Phillips, working at Princeton, created the first 'general circulation model' of the Earth's atmosphere using a computer with just 5 kilobytes of memory (plus an extra 10 kilobytes on a disk). Oceanic processes were added in the late 1960s by the US NOAA Geophysical Fluid Dynamics Laboratory, clouds started to be modelled explicitly in the late 1980s, and soil and vegetation types were first included in 1996.[28]

'Dust is one of the missing jigsaw pieces in climate modelling,' says drylands geographer David Thomas.[29] The UN's IPCC reported in its Fourth Assessment Round in 2007 that anthropogenic aerosols (here encompassing sulphate, organic carbon, black carbon, nitrate and dust) 'remain the dominant uncertainty' in modelling how much the planet was heating or cooling.[30] When Thomas and colleagues at the University of Oxford proposed their research project in 2010, there was no data on observed dust sources that matched the scale of the 'grid boxes' used for climate modelling calculations. But numerical models are 'the only tools we have to predict future weather and climate', they wrote – and so they need to include dust in order to avoid major error.[31]

Yet dust is strange and surprisingly diverse. Whether it heats or cools the planet depends on all kinds of factors. Size matters: so-called giant dust particles over 20 micrometres in size (this is big for dust, if not in the general scheme of things) are more warming than smaller dust – and also significantly under-measured, says Claire Ryder, a meteorologist at the University of Reading.[32] Shape matters too – spherical and non-spherical particles can have different reflective properties.

Alignment is also a big deal, so much so that 'Probably everything we've so far hypothesized about the impact of dust on the atmosphere might be misplaced,' climate scientist Vassilis Amiridis told *Eos* in 2020.[33] If dust isn't randomly facing every which way, but instead polarised – that is, lined up in parallel, a bit like the slats of a blind – perhaps 10–20 per cent more radiation may be able to slip through the gaps, his team's work suggests. Dust from different places can have hugely different consequences – for example, dust that enters into a monsoon system may get rained out right away, whereas dust emitted over a desert can stay aloft for much longer. As such, one study found that aerosol emissions from Western Europe had fourteen times the global cooling effect of emissions from India.[34] Another recent study in West Africa found that dust presents a net warming effect in the atmosphere and at the top of it, yet is cooling at the surface.[35]

This is true mathematical complexity. Dust isn't just complicated, operating on large scales with many moving parts, but *complex*, a system whose behaviour is intrinsically difficult to model due to the number of interactions and feedback loops between it and its environment. Dust is non-linear. Dust may have unexpected emergent properties. As such, dust is enormously difficult to summarise as just one part of a whole planet computational model.

Fortunately, IPCC climate reports aren't written on the basis of a single climate model, but rather many. For each round of reporting, climate modelling groups around the world run their climate models against an agreed set of future emissions scenarios, known as 'Shared Socioeconomic Pathways' (SSPs) – and report the results back to be aggregated and averaged. For the latest Seventh Assessment Report in 2022, 53 modelling groups applied about a hundred climate models to eight scenarios, ranging from rapid and immediate carbon reductions to runaway fossil fuel usage.

Even the latest generation of climate models have remarkably divergent ideas about what dust does, however. In 2021, Alcide Zhao at the University of Reading looked at sixteen such models and compared them against satellite data and observational measurements.[36] Some fairly fundamental divergences exist: for example, modelled global dust emissions varied fivefold, from 1,400 to 7,600 million tonnes per year, in contrast to the latest observational estimates of 5,000 million tonnes. There's disagreement about where dust comes from, with models describing between 2.5 per cent and 15.0 per cent of the world's surface as dust-emitting. Only one of the assessed models includes South Asian dust sources as significant, though it's an all-pervasive part of life in Delhi and other Indian cities. Then there's a fourfold variation in how long dust stays aloft in the atmosphere (from 1.8 to 6.8 days), plus 'very different assumptions on dust size ranges across models'.

What's more, Zhao has found that 'dust processes are becoming more uncertain as models become more sophisticated'. And, because of the ways dust is 'coupled' to many other components of the Earth system, he asserts that it is therefore reasonable to speculate even greater model uncertainties and deficiencies exist in models that take these linkages fully into account. The more scientists learn about dust, the more we realise how little we know. It humbles us. It exceeds our grasp.

***

Scientific understanding of the world's dust flows is itself in constant flux. As a major area of uncertainty in climate modelling, one of the world's most urgent areas of inquiry, dust is becoming an increasingly popular area of investigation. Writing this chapter, I had a moment of first astonishment, then amusement, when I realised that everything I thought I'd known about Saharan dust transport was now, well, not *entirely* wrong but certainly more complicated than even a few years ago.

The Amazon is a giant ecosystem held in a fragile equilibrium. Despite the lushness of the trees that grow above it, its soil is actually shallow and low in nutrients, as high levels of rainfall wash soluble minerals out of the earth. One mineral in particular was a puzzle for scientists: phosphorus, which lets plants transform the energy from the sun into a usable power source for their cells. If it kept running out of the ecosystem in groundwater and rivers, how was the rainforest staying in balance? The answer, people realised, was dust: up to forty million tons per year of the stuff, blown across the Atlantic.[37] The Sahara – harsh, dry, seemingly barren – was keeping one of the planet's most verdant places alive. Everything was connected. Dust wasn't just participating in the phosphorous cycle, but by helping forests grow, it was modulating the carbon cycle, too. 'We believe the dependence of one large ecosystem upon another separated by an ocean and coupled by the atmosphere to be fundamentally important to any view of how the global system functions,' Robert Swap and colleagues wrote in 1992.[38]

The dust was thought to come from the Bodélé Depression, a low-lying area of land in the southern Sahara. Like so many dust sources in this book, it had once been underwater, part of the ancient Lake Mega Chad, populated by diatoms, which laid down their tiny shells into a rich source of phosphorus. A quirk of topography here means that two mountain ranges funnel and accelerate the wind over this crumbly diatomite earth, giving it the power to loft huge quantities of dust into the air – as much as 0.7 million tons per day. Despite being just 0.2 per cent of the Sahara by area, the Bodélé was thought to produce 40 per cent of its atmospheric dust.[39] One ancient lakebed was feeding the lungs of the planet. 'This is a small world and we're all connected together,' NASA's Hongbin Yu told the *Independent* newspaper in 2015.[40]

In the last couple of years, however, a number of new research papers have disputed much of this narrative. 'If we really want to understand what is actually fertilising the Amazon basin, we have to track it back to the source,' NASA aerosol scientist Olga Kalashnikova told me when we spoke over Zoom. Analysing how dust plumes travel using multi-angle satellite imaging, Yan Yu, a postdoctoral fellow on Kalashnikova's team at the Jet Propulsion Laboratory, found that while the Bodélé might produce a lot of dust, most of it falls out of the atmosphere over land and over the ocean. Just 0.4 million tonnes makes the full journey to the Amazon: by contrast, El Djouf, another major dust source in

Mauritania and Mali, delivers nine times that amount.[41] Another study of lakebed sediments found that over the last 7,500 years mineral dust has reached the Amazon from all sorts of places, including the Andes and southern Africa, and the Sahara might not be the dominant long-term source at all.[42] One thing is indisputable, however: the Earth system is nothing if not dynamic.

What if the fertiliser that reaches the Amazon isn't actually mineral dust? Atmospheric scientist Anne Barkley recently found that half of the phosphorous winging its way to the Amazon actually takes the form of black carbon.[43] Across the tropical regions, in the savannah grasslands of Africa and South America and the dense forests of the Amazon and Southeast Asia, people burn vegetation to clear land for farming, and to get rid of last year's crop before planting again. Many developing countries also use wood, agricultural waste and dried animal manure for heating and cooking, as they're either free or easily available fuel sources. In northern Russia, Canada and America, the boreal forests burn. Altogether, this produces about 60 per cent of all black carbon soot in the atmosphere today – and it has continent-spanning consequences.[44] Because it's so dark, black carbon is a major contributor towards global heating as it absorbs the sun's energy so effectively – but the phosphorus it contains is also much more soluble than phosphorous from mineral dust and so it's the dominant fertiliser source for the Amazon and neighbouring oceans, too.

Further complicating matters, as they travel, mineral dust, soot and other aerosols often clump together, creating hybrid particles. This is one of the reasons why, as we'll see later in this chapter, that dust is such a profound source of uncertainty in climate modelling today.

Planet-spanning particulates of one type or another are still feeding the Amazon: the world is still intricately interdependent. But what's apparent is how much we still have to learn about this key part of the Earth system.

***

The simple fact of where it comes from remains one of dust's many ambiguities. A new investigation aims to fix this. In December 2021, I spoke to NASA's Dr Robert Green, Principal Investigator for EMIT, the Earth Surface Mineral Dust Source Investigation. This project seeks to create a new mineral map of Earth's dust-producing regions by measuring what major dust sources around the world are actually made of.

'Models are the best way to understand how dust interacts with the whole Earth system,' Dr Green told me when we spoke in December 2021. He explained how they use estimates of minerals from the UN Food and Agriculture Organisation Soil Map. 'This is a soil map where people have picked up a handful of soil and described its texture and its colour. It has soil composition for every place on Earth, but it doesn't mean people walked across Africa to get them,' Green explained. Instead, 'there's a lot of infilling. There have been less than 5,000 cases where at that location, that soil was taken to a laboratory and the minerals were determined. So basically, the current Earth system models rely on less than 5,000 measurements for the entire planet.' This is nowhere near enough for accurate modelling and forecasting. EMIT seeks to make a great leap forwards. 'In our one year of operation, we will deliver a billion direct measurements,' Dr Green said.

This has been made possible through the use of spectroscopy, which, in Green's mind, is 'perhaps the most powerful analytical method that has ever been invented by humans. It is how we understand our universe.'

'Spectroscopy began when somebody used dispersed light to answer a question,' he continued. We might think of it as the science of analysing colour. Objects have colour because they absorb some frequencies of light and reflect back others. Iron oxide appears reddish in tone because it absorbs green, blue and violet light, and reflects back just the red end of the spectrum. Black carbon is black because it absorbs almost all light. In fact, every single element in the periodic table has a unique light spectrum it reflects or absorbs, like a colour signature determined by its crystal or molecular structure. Measure the light reflected from the surface of a planet (or emitted by a star), and you can learn what elements it's made of. When NASA flew their first imaging spectrometer on its maiden test flight over Nevada in 1982, 'we found new minerals that we didn't know were there!' Green exclaimed.

Green joined NASA in 1986, and has worked on spectroscopy ever since. 'I've been involved with imaging spectrometers that have gone to the Moon and gone to Mars. Right now we're building one to go to Europa around Jupiter,' he says. He's seen a lot of change over his career: 'Once I paid $100,000 for one terabyte of disk, and now you can get them for twenty bucks or something.'

Fixed to the side of the International Space Station (ISS), EMIT will carry 1,200 parallel spectrometers, each looking down and taking a read-out every sixty metres across the Earth's surface, measuring the light spectrum they can 'see'. 'We'll cover the entire arid land regions of planet Earth multiple times in one year,' Green explained, giving them 'multiple looks: there might be clouds, there might be dust storms, but we want to see through those because we're interested in the surface.' When it works, he added: 'We can see molecules from one hundred kilometers away. And that's kind of amazing.' It is.

The project seeks to give climate modellers the ability to fore-cast future dust sources. 'Our approach is that we're going to look at the regions at risk for desertification and try to estimate their underlying mineralogy.' These might be places adjacent to known dust sources 'where there's still enough vegetation that it's not really a source, but there'll be little cleared roads, and cleared quarries and maybe cleared farmer's field': places where the land surface is being disturbed, making it vulnerable to wind erosion. The EMIT team will run their Earth system models under differ-ent IPCC climate scenarios to see what might happen next: is this place likely to get more vegetation? Or could it be desertified?

'One of our deliverables is our estimation of what we think may happen in the future 100 years from now.'

The mineral map will improve climate models by giving scientists more information about how dust will heat or cool the planet, both today and in the future. Bright white kaolinite clay dust has very different radiative properties to red haematite – which is why it's so essential to understand not just dust quantities and transport, but its composition, too. In this way, dust science does its bit to save the world.

Sometimes these studies have startlingly direct benefits, too. Green described how they had some measurements from AVRIS – another spectrometry instrument – several years ago, on asbestos on the surface in parts of California, where it can be naturally occurring. It was utilised for other purposes.

'We were deployed when the World Trade Center collapsed back in 2001, to look for where the asbestos went when those two towers fell,' he said. 'We could see the gypsum wallboard, we could see the cellulose from the paper. We flew on 16 September, and we could see the hot fires that were still burning. And so we told the people on the ground, you know, "Here's where we see the fires." We gave them the latitude, longitude: "These are the fires and this is how hot we think they are". So they could take that information to try to enhance the safety of the people still involved in rescue.'

More recently, Covid delays have caused them some problems: 'When I wrote a proposal, I didn't propose to build an instrument when people couldn't work next to each other!' Green observed ruefully. Building space hardware involves working closely in clean rooms, but the NASA JPL campus completely shut down in March 2020 – as did supplier labs – and the project hit delays. In December 2021 all the scientists I spoke to were still working remotely, and I was unable to visit in person as NASA were still strictly limiting access. But the launch was finally imminent. 'We're scheduled on SpaceX crew resupply number 25, which currently is planned for the first of May [2022], out of Florida,' Green told me.

'What do you wish more people knew about Earth systems?' I asked him.

'How interconnected it all is,' he replied. 'I think the dust cycle is an interesting way to show that because it connects to so many different parts: it fertilises tropical forests, it fertilises the ocean, it's a hazard for us, it melts snow, it helps form clouds, it heats and cools, all these things are important. So the Earth is a system of things that are interconnected. It's all connected – and yes, we can impact it. It's worth learning a little bit more about that, and thinking about that when we make our choices, so we can work to find sustainable paths.'

***

I met Dr Erica Thompson at maths camp when we were both fourteen. While I dropped out of a maths degree after a term, she stayed on track, did a PhD on climate modelling North Atlantic storms and now is Senior Policy Fellow in Ethics of Modelling and Simulation at the London School of Economics and Political Science (LSE). During her doctorate, she became fascinated by the question of how models represent the world – or rather, if they do so at all. Models, Thompson argues, operate in a parallel universe she calls Model Land, 'a hypothetical world in which our simulations are perfect'. 'Within model-land everything is well-defined, our statistical methods are all valid, and we can prove and utilise theorems' – making it a very comfortable place for researchers to reside.[45] Supporting real-world decision making – which is, after all, the main thing models are *for* – is in this way a simple matter of 'taking the output of model simulations at face value', doing some quick maths to knock out 'blatant inconsistencies' and then interpreting the frequencies the model produces as being real world probabilities and likelihoods.

Anyone with the slightest familiarity with sociological thinking can point out the many and various ways that models often fail to properly represent reality, often pointing towards the biases and shortcomings of those who created them. Thompson acknowledges all that, and then says: hang on, they aren't even mathematically accurate. She calls this the Hawkmoth Effect.

Thompson's theory complements Lorenz's Butterfly Effect, that is the idea that a small difference in initial conditions (the eponymous butterfly flap) can result in a large difference in the outcome of a complex dynamical system. Thompson argues that in the twenty-first century, the Butterfly Effect is a solved problem – instead of taking a single starting estimate, modelling should proceed from multiple initial conditions and then produce a probability distribution output rather than a single answer. When the equations of the dynamical system are known perfectly, this works well. When they're not – for example, when we don't know how much dust is heating and how much it's cooling the planet – then she says, 'the probability distribution that we arrive at by using this ensemble will grow more and more misleading: misleadingly precise, misleadingly diverse, or just plain wrong in general'. Making incremental improvements to the realism of the model might improve short-term forecasting a bit, it's true, but over time a 'slightly wrong' initial condition can still yield 'wildly wrong' outputs: this is a fundamental property of nonlinear dynamical systems, like dust and climate.

It is pleasingly poetic to be writing about a substance that is fundamentally unknowable by computational systems, I must admit – but we do need to know about dust in the Earth system nonetheless, if we're to understand the risks of our warming planet. So what is the climate modeller to do?

Thompson suggests some interesting things. First, there is a point at which it is no longer helpful to call for more research, better information or less uncertainty before making a decision: you just have to jump. Do we need better climate models in order to decide whether to upend the status quo economy into a Green New Deal of decarbonising energy production and massively retrofitting existing buildings? No: we just need to do it now – and do it fast. Thompson also stresses the need for a clear statement of the model's own limitations – and argues in favour of qualitative inputs, such as expert judgement, in order to interpret model outcomes.

In dusty terms, that might mean showing climate modelling outputs to the people on the ground who face the consequences forecast – and involving them in creating the environmental plans and policies to mitigate these problems. The Marshall Islanders and the lower Manhattanites who will lose their homes to sea level rise; the American Plains farmers whose livelihoods depend on rainfall and aquifer recharge; the Arctic hunters who can no longer take their sleds out on the sea ice. The world's most vulnerable landscapes are disproportionately Indigenous land, so connecting top-down science with their traditional ecological knowledge is particularly important here – but local, lived experience of all kinds matters. What if *that* was how governments made environmental policy decisions? What then?

I also find it useful to think with a term from the field of science and technology studies: *mess*. Sociologist of science John Law writes about 'mess' as a category of phenomena that are fundamentally resistant to simple, clear description. Alcoholic liver disease was one such 'messy' object for him, as he sought to map patients' experiences through the medical system, only to find his object of study kept slipping as he moved between different medical specialities, patients and support groups. The methods used to define the condition at each point didn't just neutrally describe its social reality, but were actively involved in creating it.[46]

In the same way, 'dust' as an object of inquiry is slippery and non-singular as we move between fields: aerosol scientists speak of mineral dust and soot, toxicologists of fine and ultrafine particles and newspaper headlines of smog, smoke, haze and – of course – 'blood rain'. Heavy metals and radioactive particles don't feature in climate research, but they're at the front and centre of human health studies. Microplastics are newly being classified as aerosols. And almost all of it operating beneath the limits of vision. Of course, dust isn't neat and tidy as an object of research. It's a reminder that the world isn't, either.

Modernity sought to make the world measurable, predictable and controllable. Dust reveals the limits of that episteme. And it might just provoke us to think and to imagine better.

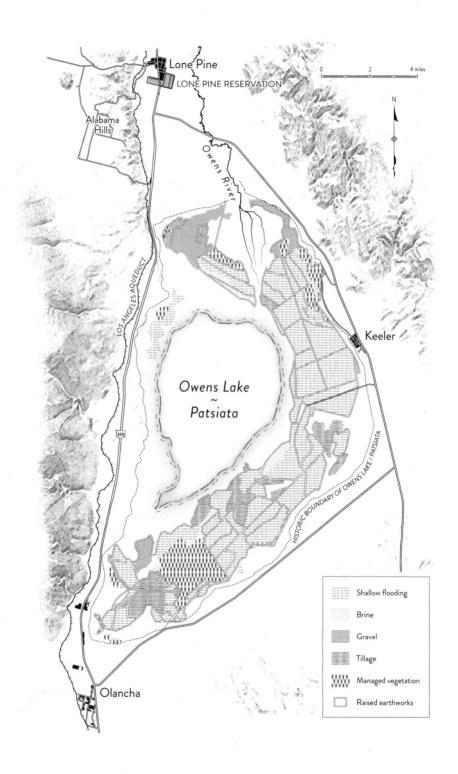

Chapter 9
Payahuunadü

The sun was slipping behind the Sierra Nevada as I drove up 395 from the south. The road was almost empty, the hills ahead of me indigo in the growing dusk. It had been a long day: 350 miles across the Mojave Desert from one drying-up sea to another, crosswinds teasing at nudging my car into the path of an oncoming truck all the way. In Johnson Valley, I'd watched dust clouds rise and haze out the hills behind as, in the distance, dirt spiralled into a twisting column that swept across the plain, then suddenly vanished. But heading into the valley of the Owens River, the land turned a little gentler, a little greener. A dozen majestic cottonwood trees bowed their canopies over a pull-in near Olancha, rich with new spring leaf. And a billboard on the side of the road proclaimed: IMAGINE PAYAHUUNADÜ. Imagine this as the Land of the Flowing Water.

And then the road crested a small rise, and there ahead of me was the valley-wide expanse of Owens Dry Lake, 110 square miles, white with salt-crust, proclaiming only its absence.

I came to the Eastern Sierra in April 2022 to find out if something could be saved from a ravaged landscape. In Chapter 2, I have already told the story of how that destruction occurred, and for whose benefit – but that's really only the first half of the story. It's taken half a century, but finally change has happened – and what change. To my knowledge, Owens Lake is now the largest dust mitigation effort the world has ever seen. The question that remains open is whether they did it the right way. That's why I want to go deep into both the politics and the design of dust control methods in this chapter. Who made it happen, and what worked and didn't work?

Owens Lake is not the only dried-up lake in America – just the first in contemporary times, and thus, the forerunner. More dust is on the horizon as the Great Salt Lake in Utah and the Salton Sea in California shrink and their shorelines lift in the wind. So there are lessons here for the world, and for the future.

This chapter also brings home the intimate interrelationship between dust and water in this book – dust and water and land. A systemic problem cannot be solved in parts: you have to heal the whole.

But back to the valley, where there's going to be a bird festival on Saturday 23 April, organised by Friends of the Inyo, a local environmental group. In a few short years, the dust bowl of the dry lakebed had become a 'globally important wetland in the making', according to the Audubon Society, America's largest bird conservation organisation.[1] Each spring and autumn, tens of thousands of shorebirds and waterfowl stop in the marshlands and rewetted areas of the lake to gorge on brine shrimp and replenish their energy before continuing on their transcontinental journey up or down the Pacific Flyway from Alaska to Patagonia.

How had this happened? Could it really be as good as it sounded, as good as I might hope? It sounded like the birds were flourishing, yet what about the human inhabitants of this landscape? And so, I made my way to the small town of Lone Pine, California – population 2,035 – and started to talk to people.

***

For much of the twentieth century, the dust from Owens Dry Lake was seen as a problem, yet one too diffuse to be anyone's responsibility. Because the source was a dried-up lakebed, and not a power plant or an ore-milling operation, the dust could be framed as 'natural': an inconvenience and sometimes a hazard to traffic on the 395 Reno highway, but not so evidently 'pollution'. But this changed. How? And why? In 1950s' and 1960s' America there wasn't just a change in science, there was a new imagination.

I asked Phillip Kiddoo, Air Pollution Control Officer of the Great Basin Air Pollution Control District – that is, the head of the dust control programme in the valley – what had happened. It required 'a change in culture, in society – a change in paradigms,' he told me.

In the 1950s and 1960s, some of America's first environmental protests were held in Los Angeles, by women in the activist group Stamp Out Smog who brought their children to rallies wearing gas masks to pressure politicians to address the city's terrible air quality. Rachel Carson's book *Silent Spring*, published in 1962, also galvanised this new way of thinking, not only igniting a wave of concern around pesticides but reframing how many Americans imagined their relationship with nature. No longer was the environment simply a base of natural resources for human exploitation, Carson revealed it as both fragile and reciprocal. That is, if humanity poisoned the environment, those poisons would come back to harm us too. Not news to Indigenous Americans, of course, for whom such reciprocity had been at the centre of their worldview for countless generations.

In 1963, Congress passed the Clean Air Act, which set national air quality standards for six air pollutants including PM10 and PM2.5 particulates. Soon after, the state of California passed stronger standards for cleaner cars and cleaner gasoline – and, in 1970, President Richard Nixon signed the Environmental Protection Agency (EPA) into existence. This 'triggered requirements for states to impose regulations to meet these air quality standards that the federal government set,' Kiddoo explained. It also set in motion the creation the Great Basin Air Pollution Control District, the body that would measure, monitor and build, Kiddoo says, 'the weight of evidence, all the supporting scientific documentation' that Owens Lake was indeed the source of the region's air quality problems.

Meanwhile, scientists were finally starting to prove definitively that mineral dust was bad for human health. In the early 1970s, statisticians analysed death rates in 117 American cities and found a 'substantial' relationship with the particulate pollution levels in those cities.[2] Harvard University environmental health researchers began a 'Six Cities' study exploring the relationship between air pollutants and heart and lung diseases – which found that tiny particle pollution was far more dangerous than gaseous pollutants such as sulphur dioxide.[3] And so the second pillar of the valley's legal case was built: there wasn't just a dust problem at Owens Lake, but it was costing residents their health and even their lives.

The 1970s saw the environmental movement come to the Eastern Sierra, too. In 1976, David Gaines, a young teaching assistant at Stanford University, camped beside Mono Lake – another saline lake 110 miles north of Owens – with a team of undergraduate scientists, counting its shorebirds, titrating its chemistry and inspecting its brine shrimp. Water levels in the lake were falling due to LADWP's water diversions, and their studies revealed just how ecologically catastrophic this would be. They formed the Mono Lake Committee to fight back, quickly winning public sympathy with their 'bucket brigades', hikers, runners and cyclists who carried water all the way from Los Angeles to symbolically replenish the lake. A group called Concerned Citizens of the Owens Valley was also set up, and a new coalition formed around the triple prongs of water sustainability, plant and wildlife preservation, and aiding the local economy. Using the new environmental legislation on air and water quality, these groups started to build a case that the DWP was responsible for the environmental problems in the Eastern Sierra – and should be held legally responsible.

***

It's one thing to say a problem needs to be solved – quite another to work out how exactly to do it. So how do you make a lakebed less dusty? There are half a dozen ways.

One is to put something on top of it: usually gravel, though a 1986 report from the US Navy also considered more exotic options, such as tarmacking the lakebed with a thin layer of asphalt, or spraying it with calcium sulphate to produce a more resilient limestone surface.[4] A ten centimetre layer of gravel will not only stay put in very strong winds, but also prevent efflorescence, the rising of salt to the surface – from where it may become salt dust. But installation, with all the gravel mining and heavy truck movements this involves, is itself a pretty dusty job.

A second set of solutions entail interrupting the wind. Sand fences were tested during the 1990s as a solution that might catch and halt the movement of dunes in sandier areas around the lake. The problem: they're only partially effective. A method called tillage turned out to work better. Ploughing deep furrows across the dry lakebed, perpendicular to prevailing winds, roughens the surface and disrupts the windflow, making it weaker and unable to pick up as many particles. Tillage's great advantage is that it's fully fifty times cheaper than any other method, coming in at around half a million dollars per square mile to set up[5] – but it offers no habitat for wildlife. And it's also pretty ugly.

A third option is to get rid of the source of the dust, specifically, the tiniest salt particles that are most easily picked up by the wind. As the water table rose and fell throughout the year with the seasonal rains, the salts it contained would repeatedly dissolve and reform, never getting the chance to grow into sizeable particles. One conceivable fix was to lower the groundwater – though the Navy report noted that 'any attempt to export more water from Owens Valley is likely to lead to public-relations problems' (a profound understatement). Another route is 'poldering', essentially washing the playa with fresh water, an approach used in Holland to desalinise land reclaimed from the sea.

But the main dust control method used on Owens Lake is instead 'brine'. This entails the controlled growth of a hard, salty crust which armours the surface and makes it more resilient against the wind. Managing this process to produce the right sort of salt crystals and not the wrong sort (too small, too puffy or otherwise flyaway) is far from easy: these salts are soluble, so it's an inherently unstable surface.

A fourth option is for the playa to be revegetated with salt-tolerant grasses and tough, hardy shrubs. The salty, boron-laced ground isn't the easiest place for plants to grow, but local species are used that have adapted to these endemic Great Basin conditions – and it's been remarkably effective. One early study found that just 17.5 per cent cover of saltgrass decreases the movement of sand by 95 per cent, as the plants' roots reach out far and wide in their search for moisture and nutrients, and hold the earth in place.[6] A bit more vegetation coverage and it reaches 99 per cent effectiveness. That said, the lake's diverse patchwork of water, salt and soil conditions mean that it's not a solution that works everywhere – and the requirement for ongoing irrigation frustrates the city, which would rather see this water going to sustain Los Angeles lawns, not Sierra scrubland.

Finally, there is the most obvious solution: bring the water back. For dust control purposes, this does not mean refilling the lake, but rather shallow flooding, a shimmering few inches of water silvering three-quarters of the designated areas in a series of pools and ponds. It is, unsurprisingly, almost completely effective against dust – and has real environmental value for birds, animals, tiny marine critters and people alike. For the Paiute communities in the area, seeing the place they call Patsiata as a lake once more has especial cultural and spiritual significance.

The plan agreed in 1999 required the City of Los Angeles to implement three of these methods: shallow flooding, salt grass vegetation and gravel. For over a decade, Owens Lake was a billion-dollar building site. Just one phase – part 7A of the ten-stage project – entailed shifting a million tons of earth and building forty miles of berm roads across the fragile, marshy salt crust to create a grid of orderly, shallow pools. Twenty-four thousand sprinkler heads, 400 miles of pipe, 900,000 tons of gravel and it became apparent that bringing dust emissions down to legal limits would require tackling a larger area than initially anticipated, and the costs kept mounting.[7]

By 2012, the LADWP had spent $1.5 billion on a place most of its customers had never heard of, and sought to argue that this was, perhaps, enough. The city was pouring 95,000 acre feet of water a year into Owens Lake – as much as San Francisco consumed in a year. They went to court to get a change of dust control tactics: less flooding, more water-minimising gravel and tillage. Despite this shift, by 2019 the city had spent over $2.1 billion on addressing the environmental crisis that it had itself created: around $50,000 per resident of the Owens Valley, or the equivalent of two months of Los Angeles residents' water bills per year.[8] Thirty billion gallons a year, fully a third of the water Los Angeles takes from the valley, is being used to solve the problems of its very removal.[9]

Was it worth it?

***

Something extraordinary has happened.

For seventy-five years, the dry bed of Owens Lake was a haunted, dead place. Less than 2 per cent of the surface was ecologically valuable habitat such as alkali meadow, saltbush or shadscale scrub.[10] Across most, there lived little bigger than a cyanobacterium or single-celled archaeon. Then, in 2001, the LADWP at long last opened the taps, and snowmelt from the high Sierra once again spread across hard-paved playa crust. Soon 26 square miles of dust control water lay in shallow inches-deep pools and deeper ponds, while on the western shore, the springs and seeps and brine pool had always kept their soggy faith.

This water brought back life – in vast and unexpected profusion. Ever-hardy saltgrass that had lain dormant for decades self-seeded. Tens of thousands of migratory birds, heading up the west coast from winter in southernmost Patagonia, saw a gleaming expanse of water shimmer in the spring sunshine and settled down to grow fat on brine flies, building up the reserves before flying on to their summer breeding grounds in the forests of Arctic Canada and on the shores of the Bering Sea.

Birds had visited the lakes of the Great Basin for 800,000 years – and now, after an eighty-year hiatus, they were back. During the first Owens Lake 'Big Day', organised by the Owens Valley Committee and the Eastern Sierra Audubon Society in April 2008, a staggering 45,650 birds from 112 different species were recorded, the largest number ever counted at this site. Delighted birdwatchers ticked off shorebirds and waterfowl ranging from the rusty-red white-faced ibis to the black-bellied plover. They counted 9,218 American avocets; 1,767 tufty eared grebes and 13,826 'peeps', the tiny, twenty-gram sandpipers that dabble at the water's edge.[11] Nature had started to reclaim this most engineered of landscapes.

In 2018, Owens Lake was designated a site of international importance for shorebirds. It now hosts 100,000 avian visitors during the spring and autumn migrations – and an annual Owens Lake Bird Festival. Tourism and recreation are now primary parts of the valley's economy. Dowitchers and whimbrels, yellowlegs and black-necked stilts, pocket mice and kangaroo rats, jack rabbits and cottontails abound. Tule elk are reported to rest in the dry alkali meadows in the valley, and bobcats, coyotes and foxes stalk through the grasses.[12]

For three or four generations – most of the twentieth century – the lake seemed a dead place. But the landscape proved more patient, and more resilient, than many could have imagined.

Just add water again: life erupts into abundance.

***

Owens Lake may have come back to life, but it remains a strange and artificial place – almost unrecognisable as a 'lake' as such. Today, dust control works now cover almost half the lakebed, forming a 25-mile curve of infrastructural engineering around the north, east and southern sides.

One Saturday in April 2022, I went out to take a look. The occasion was the Owens Lake Bird Festival, back after a two-year pandemic pause. Organised by the local environmental group Friends of the Inyo, it offered variously birds, botany and off-road driving into the mountains to admire the regional geology, a tour of the Cerro Gordo ghost town and/or some light Class 3 scrambling. My only regret was I couldn't do the lot: it had been a tough week, being based in the Lone Pine mountaineering hostel and watching groups head off to climb the neighbouring Mount Whitney without being able to join in. But I was here on dust business, and that meant heading out with photographer Martin Powell, a tall, avuncular man, teaching a photography session for beginners and men with enormous zoom lenses alike.

I hitched a ride in Martin's truck out to the lake. We headed south on 395, then turned off at Boulder Creek on to a dirt road that headed into the lake area itself. Martin just kept driving. Special access for the bird festival, I wondered? Not in the slightest. He explained how a legal quirk made it entirely public land. The lake is classified as 'submerged lands', even potentially navigable water (recall how back in the nineteenth century, a steamship once sailed across it to the Cerro Gordo mine) – and, as such, it belongs to the California State Lands Commission, quite regardless of the fact there hasn't really been much water there for a century. DWP only manage the lake, they don't own it, yet they've even created visitor trails and interpretive signs. There are Portaloos, land art, even a leaflet. Visitors are merely requested to try to keep out of the way of large earth-moving equipment and, if duck-hunting, please not to shoot in the direction of the engineers.

So we drove down Brady Highway, the main road for construction traffic on Owens Lake. A raised ridge to the side carries the 1.2-metre pipe that carries irrigation water from the Los Angeles aqueduct out to the network of pipes, pumps and irrigation sprinklers that administer 49 square miles of dust mitigation works.

Branching off it were access roads, raised up from the lake surface and dividing it into almost a hundred different dust control areas over which different dust control strategies have been applied.

In the areas where these methods have been established for the longest – about twenty years, now – the landscape resembled an estuary or salt marsh. Within its rectilinear scaffold of roadways, a more organic landscape was forming: swathes of pale gold salt-grass surrounding curving puddles of water and wide plains of mud. The water's edge was viscous, black with brine fly. California gulls paddled in the shallows, trampling their feet to stir up larvae to gorge on. Further down the lake, brine areas had grown a deep evaporated crust of salt that cracked into tortoiseshell plates several metres across. Inside each, opaque water lay in opaque pools of rust-red and dull, brickish pink, tinted unearthly by the minerals. Halophilic archaeans – strange, tiny single-celled salt-loving creatures, at home here (and perhaps, we shall find, on other planets) – are drawn to these mineralised waters, staining it further with their carotenoid pigments.

But the birds . . . I peered down the Swarovski telescope of a field biologist from the Bureau of Land Management, as Martin talked us through the different species at each location. This was a biome very different to the heathland I grew up on – but childhood memories of *I-Spy* books stirred, and I was surprised by how many names came back to mind. American avocets were the star attraction, hundreds upon hundreds of them, leggy and elegant in their blue stockings and peachy breeding plumage as they sifted their long, upturned bills through the shallow water. Swallows darted overhead and, in the distance, flocks of shore-birds rose and turned, a hawk disturbing their peace. Groups of grebes bobbed on the lake's wavelets – it was another windy day – before diving down for food, and coots dotted the water to slightly bathetic effect. (Had I travelled so far only to meet the principal occupant of the rivers and reservoirs I walked at home in north London?) I preferred the western and least sandpipers, pleasingly portly little things in tawny brown and white, punctuated by petite black legs and beaks. I looked out at the landscape, equal parts humanmade and nature reclaimed, and tried to make sense of what was in front of me.

I thought of all the other drying lakes I had seen. The Aral Sea had been sublime in the sunlit distance and yet repellent up close. In 2016, southern California's Salton Sea had been little better, ringed with dead tilapia and sea birds. I visited it again in April 2022, when it smelled much better, its beaches almost entirely corpse free. A sign at the visitor centre insisted that the water wasn't polluted and the fish were flourishing. Could it be? Is nature healing?, I asked a park ranger in faint hope. He countered that perhaps there are fewer dead fish these days because most of the fish had already died out long before. *Ah*. But compared to either, the Owens Lake felt vital, teeming and abundant; a few inches of water sustaining a complete foodweb. It was fascinating – holding an odd, austere, allure. Its hypersaline hues have caught the eye of photographers such as David Maisel and George Steinmetz, who see in it 'a strange beauty born of environmental degradation' – that is, 'a contemporary version of the sublime'.[13]

'Is it beautiful?' I asked Mike Prather, who's led countless tours of the lake in his thirty-plus years of advocacy for its avian occupants. Yes, he replied. In two ways. In part it's intellectual for him, he explained, as a former science teacher appreciating the intricate, interdependent web of life that has developed. But he added, 'For me, it's also the type of beauty where I'm just profoundly struck. I'm out there by myself, hours and hours and days and years. And I never, ever tire.'

It's not the same for every visitor, he comments. Sometimes, 'Instead of seeing the birds and going, "Holy crap! There's 10,000 avocets in front of me, there's 30,000 ducks," they say, "Well, this looks industrial."'

As well they might.

Air pollution monitoring systems stud the landscape, rendering it visible as data. Since an extended drought in 2014, Owens

Lake has been operated under what's called dynamic water management, a system of sensors, sluices and computer-controlled sprinklers that spring into action when the 'sand flux' exceeds a particular threshold or if dust is seen swirling over the lakebed on a windy day. And so this place has become a cyborg landscape, boundaries dissolved between the organic and inorganic; a hybrid space, a system equal parts eco- and engineered.

Yet, as Mike and I discussed, it is not really any more unnatural than any other waterway in America. Every river is dammed and managed and made modern as a resource for hydropower, water supply and irrigation. There is no wild water left in the Western world, and precious little elsewhere – the notion of untouched wilderness a romantic folly. As environmental historian William Cronon wrote in 1994: 'The myth of wilderness is that we can somehow leave nature untouched by our passage.' It's a desire to believe 'we can somehow wipe clean the slate of our past and return to the tabula rasa that supposedly existed before we began to leave our marks on the world'.[14] That is, it's a distinctively white fantasy that settler-colonialism did no harm. The Paiute people I talk to don't speak of 'untouched nature' – quite the opposite – emphasising instead their generations of stewardship and water management, building irrigation ditches in order to spread water, and thereby life, across more of the land.

Yet birds seem simple creatures less concerned with ecological authenticity, only water and food. The hyper-engineered landscape of Owens Lake today turns out to suit them just fine. 'Some species, there's probably more of them now than when the lake was there,' Mike told me. 'When the lake was there, there was this thin band of shoreline and then a massive 110 square miles of open water.' He asked a 'GIS fellow' (a geographic information systems expert) to map out how much shallow water bird habitat there would have been round the lake when it was full – and he told Mike 4,200 acres, 'a small fraction of what is out there right now'. I calculate that 4,200 acres is 6.5 square miles – in contrast to 35 square miles of the shallow flooding and managed vegetation today.[15] You can see why the birds are having a good time.

Yet elsewhere, the lakebed seemed lifeless still. Older areas of gravel had picked up plant life in the cracks and had naturalised into something like a shingle beach, but newer ones were monotonously flat and grey, more reminiscent of an off-road parking lot. DWP may prefer such water-free dust control, but to environmentalists and the Paiute community it's the least desirable option. 'If they're just spreading gravel out there, that isn't providing any habitat, and they're causing dust and disturbance from the mining,' said Teri Red Owl, executive director of the Owens Valley Indian Water Commission (OVIWC), the intertribal consortium for water in the valley. To Mike Prather, 'It's good for spiders, occasional lizards, and things like that, but not too much else.'

Nonetheless, when it comes to dust, it all works. Gravel, plants, brine, water: it's all remarkably effective.

'At this point in time, we believe we control about 98 to 99 per cent of the emissions,' Phill Kiddoo told me. 'To put that in perspective, we used to have days where we would have national ambient air quality exceedances when we would have levels at over 20,000 micrograms, one hundred times the regulatory limit. It's equivalent to driving 6,500 miles an hour above a speed limit of 65 miles per hour. It's hard to imagine those levels.'

'At this point in time, we see exceedances one or two days a year, at one or two or three of the monitors at the lake, not all of them, and not on several consecutive days. And the levels that we're seeing on those violation days are generally like 180–200 micrograms, you know, maybe a big day of 400. That's pretty unusual. So we're really almost there.'

It's a hell of an achievement, and testament to decades of dogged work by Kiddoo, his predecessor Ted Schade and everybody in the public agencies, tribes and non-profits who have fought to hold Los Angeles to account. The Owens Valley was a world-class air pollution disaster zone, a place where people were afraid to raise their children. Now it's not. While the conflict is still by far from over (as I wrote this chapter, Great Basin and DWP were embroiled in yet another lawsuit about inaction on a new area of lakebed requiring dust control) the horizon of the possible has, I believe, undoubtedly expanded.

Yet there is more to be done – and the path ahead is paved with conflict. Two issues in particular stand out: one is still dust; the other its antithesis, water. They are, of course, connected.

***

It was a Wednesday in April 2022. I had spent the day forty miles north, speaking with the environmental team at the Big Pine Paiute Tribe. As I drove back down 395 to Lone Pine, it was as if I was driving into a whiteout. Haze filled the entire valley, from the dry, bare rock of the Inyos in the east to the jagged summits of the Sierra Nevada in the west. It was a bright, pure, formless white, like fog or mist – but, of course, it wasn't water vapour at all: it was salt. Salt dust. Clay dust. Some non-negligible quantity of cadmium, nickel and arsenic. This wasn't good. I wondered at what point it might be worth wearing my Covid mask to go outside.[16]

This was the worst I had seen the dust – but not my first encounter. The day before I'd driven around the lake and watched particles lift up and sweep across the lakebed in ragged fringes of white, ghostly and possessed, rising perhaps a hundred metres into the air. The stronger forms coalesced almost on the verge of twisting into defined dust devils, but never quite – and then the wind would lull for a moment, and they'd vanish into thin air, as if there was nothing to worry about. The dust didn't just blow on the lake, but in skeins and flurries on the valley side, too – land that looked to have plants growing on it, desert scrub, plants that should have been holding the earth together. Their names reveal their toughness: saltbush and iodinebush and invasive Russian thistle; plants that can stand up to a bit of a chemical beating. Yet they can't live without water. The fact dust was blowing on the valley sides was a sign that water in the Owens Valley was still out of balance.

'Who's the most radical voice for water rights and land rights in the valley?' I'd asked part way through my conversation with the Big Pine Paiute Tribe environmental team. 'It's probably Sally!' said Paul Huette, the utility operator for the tribe, and everyone around the table laughed. 'My hands are tied with some situations because of the position that I'm in,' he explained. As an enrolled member of the Big Pine Paiute Tribe, he sits on myriad public service boards for everything from environment and water to housing, schools and the volunteer fire department.

Sally Manning, by contrast, can speak more boldly. An ecologist by training, she wrote her PhD on water usage in Owens Valley plants while working for Inyo County as an ecologist, while also being highly involved in environmental activism in the region. She became Big Pine's environmental director in 2009. As a white American, Sally does not speak as a tribal member, but she's a deeply informed advocate for the tribe's rights and interests nonetheless. We were joined by water programme coordinator and member of the Bishop Paiute Tribe, Noah Williams, and Cynthia Duriscoe, who manages the air programme.

Dust is a continuing concern for this community. 'Here in Big Pine, we are about eight miles north of the socalled non-attainment area for the Owens Lake – the boundary stops at the Poverty Hills just south of here,' Sally explained. 'There really wasn't any long term monitoring of weather or air pollution in the Big Pine area, even though we have significant population – well, for Owens Valley anyway,' she qualified. Everyone laughed: Big Pine has just 1,563 residents and yet it's still the third largest town in a sparsely populated county. 'And we could sit here and see dust coming off, and it's not Owens Lake, it's out there,' she said, gesturing out the window. The culprit? Groundwater pumping.

The water that flows to Los Angeles isn't just natural run-off from the mountains and the flow of the Owens River. Surface water flows weren't enough for this thirsty city: the DWP pumps groundwater from underground aquifers in the valley, too. 'There are some hundred-year-old holes around here', Sally told me, installed by private ranch owners before the aqueduct's construction in 1905 – but the DWP's groundwater pumping really expanded in 1970, when they enlarged the aqueduct to carry 60 per cent more water. They have about a hundred pumps in the valley now, mostly located on alluvial fans on the valley sides, to capture the water coming down off Sierras. These sedimentary soils had for hundreds of years supported alkali meadows, a grassland interwoven with irises, lilies and native shrubs and home to meadowlarks, hawks and the Owens Valley vole – all fed by a high-water table. In 1859, US army captain John W. Davidson mounted an expedition to the valley, declaring it some of the finest country he had ever seen: 'a vast Meadow, watered every few miles with clear, cold mountain streams, and the grass (although in August) as green as in the first of spring'.[17]

'So that's what they really destroyed,' Sally said. 'The surface just lifts up and blows away in the wind.' She'd go out on rare-plant monitoring surveys and find herself tramping through dead vegetation as dust rose about her ankles.

'You gotta pump 5,000 more acre feet a year and drop a natural spring, because you want to see more water going somewhere else that's 250 miles away?' Paul, Sally's colleague, commented, angry and incredulous at the waste of it all.

'The dust affects a lot of health issues here,' he continued. 'A lot of people have asthma or need an inhaler, because of the dust and the particulate matter in the dust.'

Noah is one of them. He recounted how his father, Harry Williams, a tireless activist for Nüümü (Paiute) cultural heritage and water rights, would tell a story of when I was just a small baby, back in 1996, before a lot of these dust mitigation measures were in full effect. They were coming back from Los Angeles, my mom and my dad, me in the backseat, and they were driving through the Lone Pine area when one of our crazy dust storms was coming through. He couldn't even see the surrounding area:

it was just completely covered in dust. He said that he could taste it in his mouth, a metallic sort of taste – and then he was just looking back and seeing his new child, myself, in the backseat and thinking, "Oh, that can't be good for him." And then just thinking to himself, "How the heck do people live like this? How do people live?"

'He unfortunately developed interstitial lung disease,' Noah continued. 'There's probably a whole slew of things that might have contributed to that, but a big contributor would have been living in this area. There's a lot of elders that have respiratory issues, especially some down closer towards the lake, who have it even more. Myself I dealt with childhood asthma.'

Harry Williams passed away in 2021. His life's work had been *paya* (water), and so it was that the day he departed, thunder reverberated between the mountains and the heavens opened, there, in the shadow of the Sierras, where rain hardly ever falls.

***

It's tempting for an outsider to write about the valley as something 'ruined', sucked dry and left for dead by the ceaseless thirst of an ever-expanding Los Angeles. But from the moment I drove up 395, it was evident that wasn't right at all. There was more life here than that – from springs and streams on the mountainsides to the quarter-mile wide band of green fed by the meandering progress of the Owens River on the valley floor, not to mention the decades of legwork (and legal action) by Indigenous people, environmentalists and public servants alike. To speak of ruination is to deny their agency and erase their power.

'How might we think of this place instead?' I ask the people I speak with in the valley. 'How might we imagine what it could become?'

'I wouldn't call it ruined,' Teri Red Owl responds. 'I would call it damaged. But, you know, damages can be repaired. Maybe it's not going to be exactly the way it was. But they can often be repaired. And so that's how I look at the land, like it's been damaged, it's been hurt, it's been injured. But if we take care of it before it's actually dead, then we can get it back to a much better place than where it is today.'

The day before, Noah Williams had shared the example of the Fish Springs hatchery just down the road from Big Pine. There, an enormous water allocation of over 20,000 acre feet – ten times the usage of the town of Bishop – sustains a trout hatchery before it's siphoned off by the two largest pumps in the valley into the aqueduct to Los Angeles. In 2020, however, there was a bacteriological outbreak so severe that operations stopped and the ponds were emptied.

'In that brief, brief time, about a four month period, we did see recovery of the groundwater table,' Noah said – and not just a little bit, but an enormous 8.5 feet. As water coordinator for the Big Pine Paiute Tribe, this is the data he monitors day in, day out. Even when they restarted operations with just one pump running, the groundwater was still rising. 'And that just goes to show the hydrology in the area: just how able it is to recover,' he said. 'You could heal the land, and you can heal the groundwater.'

Sally chimed in, saying she's seen exactly the same in the grass-lands she monitored in her previous role at Inyo County Water Commission. 'They pumped the water down so fast that these things just croaked. One year to the next, things went from beau-tiful green to the point people thought someone must have come and sprayed herbicide out here. My first impression was this place is done for now. We wrecked it. But they're perennial native grasses in Owens Valley. A few years later we had a wet year and they turned the pumps off – and the water table did come up. And we went out there and said, "Thank God, you're coming back to life." And so there is resilience, and it does give me hope.'

As with the birds arriving back on Owens Lake, it doesn't even take years – just months. Payahuunadü is a place so very close. Yet this very resilience is a property that can be used against the land. Sally played out a typical conversation with DWP, where admit-ting that vegetation has the ability to regenerate produces only dismissal. *Well, then, the damage isn't permanent, and so we don't have to do anything differently*, they say. 'That's where you get into the politics,' she sighed.

'The white man's idea of mitigation is very different to the Indigenous idea of mitigation,' Teri Red Owl tells me the next day. 'In our opinion, you don't "mitigate", you just don't cause the problem to begin with – because I don't believe you can ever get something back the way that it once was. You can make things better, but you don't ever get everything back. There's still that trauma.'

Her colleague, Kyndall Noah, communications lead at OVIWC, agrees: 'The mitigation efforts that they have done for the dry lake doesn't talk about the history of health disparities, it doesn't talk about the cultural effects that had on people and their relationship to the water. That's a loss of whole generations of cultural ties to that water.'

Teri talks about how Patsiata (Owens Lake) used to be a major food source for Indigenous people in the valley. 'We would collect brine flies off of the lake, that was a huge source of our protein – and of course all the birds that were there, too.'

Large bodies of water are always ceremonial places too, she says. This one is also a place of remembrance. Teri speaks of a massacre in 1863 during the Owens Valley Indian War, where a militia of soldiers and settlers pursued Paiute men on accusations of livestock raiding. 'The men ran into the lake to try and escape the muskets, and a lot of them were killed right there in the lake,' she said. Thirty-five men died – some say women and children, too. 'You can start to begin to heal, but that's going to take generations.' For people in the community this is still within family memory.[18]

Yet can space be found – in 110 square miles of lakebed – for such history to be remembered? Efforts have been made by Great Basin and DWP to preserve this particular site from disturbance, but elsewhere in the lake it's still a contested matter. For years, Teri Red Owl told me, DWP employees and consultants would pick up artefacts they found on the playa surface and pocket them, bragging about their finds in the bar that evening. (Cultural monitoring has since been introduced to try and stop this.) Remnants of a historic dry period 800 years ago face being ploughed into oblivion by tillage.

The natural landforms are themselves culturally important, argues Kathy Bancroft, tribal historic preservation officer of the Lone Pine Paiute–Shoshone Tribe. 'Our family's history is in the landscape,' she wrote in 2013. 'Not only are they destroying the proof of historic events and our prehistoric way of life, but they are changing the landscape and geology that our stories are built on.'[19]

***

'We've been a water colony for Los Angeles DWP for the longest time,' Kyndall Noah says. It's a phrase that brings the power dynamics in this landscape into sharp focus.

'A lot of the important decisions that are made in Los Angeles, we don't have any kind of say,' he continues. 'Any decision that's made 300 miles away affects us here – but we can't even get on the agenda to make our reasons why the groundwater shouldn't be pumped.'

Los Angeles literally owns the Owens Valley. Half the valley is public land, either federally owned and managed by the BLM (33 per cent) or belonging to the State of California (14 per cent). [20]Over 95 per cent of what's left is property of the City of Los Angeles. Other landowners hold just 2.9 per cent – and the four Tribal reservations, just 0.4 per cent, or 1,700 acres. They might have had more, Noah Williams explained. Back in 1912, President Taft approved a plan for 2,800 acres of reservation land, alongside 67,000 acres put in trust for tribal usage. But Los Angeles lobbied the federal government to take this away and assign it as Forest Service and federally owned lands. Why?

'To, quote, "protect the water resources of Los Angeles",' Noah Williams said.

'Everybody likes a spotlight on the fall of Mulholland and the dam debacle,' he continued. 'But, shoot, this is our own dam debacle, too.' His words are met with a chorus of assent from his colleagues. In 1932, the DWP's land agent made another attempt to eject the tribes from the valley entirely – which resulted in a land exchange and the tribes losing their legal water rights. In a desert, this is akin to being dependent on Los Angeles to have air to breathe.

Now, the first peoples of this valley find themselves excluded from having a real say in its future. Even the most prosaic of demands get snubbed. Paul Huette described how, a few years ago, a DWP irrigation pipe sprung a leak due to overgrowing tree roots. For two years, during a period of major drought, 'more than half of the water that we were supposed to be getting was running down the road'. Attempts to straightforwardly meet round a table, one water utilities operator to another, got nowhere. In return for fixing it, DWP demanded the tribes sign away all historic water rights disputes as settled, a position Paul described as 'pretty much signing over your firstborn children'. The tribe had to refuse.

'We're sovereign,' Paul explained – a point I want to emphasise for readers who may not be familiar with the literal truth of that word. During America's colonial expansion west, the federal government recognised Indigenous tribes as independent nations and came to treaty agreements with them on these terms. Today, tribes are legally recognised as 'domestic dependent nations', and the US government interacts with them on the principle of 'government-to-government relations'.[21] Some tribes may today be very small, but then so are Monaco or the Vatican.

'Our people are elected by the people here on the reservation,' Paul continued. 'Our chairperson is equivalent to the President of the United States in tribal aspect. And they wouldn't put our chairperson on the agenda to talk about this issue.'

On 21 March 2017, a group from the reservation travelled down to the DWP Board of Commissioners meeting in Los Angeles and seized the floor during public comment time in order – finally – to be heard. Tribal members spoke about the impact of the lack of irrigation water on their farms, their crops, their ability to feed themselves. Moved by these stories, Commissioner Christina Noonan offered to cut the tribe a personal cheque. 'I, as an individual and mother, could no longer withstand the idea of Big Pine Tribe not having available water to bathe their children nor grow food for their families,' she told KCET local news.[22] The next day, the DWP was back on the phone to Big Pine Paiute Tribe, their tone suddenly very different.

'They had the pipe fixed within two weeks after that,' Paul said bitterly.

***

Why tell a story about water? Because water and dust sit on opposite sides of a delicate ecological equilibrium – in large part, this is a book about what happens when the water is gone. Out in the desert drylands, all conflicts about water end in dust, given time. Often it doesn't even take very long.

Dust storms on the dry bed of Owens Lake and the Big Pine Paiute Tribe's fight to fix a leaky pipe are two sides of the same problem: the fact that the valley is, as Kyndall Noah said, a colony. It's a colony twice over: first came white settlers with guns in the 1860s, who took the land – and then came Los Angeles, forty years later, with lawyers and lawsuits, and took the water.

It's not just Kyndall who uses the term. Writing for the *Los Angeles Times* in 2013, the historian William Kahrl – author of the definitive history *Water and Power* – called the Owens Valley a 'virtual colony'. He outlined how Los Angeles owns most of the land, disallows any economic development that might compete with or threaten their water supply (at one point deliberately flooding a chemical plant sited on the dry bed of the lake), and silences dissent, once forcing a radio station critical of the city to fire its reporters or be put out of business.[23] Economically and ecologically, the Owens Valley is part of Los Angeles in that it is a central part of how the city operates as a system – yet not one person in the Owens Valley has a vote for the municipal authority that determines their fate. That doesn't seem fair.

This is not to claim that opposition towards Los Angeles and the DWP is a universal sentiment in the valley. Many people work for them – I was repeatedly told that these are the best paying jobs in the valley, good unionised jobs that can command a six-figure salary. There's also gratitude for the way the water authority's ownership has saved the valley from development and suburbanisation, as is happening at astonishing pace elsewhere in the West. This is not in any way a pristine, untouched landscape – but its beauty and emptiness nonetheless offer an impression of wilderness that attracts many thousands of tourists on whom the valley's livelihood now depends. Despite decades of energetic disagreement with DWP leadership, environmentalists like Mike Prather and Friends of the Inyo collaborate with DWP biologists on bird counts and lake visits, and call them friends.

There is a growing coalition for change, however. Emboldened by considerable success in the battle for dust control, people in the valley are starting to imagine more. What does justice mean for a water colony? It means getting the water back.

And not just water – land.

***

'I think that Los Angeles has the means to supply its own water without the need of Owens Valley water.' Noah Williams' voice is soft but firm.

Mulholland's greatest success was making the Los Angeles Aqueduct look like an inevitability. It's a conclusion accepted by even the critical historians: 'Los Angeles could not have developed without the water supply of the Owens Valley,' William Kahrl concludes.[24] This is part of the tragedy of it: it was a rational and efficient use of resources, a utilitarian maximisation of the greatest good for the greatest number. If enough millions benefited, then to hell with the few.

Noah had been reading up. A 2018 study by researchers at

UCLA's Institute of the Environment and Sustainability makes the case that '100 per cent local water' is in fact possible for the city by 2050.[25] At present, 60 per cent of Los Angeles' daily water usage is treated to near drinking water standards before being swept out to the ocean every day.[26] What if it wasn't? The researchers quantify how four strategies could make the city completely self-sufficient, namely stormwater capture, groundwater remediation and storage, making more use of recycled water, and, of course, reducing demand through water-saving measures. Through these measures a fully local water supply for the city would be possible, no 338 mile trans-basin aqueduct required. That is, it's possible for Owens Valley water to stay in Owens Valley in its entirety.

After one hundred years of extraction, this may seem an almost unimaginably bold proposal – not even all of Noah's colleagues could fully imagine that LA could really, truly, stop taking *any* water from the valley. But there is real, growing evidence that Los Angeles is starting to think differently about water.

As at Owens Lake, this change of heart arguably isn't voluntary so much as forced – this time, by climate change, and what's now a twenty-year megadrought cycle across the American southwest. The Colorado River is drying up, its reservoirs such as Lake Mead and Lake Powell hitting record lows. Snowmelt from the Sierra Nevada is the prime source of water for the Owens Valley – but after a record-breaking 210 inches (5.3 m) of December snowfall, winter 2021–2 saw almost no precipitation at all. Some years, historic water rights allocate more water than actually flows.

In response, LA's Green New Deal Sustainable City pLAn [sic], agreed in 2019, details an approach to source 70 per cent of Los Angeles' water locally by 2035, compared to just 15 per cent in 2013–14.[27] The new Hyperion Water Reclamation Plant will set the city up to recycle 100 per cent of its wastewater for reuse, and 'green street' refits will let stormwater soak into groundwater reserves rather than running off into drains. Just one storm dumping an inch of rain deposits the equivalent of 5 per cent of

LADWP's total annual supply, so capturing even a fraction of this can make a significant improvement.[28] The city has also declared its intention to reduce water use by 25 per cent by expanding household metering and incentive programmes: buy a high-efficiency washing machine, and receive a $500 rebate.

The press coverage may be glowing – 'LA Is Doing Water Better Than Your City', *WIRED* magazine reported – but over in the Owens Valley, there's a sense of scepticism.[29] Improvement is noted – in recent years, the region supplies about one-third of the city's water, down from two-thirds – but so's the fact that none of the city's plans suggest actually dropping this figure any lower.

Yet it seems eminently possible that they could.

The UCLA researchers write that: 'The City can come very close to full self-sufficiency if conservation performance mirrors per capita consumption in other parts of the globe, including Australia and numerous European countries.' Yet in spring 2022, LADWP customers used an average of 424 litres of water per person per day, nearly three times the volume of much of the Western world.[30] People in the equally sunny and suburban city of Melbourne only a little more (161 litres) – while the UK averages 152 litres and Germans get by on 115 litres per day without evident hardship.[31] This makes Los Angeles's 25 per cent reduction target look frankly a little wet.

Mayor Eric Garcetti boasts of how far the city's already economised its water usage (an admittedly pretty respectable 40 per cent or so), but it remains extravagantly careless still. Thirty-five per cent of urban water use goes towards the irrigation of lawns and gardens.[32] Utilitarian arguments about the 'greatest good' rather fall apart when you realise the Owens Valley has been devastated for the sake of rich homeowners' Anglo-pastoral fantasies of rolling green lawns.

Xeriscaping is exceptionally chic.

Anything else should be illegal.

***

The way Noah Williams sees it, Los Angeles has grown rich on the back of its extraction from the Owens Valley, the water from its rivers, the silver from its mines. 'They wouldn't have been able to grow the way they have if it wasn't for Owens Valley water,' he asserted. 'You know, they always like to pat themselves on the back for the billions of dollars that they spent down at the Owens Lake. But that's just a drop in the bucket of how much economic prosperity that they've had over the last hundred years.'

The city owes the valley a debt. But what might happen if they at last gave the Owens Valley its water back? How realistic is it?

'Environmental justice is like, we want to restore the water we have,' Paul Huette told me. 'We want the naturalness of *this* creek that comes from *this* glacier, the southernmost glacier in California. We should have these things flowing the way that it was supposed to.'

Stopping pumping groundwater is a priority for tribes and conservationists alike, the first stage in healing the hydrological system as a whole. Wendy Schneider, executive director of Friends of the Inyo, concurred. 'The first priority would be to stop the groundwater pumping, and let the seeps and springs recover. The second priority would be to cut back all of the stream diversions that they're doing, and see what happens. Filling the lake is something that will hopefully happen as a result of those things. But the conservation community isn't really focusing on that.'

Nonetheless Teri Red Owl dreams of it, as proof of an ecosystem made whole. 'My ideal vision would be no more exporting our water and letting it all flow back into Patsiata (that's Owens Lake) and bring life back there. If I could see in my lifetime water going back into that lake and it becoming a real lake again, that would be my ultimate dream, because I would know the rest of the valley up here would be healthy if the water was flowing naturally down into the lake.'

But this isn't a call to artificially refill the lake. That would require running the entire Los Angeles aqueduct into it for ten years, Mike Prather told me (he'd asked the Inyo County Water Department hydrologists to work this out). And that's just the best case scenario.

'Then it would get up to the shoreline again, if we had normal years of rain and snow,' he said, 'which we don't seem to have any more. And then of course, if you want the lake to stay relatively the same size, you're going to have to keep the water going out there.' It seems impossible to Mike: 'I just don't know if you can undo things like that.'

Teri dreams of naturally flowing water – that is, a vision of making the whole system of the valley whole again. Indigenous ecological knowledge and complex systems theory once again align. You shouldn't try to brute force an outcome, they say, but rather address the leaks and blockages and diversions in the system that prevent it from functioning as it should. Turn off the water pumps and let the groundwater levels rise. Let the springs re-spring, and streams burble back into life. Let the water flow and the basin recover its equilibrium – and then whatever level the lake might settle at, you know it's the right one for the landscape today.

Yet right now the water on the lake comes through a 1.2-metre pipe off the Los Angeles Aqueduct, while the Owens River itself is choked with reeds and barely has enough water to flow. The lake-bed dust may have been substantially resolved, but now other parts of the valley are instead starting to blow. The ecosystem here is no way in balance. The last twenty years of dust control produced such an artificial landscape because they tried to fix a single problem in isolation, without acknowledging that dust and water are inextricably bound together in a single hydrological system. The problems caused by one grand modernist engineering project were addressed by simply installing another – one more alert to shifting environmental dynamics and feedback loops, it's true, but fundamentally still a vast brute-force imposition on how this landscape naturally works.

It's all upside down: to DWP, water that's allowed to flow back on to the lakebed is 'wasted', whereas water that flows to Los Angeles to fuel more economic growth (and thereby air pollution) is 'productive'. Mike Prather described how DWP and others express 'concerns' that even the existing, partial levels of water and natural vegetation on the lake are not 'sustainable'. But what does that really mean?, he asked.

'If you look at it historically, the water came down the river and all of it went out on Owens Lake. Well, right now, a fraction of that goes out to Owens Lake: it's water from melted snow and it's going back to the lake where it was, and the birds that were once there are back again. So how is that not sustainable?'

***

Environmental justice in the valley isn't just a matter of where water flows, but also power.

When Inyo County and Los Angeles set up the Standing Committee for water governance in 1982, 'The tribes weren't in at the table. They excluded the tribes from that,' Paul Huette said.

Sally Manning concurred: 'I'm not a tribal member but I've worked here long enough to internalise this and it's just so frustrating. LA is planning ahead 35 years for its own water supply, but nobody here can do that. We have no control.'

'What we've got is one government agency and another,' she continued. 'The lawyers argue in the courtroom but then they go to the bar afterwards and drink together, it's kind of the same culture.'

They say in the American West that 'whiskey is for drinking and water is for fighting over'. In a too-dry valley, what sounds like tedious municipal bureaucracy is probably the most consequential political space there is. As such, a whole range of people are coming to think that the obvious way to improve management would be to add a tribal representative to the Standing Committee.

'Adding any third party to the Standing Committee would break the 1:1 gridlock that so often happens,' the Owens Valley

Committee write. And, 'Of all possible third parties, Owens Valley Paiute have the strongest claim as they successfully managed Owens Valley water for centuries before Los Angeles or Inyo County existed.'[33]

Or as Teri Red Owl puts it, 'They haven't done a very good job taking care of it. So why not let us have a chance to see what we could do?'

The fight for water back is part of a wider Indigenous movement across America – and around the world: the fight for land back.

Owens Valley Paiutes lost control over their land and water in the 1930s. Through a series of congressional Acts and Presidential orders, their initial allocation of 67,000 acres was forcibly exchanged for just 1,747 acres at four sites spread out across the valley – land that mostly isn't even owned by the tribes directly, but rather held in trust for them by the federal government's Bureau of Indian Affairs. They're the fifth largest tribe in California, but their land base is miniscule – Teri contrasts it with the Pyramid Lake Paiute Tribe in Nevada, who have 475,000 acres for a similar number of enrolled tribal members.

Meanwhile, the water rights to their original land were never traded, the Owens Valley Indian Water Commission have argued: there wasn't a vote, and the proper procedure was never followed. Eighty-plus years on, the legal matter remains unresolved.[34] The irrigation ditches dug by tribal members' grandparents and great-grandparents now funnel water into pipelines and canals for the DWP.

'Tribes are not able to really thrive because they're limited,' Kyndall Noah explained. 'They're landlocked, so they can't develop themselves economically.' Their reservations are mostly taken up by housing, leaving little room for people to set up businesses or grow food at scale. Land back would mean gaining access to places of cultural and spiritual significance – one of which is unfortunately located within the China Lake Naval Weapons Center. Visiting requires scheduling far in advance, running a gamut of background checks and clearances and then being accompanied by a Navy escort at all times – with the result: 'We're not able to practice our religious customs and ceremonies freely.'

Above all, land back – and water back – would mean healing. The groundwater pumps would stop and water could flow through the old irrigation ditch networks again. Native plants could be re-established and biodiversity given a chance to recover.

'You're not going to turn it back to how it was, but it could be a new, beautiful place,' Sally Manning said. 'It could be a working landscape for agriculture, for flowers – a wetlands, a diversity of animals and fish and birds. It could happen.'

'And you're not going to have dust where there's a wetland,' Noah added.

Teri is hopeful – though she also acknowledges that she has to be: her people can't afford the luxury of giving up. 'There are opportunities for the tribes, I think more now than there ever has been since OVIWC were established in 1991. A lot of that has to do with the political climate in Washington, DC, with so many Indigenous people in these higher-up positions' – such as Secretary of the Interior Deb Haaland, who is a member of the Pueblo of Laguna.

What's also striking are the alliances being formed. The Owens Valley Committee – whose board are mostly current or retired public servants, and many or all of whom are white – do not pussyfoot around. 'The only long-term solution to Owens Valley resource management problems is to solve the underlying political injustice of colonial rule,' they write.[35]

The conservationists are on the same page. Friends of the Inyo's Wendy Schneider told me how they were working on two permanent protection campaigns in Conglomerate Mesa and the Bodie Hills. 'And in both of these cases, we want to help the tribes get some land back. That's just something that communities have realised is the right thing to do.'

I'm surprised, I say – I would have thought that was seen as quite a radical stance. Not really, she explained. 'You know, in the American West you have so much public land. Nobody is saying, "Let's take people out of their houses and give that back to the tribes" – which is what we did to them. But land that has been managed by BLM or the Forest Service, I think people can get on board with transferring that to the tribes pretty easily. It's not super controversial.'

Just to the north in Mono County, Friends of the Inyo is running a campaign called Keep Long Valley Green. As Wendy told it, 'In 2018, the Los Angeles Department of Water and Power said, "Hey, we're not going to allow irrigation anymore on all those lands that we've been allowing it on for the last 80 years." And everybody kind of went "What?!" In the end, a coalition was formed – and it's a really diverse coalition: we have environmentalists and ranchers and tribal members and recreationists from the area who have all come together – and we probably don't agree on anything else except that the Los Angeles Department of Water and Power needs to continue to allow historic levels of irrigation in Long Valley.'

This gives me so much hope. City dwellers might not think of America's rural fringes as radically progressive places – but listen, and pay attention, and your stereotypes might just get annihilated.

Still, it's painfully slow work. The Owens Valley Indian Water Commission was founded in 1991 to fight for water and land rights, but in thirty years, it's a fight they haven't yet won.

'Sometimes it takes a toll on you,' Teri told me frankly. 'A lot of times I don't feel like I'm doing a good enough job, or I'm making good enough progress. And then that hurts, it hurts me, because that's what my life work is about. But we're not where we need to be. And so it means having to balance those feelings of optimism with feeling like you haven't done enough to take care and protect the water. It's a job that's never going to end and it isn't going to end with you. It's something that has to be taught to all the generations that are going to come behind you.'

'It was an inherited fight,' her son Noah had said the day before. 'I learned closely from both of my parents, and so I think it is almost my duty to pick up the work my father had left off, and continue this fight to restore our lands and reclaim water, in this area that I view as rightfully ours. I want to see restorative action take place, for the benefit of our people. This is my home. When I take my last breath, I want to be put into the land here: this is home for me. I grew up playing in the water in the canals, my grandma's creek: I really got to enjoy the natural world and being connected with it in that way. I want to see that sort of life – if not better – for our future youth, whether it be my own children or relatives or other community members in the valley. This is a very important place to me. It's home.'

Noah's voice is quiet, measured, unhurried, and yet there's a vision and a boldness to his thinking that's thrilling. He continued. 'I believe that Los Angeles is temporary. They have done a lot of damage – in some areas, probably irreparable – but I wouldn't say that the land is ruined. I think that they're just guests in this valley, and that we still have a long-standing history to come.'

There is laughter around the table because on some level it seems as though he's not just talking about the city's presence in the valley, but about the city itself. Los Angeles is temporary.

In Indigenous thinking, Noah said, 'we've got to keep the seventh generation in mind.' Los Angeles has been present in the Owens Valley for a little over 120 years. In 1863, it was his great-great-grandparents who were forcibly displaced from their lands by the US Cavalry – that is, the valley has only been 'America' for five generations.[36] Noah can see beyond that. He can imagine otherwise.

'From the Native perspective, this is always going to be our homelands. Our people, our ancestors, are buried and tied into this land. So we're not going anywhere.'

Before Los Angeles built the aqueduct, this was Payahuunadü: the land of flowing water. What Noah is saying is that it can and will be again.

Coda

Dust, I have argued through the course of this book, is a means of making modernity's mess and destruction visible – of turning our attention towards the costs and the consequences of so many grand schemes for human progress, from the growth of world cities such as London and Los Angeles, to the rise of industrial agriculture and the mastery of nuclear power.

Progress, though, for whom, dust reminds us to ask, with its sympathies for the subaltern: the people doing the dirty work, be they the urban poor, rural people, women, ethnic minorities and Indigenous peoples.

And progress for how long? Farmers prospered on Dust Bowl land for barely more than a decade before it started blowing.

'The driving cultural force of that form of life we call "modern",' the German sociologist Hartmut Rosa writes, 'is the idea, the hope and the desire, that we can make the world *controllable*.'[1] Yet through a particulate haze, we might come to see more clearly that it is not. That our place on this planet has to be one of partnership and care, not high-modern domination, if we wish for there to be a place for the human on this planet at all.

Sociological theorists would say that high modernism in its most muscular sense ended some fifty or more years ago – but I'm less sure that its habits of thought have entirely left us Westerners' minds. We live in another time, now – the 'Information Age', some call it, in which brute force mechanistic notions of

325

domination over nature have been superseded by a networked notion of (eco)systems, cybernetics and (climate) modelling, the metaphors computational in line with the preeminent technologies of our time. Dreams of mastery become more complex, increasingly sensitive to feedback loops and uncertainty. But yet we're by no means beyond what the sociologist Max Weber called the great Western 'process of rationalisation': as much now as a century ago, the belief sustains that 'one can, in principle, master all things by calculation'.[2] The yearning for technofixes remains.

Yet dust complicates all that. To think seriously with dust is to have one's faith in governing the world purely through 'technical means and calculations' shaken. As I have argued in Chapter 8, dust troubles even the cutting edge of climate modelling, bringing vast variability into nonlinear dynamic systems highly sensitive to differences in initial conditions. It suggests other approaches: incremental and iterative learning rather than big leaps. The precautionary principle. The importance of local, traditional and indigenous ecological knowledge alongside science. Qualitative interpretation, not just quant. Paying attention to tangible, material realities to ground our theoretical models in the world.

Dust may actually point to the fact that we have never really been modern at all – that control was illusory, humanity's separation from nature nonexistent. Modernity was meant to be a process of 'build[ing] a brave new world out of the ashes of the old', as David Harvey has asserted, but what that really means is that the ash stays with us into the present, even long into the future.[3] So much for boundaries and separation: the environmental consequences of a century of modernity are written into people's lungs and brains and blood.

In 2008 in Kingston, Tennessee, 4.2 million cubic metres of toxic 'fly ash' sludge – a waste product of coal power plants – broke out of its containment pond, swamped people's houses and ran into the Emory River. It was a spill one hundred times larger than Exxon Valdez, with arsenic levels one hundred times over safe levels and 'ashbergs' the size of houses.[4] 'In what I call late industrialism, the levee has broken, retention walls have failed,' Kim Fortun, an environmental anthropologist at University of California, Irvine writes.[5] 'Modern ontology . . . works by maintaining binaries and boundaries,' she continues: 'the sludge is supposed to stay in the sludge pond, out of the rivers, air, and human bodies.' But it doesn't. The logic that invented coal power plants also invented coal power plant disasters, by only doing so much to prevent them. At some point, industrial safety becomes unprofitable, and trade-offs are deemed acceptable – with certain people and places more expendable than others. As such, Fortun writes, 'even if we have never really been modern, we still have a modernist mess on our hands.'

Such an enormous mess. America produces some 77 million tons of coal ash per year.[6] There are so, so many place-histories I haven't had the space to tell. So many dusty problems.

***

I have come to think that these 'dusty problems' are a distinct type of modern tragedy, environmental horror stories of a particular hue. Such similarities exist between the places I've visited in the course of this book, so many common patterns beyond the mere material fact of dust. So what might a dusty problem entail?

A problem of one sort (a challenge to economic growth) has been solved by creating another problem – one that can be more easily brushed under the carpet: denied, ignored or displaced as somebody else's responsibility. It's slow, taking at least decades to build and decades to resolve – if this happens at all. The problem operates on difficult-to-govern scales, probably very small and very large at the same time, and it contains positive and interlocking feedback loops, so it's a dynamic and unpredictable problem. There's the possibility of tipping points and phase-change consequences, where the problem goes from bad to calamitous in a blink. A dusty problem definitely does *not* operate within standard political boundaries or election cycles, making it particularly tricky to act upon and sort out. And above all, a dusty problem is an environmental injustice, where the people who have caused the problem aren't the one dealing with its consequences – separated as they are by distance and time, but also wealth, race, class and power.

Harm is caused to both people and places – and the resolution to dusty problems, I suggest, has to involve people as much as place-centred solutions too. We cannot heal damaged landscapes without addressing the social, political and economic tensions that have produced this despoilation. As environmentalists are increasingly coming to recognise, it's impossible to implement real ecological stewardship without thinking about governance, too. And so that means giving control and sovereignty back to local communities and Indigenous people – making restitution for old thefts, and returning the land to people embedded in a place, not those who seek only to extract from it. Colonial models of land expropriation and ownership are what got us into this mess. But they can be undone. And yet there is no turning the clock back to a past, pristine state. The fantasy is a romantic one, but that's all it is: a naïve pastoral dream. The reality is that ecosystems are complex, dynamic systems that can't run backwards in time, only forwards, starting from the conditions in which we find them now.

But people and authorities talk of 'repair', 'renewal', 'restoration'. When we're talking about nuclear mining waste, the phrase is 'remediation', a word of implicitly low expectations: remedial work for land that, like a school child, hasn't met the necessary level of 'attainment'. I'm cautious of these '*re-*' words – too often the promise of renewal is simply cover for simply shipping the problem elsewhere. The sludge from the Kingston coal ash spill was 'remediated' by moving it to 'another, more marginal place, out of sight and mind', Kim Fortun wrote. Four million tons was shipped off to landfill near Uniontown, Alabama – where ninety per cent of the population is African American, and almost half of residents live below the poverty line. Ash was piled in uncovered mounds just thirty metres from people's homes. When the wind blows, 'we have to breathe the toxic coal ash dust,' Uniontown resident Esther Calhoun said. 'The odor is sickening.'[7] There's no remedy in that.

Similarly in the Owens Valley, the LADWP talk of 'mitigation' but that sounds a lot like excuse-making to the people who live there: they talk about justice and they talk about healing. They know that there's no return to the way things once were – but it could become 'a new, beautiful place'.

And so in the uranium mining-scarred hills of New Mexico, in the Aralkum Desert in Uzbekistan and Kazakhstan and in the Owens Valley, California, something is happening. None of the people in those places believe the land they live on is 'ruined'. It's not over; it's not past tense. What's happening might be limited, deeply compromised and massively insufficient, but there are green shoots, from which a new way of being may grow.

I think of the Aral, where a state programme of 'revegetating' has been met with nature's own processes of primary and secondary succession, as the toughest, most stubborn and salt-loving grasses and plants provide the biological foundations for a new ecosystem to reform. That's the hope we must have in mind. Not 'restoration', no turning back the clock to some innocent past. The scars on the land are too big; they will never disappear. And so much better *is* possible.

The word I've come to think with is 'salvage' – to save something valuable or important from disaster. A rescue carried out in difficult conditions, a rescue that may not completely succeed but you do what you can anyway. To salvage a difficult situation isn't to solve it but to fight to avert failure and strive for even a crumb of something better. An architectural word for reusing and repurposing – for bricolage, jugaad, hacks. It's a word halfway between salvation and garbage, as the writer China Miéville told the *Boston Review* in 2018: 'This shit is where we are. A junk heap of history and hope.'[8] It's a word for difficult times, Miéville explains, times when we fear 'It's too late to *save* but we might repurpose. Suturing, jerry-rigging, cobbling together. Finding unexpected resources in the muck, using them in new ways. A strategy for ruination.' A lesson that 'it's worth fighting even for ashes, because there are better and much, much worse ways of being too late.' It is worth fighting even – perhaps especially – amid the dust.

Salvage is pragmatic and practical – a word that reminds us that solutions don't have to be perfect to be better than the problems they address. This isn't a romantic environmentalism but a hybrid Anthropocenic one, as acres of riprap and gravel keep the dust down on the nuclear mine tailings site and the dried-up lakebed. It's not always 'pretty' and it's not 'natural' in the familiar sense – but perhaps we in the West need a new understanding of nature. One that reminds us we are always already part of it. Which is why I write about dust.

Dust is part of the earth's metabolism, and part of its pathology. As are we all – some of us more so and some of us less, but no-one standing apart, every last one of us enmeshed with the world from the molecular scale right on up. Dust makes the idea that we ever could be separate from the world – let alone its master – seem ridiculous and infinitely futile. We are intimately and intractably entangled – and recognising this, finding a way to live with our inherent interdependence, will be the only way through the multiplying environmental crises we have created.

And the crises do proliferate.

***

In January 2023, scientists from Brigham Young University, Utah, warned that the Great Salt Lake might have just five more years left on this planet.[9] It's a crisis akin to drying-up of Owens Lake (Patsiata) but on some twenty times the scale: at its highest water periods in 1873 and 1987, the Great Salt Lake (or Ti'tsa-pa, 'Bad Water' in the local Shoshone language) covered 2,400 square miles. But the lake has since lost 73 per cent of its water and 60 per cent of its surface area, and without a coordinated rescue plan and a million more acre-feet of water per year, it's about to evaporate into nothingness. Or rather, into a cloud of terrible, toxic dust.

Like Owens Lake, the Great Salt Lake is a 'terminal lake' in an endorheic basin, meaning it gathers all the toxins and pollutants in the watershed around and stores them in the silt of the lakebed. Heavy metals were the legacy of mining in the region – arsenic and antimony, copper, mercury and lead – while organic contaminants and cyanotoxins come from agricultural run-off. There they sat, covered by the water, until the water was gone and the lakebed dirt started drying out. For now, most of the exposed soil is still protected by a hard salt crust. But as wind erodes the crust over time, those contaminants will become airborne.

So much dust.

Recall that Owens Lake was, for decades, the largest source of dust in the United States. Now multiply the size of its lakebed by twenty and put it right next door to a major city. Some 2.6 million people live in Salt Lake City and the wider Wasatch Front region around the lake. Already the air quality is poor. Writing in the *Salt Lake Tribune* in January 2023, paediatrician Hanna Saltzman pointed out: 'With our smog, wildfire smoke, high ozone levels and worsening dust, children here often breathe unsafe air.'[10] Salt Lake City is the nineteenth worst metropolitan area in the United States for 24-hour PM2.5 particle pollution; and has never met federal attainment levels.[11] An estimated 75 per cent of Utahns lose one year of life or more as a result of state air pollution levels. Some 23 per cent lose five years or more.[12]

The cost to human life of the lake vanishing entirely is hard to contemplate. 'We can't risk finding out,' Saltzman wrote in her op-ed. On a day off from the hospital she went to a demonstration outside the Utah capitol, where people gathered to press legislators for decisive action. They carried placards: '*Save Our Lake*', '*Defend Our Future*', '*No to Toxic Dust Bowl*'. 'The health of the lake is the health of our kids,' Saltzman emphasised. It's the health of everybody – every body and every thing.

Over a million people in the region are particularly vulnerable to the effects of air pollution, be that through age (old as well as young), asthma or lung and heart diseases.[13] Ten million birds across 350 species depend on Great Salt Lake at some point in their migratory journeys up and down the Pacific Flyway, feasting on the lake's brine shrimp that are starting to struggle as salinity levels in the North Arm reach 28 per cent.[14]

Nature writer Terry Tempest Williams described watching adolescent white pelicans die out on the salt flats, spooked by coyotes newly able to reach their breeding island due to the low water. 'Their wings were not strong enough to fly the miles to fresh water for fish,' she wrote in the *New York Times* in March 2023.[15] 'Forced down by fatigue, they were dying from hunger and thirst. Walking behind them at a respectful distance felt like a funeral procession. I passed 60 salt-encrusted bodies stiff on the salt flats, hollow bones protruding from crystallized clumps of feathers, wings splayed like fans waving in the heat.'

Great Salt Lake is drying up not solely due to the regional megadrought but because of water consumption. Utah State University researchers estimate net inflow to the lake has reduced 39 per cent since 1850, partly for urban uses (per capita usage in Utah is the second highest in the United States), but mostly due to agriculture.[16] The *Salt Lake Tribune* reports how 'growing alfalfa and other kinds of hay sucks up 68 per cent of the 5.1 million acre-feet of water diverted every year in Utah,' but produces just 0.2 per cent of state GDP in return.[17] Almost a third is exported to China; if I wrote of 'ghost acres' in Chapter 1, then this is a kind of 'ghost water', its beneficiaries completely severed from the consequences of its extraction. Only when it's almost too late is Utah now waking up to its true value.

'We have this potential environmental nuclear bomb that's going to go off if we don't take some pretty dramatic action,' Joel Ferry, a Republican state lawmaker and rancher told the *New York Times* in June 2022.[18] Activist groups such as Physicians for a Healthy Environment are calling for the city to buy out alfalfa farmers so more water will return to the Great Salt Lake. The

Tribune's editorial board call for no less than an end to most farming in the state as the population grows and drought conditions become the 'new normal', saying, 'The simple fact is that agriculture . . . is just not the future of Utah.'[19]

State leaders are reluctant to act, but their hands may yet be forced. 'Examples from around the world show that saline lake loss triggers a long-term cycle of environmental, health, and economic suffering', the scientists from Brigham Young University write – citing studies from the Aral Sea as evidence of the harm, followed by the Owens Lake dust control study as evidence of what can be done to mitigate.[20] It's all connected. How much of another catastrophe will lawmakers need to witness before they realise disaster must be averted? That's the open question.

***

The future is a dusty one in so many places: both drylands and desertification will increase drastically in a warming world. Drylands are places where at least 1.5 times more water evaporates from land and plants into the atmosphere than falls in rain. Some 41 per cent of the planet is classified as dryland, from grasslands and savannahs (9 per cent) to sparsely vegetated semi-arid land (15 per cent), and arid and hyper-arid lands recognisable as desert (17 per cent).[21] As the planet warms, places on the boundaries of existing climate zones will tip over into drier aridity regimes, implying profound consequences for how people and animals can live in these regions. Static farming is impossible on semi-arid land: livestock must move long distances to find water. On fully arid land, even that becomes impossible. In 2018, scientists estimated that if the world warms by 2°C by 2050, as seems very likely, then between 24 to 32 per cent of the entire land surface will aridify.

'Desertification' is a separate process. In scientific usage it doesn't refer to increasing aridity so much as land degradation: the loss of soil, soil fertility, vegetation and biodiversity as fields and grasslands are scoured down to bare dirt. Desertification is caused by deforestation and poor agricultural practice, as over-cultivation and overgrazing break up plant roots and the biotic crusts that hold the soil together in dryland regions. Poor irrigation can fill the earth with salt, rendering it infertile. In many places, the soil is just being washed away: erosion from conventionally ploughed fields is over 100 times the natural rate of soil formation.[22] Land degradation also carries climate consequences. If vegetation is lost, less carbon dioxide is sequestered from the atmosphere, while soil degradation also sees carbon released from dryland soils, which are a major global store. The feedback loop feeds on itself. The temperature rises. Drylands get drier.

Already between 20 and 35 per cent of drylands are impacted by land degradation, with 250 million people affected and 1 billion livelihoods at risk.[23] By 2050, some regions are forecast to see a 50 per cent reduction in crop yields, leaving people poor, vulnerable and hungry. A 2018 report from the IPBES, a UN-founded biodiversity organisation, laid out the consequences in stark statistics: 'Years with extremely low rainfall have been associated ith an increase of up to 45 per cent in violent conflict.'[24] By 2050, they add, 'it is likely that land degradation, together with the closely related problems of climate change, will have forced 50–700 million people to migrate'.

The Gobi Desert, in Mongolia and northern China, is the second largest source of dust on the planet – and it is growing. Each year, another 6,000 square kilometres of grasslands turn to sand and gravel.[25] Animals starve. Roads and railways are regularly closed due to blowing sand. Villages have to be abandoned. And sandstorms race southward, deluging Beijing in dirt.

Some 1.7 million square kilometres) of China is sand or gravel desert, and in 2006, the country reported that 27 per cent of its land area (2.6 million square kilometres) was degraded, compared to 18 per cent barely over a decade before in 1994.[26] The causes encompass climate change (rising temperatures, drought and declining glacier snowpacks) and more local human action: over-harvesting of trees for firewood, over exploitation of water resources for agriculture and industry, as population of China's northern provinces has quadrupled in just a couple of decades, and possible overgrazing by pastoralists, though this latter is disputed. [27]

In response, China has created one of the largest ecological engineering projects in the world: the Three-North Shelterbelt Program. The vision: a 4,500 kilometre line of trees across the northeast, north and northwest of the country (hence the name), to stop the desert in its tracks. By 2014, some 66 billion trees had been planted – by 2050 the vision is for 100 billion trees, covering fully one-tenth of the country: a Great Green Wall against the 'yellow dragon' of Gobi sand.

In 2016, environmental journalist Vince Beiser visited Duolun county in Inner Mongolia province.[28] The view 'could be described as either profoundly inspiring or deeply strange,' he wrote. Duolun sits on the southern edge of the Gobi Desert, and 'for miles around, the terrain is dun-coloured, dry, sandy desert stubbled with yellow grass', except for a group of hillsides vivid with green. Trees, countless thousands of them. He described how they were planted to form geometric shapes: 'a square, a hollow-centred circle, a set of overlapping triangles. The flatland below them was striped with ruler-straight bands of young pine trees, all the same height, standing in formation like soldiers ready for battle.' The deputy director of the local 'greening office' of the State Forestry Administration showed off satellite photos illustrating how much the landscape had changed in just fifteen years. The government claims 31 per cent of the county is now forested, from almost none.

The pine trees planted in Duolun may have survived, but this is the exception, not the rule. The Three-North Shelterbelt Program is a high modernist megaproject and as such has sought standardisation, not adaptability to local conditions. Vast monocultures of poplars were planted in the scientifically recommended 'straw grid' method to stabilise the sand and watered by dripline irrigation. And then the irrigation pipes clogged up with silt, and monocultures failed to regenerate the soil very deeply, and proved vulnerable to pests and diseases. In their early years, poplar saplings may be able to push their roots deep enough to tap whatever soil moisture is available – forcing out other species – but then along comes a real hard, dry year and the trees perish en masse. In Ningxia Hui Autonomous Region, in 2000, one billion poplar trees died all at once, from the same pathogen, because they were clones from a limited set of cultivars and so shared the same vulnerabilities. According to some studies, only fifteen per cent of the billions of trees planted since 1978 have survived.[29] The plantations that remain are 'green deserts' only a little more hospitable to biodiversity than the sand they replace. Landscape architecture scholar Rosetta S. Elkin describes the poplar plantations as enacting 'slow violence on the social and biotic structure of the land'.[30]

But have they stopped the dust?

Maybe.

'Although numerous Chinese researchers and government officials have claimed that the afforestation has successfully combated desertification and controlled dust storms, there is surprisingly little unassailable evidence to support their claims,' X.M. Wang and colleagues wrote in the *Journal of Arid Environments* in 2010.[31] Tree cover in northern China has increased, though whether to credit this wholly to the tree planting project is debated: wetter years in the 2010 nurtured natural greening, too, while warmer temperatures boost plant growth as well.

Yet the number and duration of sandstorms did fall markedly that decade, and scientists expressed tentative hope that the state tree planting programmes had contributed at least a little to this mitigation. Though not necessarily much: one study attributes just 4.6 per cent of the reduction in dust 'to grassland vegetation increase' which is in turn only 'partly ascribed to the ecological restoration'.[32] Natural climatic factors – namely slower windspeeds, and greater soil moisture suppressing dust formation – are also influences. For a time, however, even NASA were willing to acknowledge the possibility that 'the new greenery may be reducing the impact of dust storms in the region.'[33]

Then on 14–16 March 2021, the biggest sandstorm in a decade rose, as a cyclone picked up dust from loose ground in Mongolia and dumped it on Beijing. The air turned thick and orange. The city choked. (Then shared *Bladerunner 2049* memes on social media.)[34] Fine particle pollution in the city exceeded 7,400 μg/m^3, thirty times higher than 'very high' levels.[35] That Monday, the Inner Mongolia edition of *The People's Daily* newspaper carried coverage of regional efforts to tackle desertification. 'Yellow sands are going away and green trees are flourishing,' the headline read.[36]

In her 2022 book *Plant Life: The Entangled Politics of Afforestation*, Rosetta Elkin discusses the Three-North Shelterbelt Program alongside two other tree-planting megaprojects: the United States' Prairie States Forestry Project, launched by President Roosevelt in 1935 in response to the Dust Bowl – and the Great Green Wall, a plan to plant a 7,775 kilometres belt of trees right the way across Africa, from Senegal in the west to Djibouti in the east, to hold back Saharan sands. She is immensely critical. Tree planting programmes don't just fail to do what they promise and aren't just a waste of time and effort and money, she argues: they're much worse. They dispossess people from their land, replace dynamic ecological webs with monoculture, and wreck landscapes with fallow fields, sinkholes and abandoned irrigation pipes. The continuing use of evidently unsuitable poplars in the Three-North Shelterbelt Program demonstrates that its real motivations are less about dust control than providing a ready source of timber, she argues. Tree planting is solutionism, a techno-fix that distracts attention away from addressing the real causes of land degradation: the over-extraction of water, industrial agriculture and short-term profit-chasing. It produces not forests, but plantations. As seen in Chapter 6, it's all sustained by the colonial fantasy that the desert or indeed drylands, are empty places and nothing there counts. *That isn't true.* There is just one way to stop the dust in these degraded landscapes: protecting the native ecosystems that remain.

In the Sahel, as in northern China, the megaproject model wasn't working. 'If all the trees that had been planted in the Sahara since the early 1980s had survived, it would look like Amazonia,' Chris Reij, a sustainable land management specialist, told journalist Jim Morrison in 2016. 'Essentially 80 percent or more of planted trees have died.' By 2017, just 4 million hectares of land out of a target 100 million had been restored, mostly in Ethiopia.[37] The 2030 completion date seems unlikely.

But the Great Green Wall is changing. The UN Food and Agriculture Organisation (FAO) describe it now as 'Africa's flagship initiative to combat climate change and desertification and address food insecurity and poverty'.[38] The name is now disavowed: 'The Great Green Wall must not be seen as a wall of trees to hold back the desert,' they write – instead, the vision is a 'mosaic of sustainable land use practices', a patchwork of things that work, where they work, not one great big overarching masterplan.

While the mass-planted saplings were dying, other trees were sprouting green shoots. Under French colonial laws (which remained on the statute book into the post-independence period), trees were the property of the state – making them a bit of a hazard to own: to cut one for firewood was to risk jail.[39] The intent was conservation, the result quite the opposite. But the laws were finally changed, and since the 1980s, farmers in Niger have started tending trees with a little more care, as a source of useful fuel, animal fodder and to hold down the fragile topsoil which threatened to blow away.

For years there was almost no international recognition of what was happening. 'This regreening went on under our radar, everyone's radar, because we weren't using detailed enough satellite imagery,' Gray Tappan, a geographer with the US Geological Survey's West Africa Land Use and Land Cover Trends Project, told Jim Morrison. 'We were looking at general land use patterns, but we couldn't see the trees.'

Niger had in fact turned green – and farmers in Burkina Faso were seeing significant changes too, by using *zai* planting pits to concentrate nutrients, and stone barriers to contain water run-off from fields, increasing the amount of rain that can soak into hard ground. Cereal yields increased between 40 per cent or even doubled, and rehabilitated fields gained 22 per cent more trees per hectare.[40] 'That doubling of yield is due to trees,' scientist Patrice Sawadogo of the World Agroforestry Centre (ICRAF) told *Mongabay* conservation news.[41] What worked was not one thing or another, but the ecological synergy: the trees add biomass to the soil and build up soil carbon (as does animal manure), while producing a beneficial microclimate for crops such as millet. Farmers even 'regenerate trees by feeding seed to livestock,' Sawadoga added, 'some germinates better if it goes through the gut'. Though farmers were mostly not planting new trees at all but rather regenerating them from stumps: the trees may have been killed in the 1970s by drought or over-harvesting of fire-wood, but their roots kept living underground. As in the Owens Valley, the land is ready and waiting to regreen if people will just give it the chance.

Farmer-managed natural regeneration (as it is called) has to remain farmer-led, however. Rosetta Elkin recounts examples of the *zai* pit technique failing when it's viewed as a 'climate-smart' technology that can be appropriated by industrial forestry and rolled out at larger scale. Extracted out of context, without the farmers' care and the manure from their animals, the plants struggle to get established. The success stories are smaller, simpler: an old man in his seventies who's spent his whole life digging pits and moving rocks and planting saplings until he has turned what was open desert into a forty-hectare oasis of spiny acacia and yellow-fruited saba trees.[42]

It's the opposite of a high-modern environmental megaproject – it also might work. And so the Great Green Wall evolves: as Mohamed Bakarr of biodiversity funder Global Environment Facility puts it, 'It is not necessarily a physical wall, but rather a mosaic of land use practices that ultimately will meet the expectations of a wall. It has been transformed into a metaphorical thing.'[43] It has become a metaphorical forest.

***

What sort of metaphor might dust be for us, now, as we face a deeply uncertain future?

We have to find ways of living with dust – of living not as masters but participants in a world of complex systems and environmental limits. What kind of people might we moderns need to become in order to do this?

It is tempting to despair and see the task as impossible. How can the same culture that created the climate crisis possibly become capable of salvaging some kind of future from it? There's an image that stays with me, from an eighth-century Anglo-Saxon poem called 'The Ruin'. An anonymous narrator stands looking over the ruins of the Roman city of Bath, four hundred years after the empire in Britain had collapsed – and marvels over how the work of 'giants' is decaying. Stone towers are now fallen, roof beams have snapped, and frost eats away at the mortar. The builders were mighty and brave men, the poet muses, yet now 'earth-grip holds them – gone, long gone, fast in gravesgrasp while fifty fathers and sons have passed'.[44] There is a tremendous sense of lateness: the author understands themselves as living in a time when the great deeds of mankind are all in the past. There is no future. Everything, in the end, turns to dust. The Anglo-Saxons called this *dustsceawung*, or the contemplation of dust – the reflection on the transience of things and the inevitability of loss. For some reading this, it may be a familiar feeling.

I have spent more time contemplating dust than most, yet I do not feel this as melancholy. It is, in the end, an ethics.

I went to Owens Valley in 2022 to learn more about the dust – but what people really wanted to talk about was water. Dust is an environmental disaster, but water: that's the thing you're fighting for.

So I spoke with Teri Red Owl and Kyndall Noah of the Owens Valley Indian Water Commission about what it means to be a water protector. 'It's about helping people realise that water is not a commodity, it's a life-giving force,' Teri said. The term is both new and old at the same time. People started using it in 2016, during the protests at the Standing Rock Reservation against Dakota Access Pipeline, a major trans-American oil pipeline that threatened the Missouri River – the reservation's water source – with contamination. Indigenous people, allies and environmental activists came from across America to join the resistance, but sought a word beyond 'protest' that captured not just what they were against but what they were fighting *for*: 'Mní Wičóni', or 'Water is life.' But the knowledge that it encapsulates is as old as time: water brings life and water sustains it – and in this way water is sacred in Indigenous cultures (and beyond).

'I believe that our Creator put the water there for us to take care of us and to take care of all the other species that rely on the water,' Teri comments. Embodying that duty of care is at the heart of being a water protector: 'It's this responsibility that you have to help sustain life, not just for us but for the wildlife, for the plants and for Mother Earth,' Kyndall Noah adds. 'So it's a huge responsibility.'

'To be a water protector is to know that I have a very close relationship with water where it's almost like it's another being,' Teri explains. 'It's like your child that you take care of, and that you nurture, and that you protect – and that you want to see grow, and flourish and be respected.' It's to be in relation to water and everything that water sustains – a model of self not as sovereign individual, but as deeply embedded and interconnected with a place.

'It's a job that's never going to end and it isn't going to end with you,' she says. 'And it's something that has to be taught to all the generations that are going to come behind you.' Noah Williams, her son, carries this responsibility deep in his bones. Despite his youth, he speaks already of the next generation coming into adulthood and the hope they give him for change.

To be a 'water protector' is an Indigenous way of being, then – but it's not an exclusive one. As Kyndall Noah puts it: 'Everyone should be a water protector: everyone should be concerned about where their water comes from or where it's going.'

Other Indigenous thinkers concur. Standing Rock wasn't an exclusively indigenous project, comments Nick Estes, a citizen of the Lower Brule Sioux Tribe and author of *Our History Is the Future*, a book about the protests. 'Everyone who walked through those camp gates became a water protector by act of merely being there, with the intention of protecting the Missouri River for the millions of people who depend on that river; for the countless non-human relations that depend on that river for life.'[45] The diversity and our multiplicity this enabled was its strength.

While the term may have arisen out of Indigenous struggle, water justice is a matter for us all. When you turn on a tap in Los Angeles, a thread of blue traces a path 233 miles back to the Owens Valley. We are all intimately entangled and interconnected with the world, and it with us. The ethical challenge is to recognise that, and act on it. So it is to be a water protector, yet so too with its material opposite: dust.

We must act in a world that is a mess, and only getting more disorderly. Our planet is a mess ecologically – carbon dioxide levels climbing, the boundaries of planetary systems being breached and temperatures ever-rising. We're in a mess politically: whether you call it a polycrisis or a clusterfuck, you don't need me to elaborate. We're even in a tangle theoretically, the systems scholars argue. 'Every problem interacts with other problems and is therefore part of a set of interrelated problems, a system of problems . . . I choose to call such a system a mess,' Russell L. Ackoff wrote in 1974.[46] High-modern rationalisation got us into this predicament, but it doesn't get us out. We have to 'stay with the trouble', as Donna Haraway has put it, to think through the mess *with* the mess if we want a better view.

Dust is such a gift to think with: rich in lessons for a complex world. Each microscopic particle still carries its own histories and particularities: a blanket of dust does not erase information, only form. In following the stories of dust, we are forced always to reckon with feedback loops and dynamic, nonlinear systems; vast diversities of scale; the deep reaches of geologic time. Far from obscuring, dust turns out to be a kind of magnifying lens for seeing these things a little clearer. It helps us understand situations and systems more completely, recognising waste and contamination – air pollution, radioactive ore piles, toxic lake beds – not as forgettable externalities but intrinsic parts of the whole. It helps us both take responsibility for our actions and recognise the limits of human control.

The ethics of dust unites both new and old ways of thinking: from an Indigenous ethics of 'being a good ancestor' to the generations of the future, to the most sophisticated systems theorising and climate modelling. It reminds us to ask, who's living with this dust? Who's breathing it deep into their lungs and into their blood? Is that justice – or is it anything but?

Global heating of 2°C is baked into the century ahead, and very likely more. Two degrees Celsius means chaos on a scale it is still hard to imagine, as much as it comes closer every year, as much as it is happening right now around the world and also for me at home, as London summers top 40°C . Technosolutionists punt the possibility that nothing in our modern ways of life may have to change, if only renewables, carbon capture, geoengineering. They say we should sow the ocean with volcanic ash or iron particles to feed the phytoplankton; they say we should seed the sky with diamonds to block out the sun.

But enough: enough high modern heroics. This is a salvage job at best, our hero less the entrepreneurial innovator than the janitor. The ideology of 'innovation', novelty and disruption that guided the twentieth century urgently needs replacing with an ethics of maintenance and care if we are to make it through the twenty-first.

343

Thinking with dust is one tiny way to do that. It might just help us have a future.

***

As I write these last words, in April 2023, I get the news that Patsiata (Owens Lake) has refilled. Mike Prather shares pictures on Facebook, the water shimmering like a mirror, reflecting the mountains behind.[47] In mid-March, an 'atmospheric river' – a plume of concentrated moisture in the atmosphere – brought heavy rain to central California, and a record snowpack has built up in the Sierra Nevada this winter as well. As this melts in May, June and July, there will be even more water for the lake – a lake which gets to be a lake again for a moment, for all that had seemed so unlikely for so long. Of course, it's complicated: the floodwaters risk the dust control infrastructure, Mike explains, such that when the floodwater evaporates later in the summer, there could be more dust, not less. 'Nature will make the call,' he writes.

He's right, but still: what a thing to see again. What a heart-filling sight. Getting the water back isn't a pipe dream for the Owens Valley. It's there, ready, waiting to return.

Acknowledgements

This book was born from my encounters with land and place, but it stems equally from many years of encounters with people: days and nights, walks and car journeys, talking at length about the places that moved us and how we might think and write about them. Dust is an absurd topic to have lit upon for what's now been eight years' work, but it's proven anything but dry or lifeless. I was writing during a strange and isolated time (2020–22) but it always felt like I was part of a much wider conversation. I am in this way indebted to more people than I can possibly mention in these acknowledgements, delightedly so.

Much of my research has been as a visitor on Indigenous land. I honour with gratitude the land itself and the people who have stewarded it throughout the generations into the present day, and commit to supporting their work for reconciliation and reparations. In California, I spent time on Nüümü (Paiute) land and that of the Newe (Shoshone), Cahuilla and Tongva tribes. Pesa mu – thank you, all of you – in Payahuunadü, including Teri Red Owl and Kyndall Noah of OVIWC; Noah Williams, Sally Manning and Paul Huette of the Big Pine Paiute Tribe environmental department; Phil Kiddoo of GBUAPCD; Wendy Schneider and Louis Medina of Friends of the Inyo; and Mike Prather and everyone involved in the Owens Lake Bird Festival. I hope that my work can in small part aid yours.

In New Mexico, I visited Laguna Pueblo and also spent time on

Diné (Navajo) and Apache lands. My sincerest thanks to Gregory Jojola for sharing his knowledge of Laguna, alongside everyone at the UNM METALS Superfund Research Program Center, in particular Johnnye Lewis, Thomas De Pree and Chris Shuey, alongside the New Mexico Environmental Law Center and everyone at the Science for Health of Indigenous Populations seminar. You are doing powerful and inspirational work, and I wish I could have written more.

In Uzbekistan, thank you to Sarsenbay and Max for guiding, and Otakbek Suleimanov and the Stihia crew for a festival I'll never forget. In Greenland, my gratitude to Nivi Christensen who hosted us in Nuuk, came hiking, and taught us the proper Inuit way to relate to a boulder.

At NASA, I am grateful to Rob Green, Olga Kalashnikova and Bart Ostro for explaining the subtleties of aerosol science, Jane J. Lee for coordinating, and Joshua Stevens of NASA Earth Observatory who created the extraordinary 'aerosol earth' image on my UK cover. At British Antarctic Survey, thank you to Liz Thomas and Eric Wolff for ice core insights, plus Emily Neville and Amanda Wynne for coordinating. I'm grateful to Peter Neff for bringing the experience of ice drilling to life – and to Tom Gill, for all you do to coordinate the dust science community. Will Cavert, Rachele Dini and Claire Ryder shared their expertise and sharpened my thinking.

This book also could not have happened without my freelance clients whose projects afforded me the deep drifts of time this book has required. Thank you to Giuseppe Polimeno, Rob Blackie, and the Black Swan team – and the 'Holes, for the courage to take the leap.

I started writing my email newsletter, Disturbances, to see what I might have to say about dust – but it turned out people wanted to read it, too. In no small part this book exists because of you, those early readers. Clive Thompson and Robert Macfarlane supported this work very early on, and vastly expanded its scale by so doing.

An especially effusive thank you to the irrepressible Max Edwards at Aevitas Creative Management: your belief in this book – your belief in me – means the world. Tom Lloyd-Williams, Gus Brown and Vanessa Kerr at ACM have also done vital work. Huw Armstrong at Hodder: your attention has been a gift. Thank you also to Kate Keehan, Naomi Morris Omori, Christian Duck, Inayah Sheikh Thomas and Helena Caldon. John Plumer gave me the best gift a geographer can get: really good maps.

But ultimately I have written this book for my friends. Thank you to Daniel Trilling and Alexandra Parsons for our miserable mizzling lockdown walks talking writing, and to Rishi Dastidar for same except over lunch (you wise and civilised man). Alex also shared her academic journals log-in over years of password changes: a sainthood awaits. Hannah Gregory helped make my book proposal work, and Beth O'Brien's love, pep talks, and late-night texts held everything else together. Huw Lemmey, Deb Chachra, James Butler and Jesse Luke Darling.

Thank you to Brad Garrett for an invitation that changed everything. To Joel Childers, who helped me learn to see (and love) the Great Basin. To Wayne Chambliss, a catalyst.

Notes

Introduction

1. Marder, Michael 2016, *Dust*, Object Lessons series, Bloomsbury Academic, New York.
2. Models vary. Specifically, 'The estimated global dust burden ranges from 2.5 to 41.9 Tg; it ranges from 8.1 to 36.1 Tg when [climate models] HadGEM2-CC, HadGEM2-ES, and MIROC4h are excluded.' In Chenglai Wu, Zhaohui Lin, and Xiaohong Liu, 2020, 'The global dust cycle and uncertainty in CMIP5 (Coupled Model Intercomparison Project phase 5) models', *Atmospheric Chemistry & Physics* 20: 10401–25, DOI: 10.5194/acp-20-10401-2020.
3. I acknowledge that is only a theory of the painting's origins, put forward by Olson, Donald W., Doescher, Russell L. & Olson, Marilynn S., 2004, 'When the sky ran red: The story behind the "Scream"', *Sky & Telescope*, 107(2), pp. 28–35. There is also a competing nacreous clouds theory that is not so dust-based.
4. Black carbon emissions of 8.54 Tg in 2017 with 70% (6.2 Tg) anthropogenic, from Xu, Haoran et al. 2021, 'Updated Global Black Carbon Emissions from 1960 to 2017: Improvements, Trends, and Drivers', *Environmental Science & Technology* 55(12), 7869–79. DOI: 10.1021/acs.est.1c03117.
5. Domestic wood burning represents 21 per cent of total PM2.5. emissions in 2021, vs. 13 per cent from road traffic. DEFRA 2023, *Emissions of air pollutants* annual statistics publication, Department for Environment, Food & Rural Affairs, 22 February 2023.
6. Schraufnagel, Dean E. et al. 2019, 'Air Pollution and Noncommunicable Diseases: A Review by the Forum of International Respiratory Societies' Environmental Committee, Part 1: The Damaging Effects of Air Pollution'. *Chest* journal 155(2): 409–416, DOI: 10.1016/j.chest.2018.10.042.
7. Medina, Sylvia et al. 2011, *Summary Report of the Aphekom Project,*

2008–2011, Institut De Veille Sanitaire, Saint-Maurice, Paris. (Aphekom. org).

8. Mavrokefelidis, Dimitris 2021, 'Ella Kissi-Debrah: The story of a canary in a coal mine', *Energy Live News*, 17 June 2021.

9. BBC (no byline) 2021. 'Air pollution: Coroner calls for law change after Ella Adoo-Kissi-Debrah's death', *BBC.com*, 21 April 2021.

10. Stone, Richard 2002. 'Counting the Cost of London's Killer Smog', *Science*, 298(5601): 2106-2107, 13 December 2002. DOI: 10.1126/science.298.5601.2106b.

11. Hodgson, Camilla; Hook, Leslie and Bernard, Steven 2019, 'London Underground: the Dirtiest Place in the City', *Financial Times*, 5 November 2019.

12. OECD 2020, *Non-exhaust Particulate Emissions from Road Transport: An Ignored Environmental Policy Challenge*, OECD Publishing, Paris, DOI: 10.1787/4a4dc6ca-en. See section 3, 'The implications of electric vehicle uptake for non-exhaust emissions'.

13. Evangeliou, N., Grythe, H., Klimont, Z. et al. 2020, 'Atmospheric Transport is a Major Pathway of Microplastics to Remote Regions', *Nature Communications* 11, 3381. DOI: 10.1038/s41467-020-17201-9.

14. 28 per cent of ocean microplastics are from tyre dust and 7 per cent from brake dust, vs 35 per cent from synthetic fibres. Mixed 'city dust' contributes another 24 per cent of microplastics. Boucher, Julien & Friot, Damien 2017, *Primary Microplastics in the Oceans: A Global Evaluation of Sources*, IUCN (International Union for Conservation of Nature). DOI: 10.2305/IUCN.CH.2017.01.en.

15. Tian, Zheynu et al. 2020, 'A Ubiquitous Tire Rubber–Derived Chemical Induces Acute Mortality in Coho Salmon', *Science* 371(6525):185–189. DOI: 10.1126/science.abd6951.

16. Ginoux, Paul et al. 2012, 'Global-scale attribution of anthropogenic and natural dust sources and their emission rates based on MODIS Deep Blue aerosol products', *Review of Geophysics*, 50, RG3005. DOI:10.1029/2012RG000388.

17. NB older American sofas! From Rodgers, K.M., Bennett, D., Moran, R., et al. 2021. 'Do Flame Retardant Concentrations Change in Dust After Older Upholstered Furniture Is Replaced?', *Environment International* 153:106513. DOI: 10.1016/j.envint.2021.106513.

18. Patel, Sameer, et al. 2020, 'Indoor Particulate Matter during HOMEChem: Concentrations, Size Distributions, and Exposures', *Environmental Science & Technology* 54(12): 7107–16. DOI: 10.1021/acs.est.0c00740.

19. Caillaud, Denis, et al. 2018, 'Indoor Mould Exposure, Asthma and Rhinitis: Findings from Systematic Reviews and Recent Longitudinal Studies', *European Respiratory Review*, 27(148) 170137. DOI: 10.1183/16000617.0137-2017.

20. Head hair is approximately 70 micrometres in diameter if you're white, African American or Caribbean, and thicker for Asian people – up to an average of 89 micrometres if you're Chinese. So it's a variable thing to use as a yardstick – but this ~25 per cent variation is pretty inconsequential given the 10,000-fold size variation we see with dust. Loussouarn, Geneviève

et al. 2016, 'Diversity in human hair growth, diameter, colour and shape', *European Journal of Dermatology* 26. DOI: 10.1684/ejd.2015.2726.

21. Morton, Timothy 2013, *Hyperobjects: Philosophy and Ecology after the End of the World*, University of Minnesota Press.

22. Nixon, Rob 2011, *Slow Violence and the Environmentalism of the Poor*, Harvard University Press.

23. Liboiron, Max & Lepawsky, Josh 2022, *Discard Studies: Wasting, Systems, and Power*, MIT Press.

24. Four deaths after the Foxconn plant explosion, May 2011, in Chengdu, China. Fifty-nine injured in a December 2011 explosion at a Shanghai factory run by RiTeng Computer Accessory, a subsidiary of Pegatron. Seventy-five deaths at a Foxconn factory in Kunshan, Jiangsu province, August 2014. See news reports, e.g. 'Apple confirms aluminum dust caused Chinese factory explosions', *Computerworld.com*, 15 January 2012, and 'China factory explosion expected to affect iPhone 6 production' *ZDnet.com*, 4 August 2014. Other workers' rights monitoring reports have been removed from the internet.

25. Henni, Samia (ed.) 2022, *Deserts Are Not Empty*, Columbia Books on Architecture and the City, New York.

26. Tsing, Anna 2015, *The Mushroom at the End of the World: On the Possibility of Life in Capitalist Ruins*, Princeton University Press.

27. Vohra et al. 2021, 'Global Mortality from Outdoor Fine Particle Pollution Generated by Fossil Fuel Combustion: Results from GEOS-Chem', *Environmental Research* 195 (110754). DOI: 10.1016/j.envres.2021.110754.

Chapter 1

1. In Letter CIV (104) of Seneca and Campbell, Robert (ed.) 2004, *Letters From A Stoic: Epistulae Morales ad Lucilium*, Penguin Classics.

2. Turner, B.L. & Sabloff, Jeremy L. 2012, 'Classic Period Collapse of the Mayan Lowlands,' *PNAS* 109 (35) 13908–13914, DOI: 10.1073/pnas.1210106109, and Stromberg, Joseph 2012, 'Why did the Mayan Civilisation Collapse?' *Smithsonian Magazine*, 23 August 2012.

3. Cavert, William M. 2016, *The Smoke of London: Energy and Environment in the Early Modern City*. Cambridge University Press, p. xviii.

4. Dodson, John, et al. 2014, 'Use of Coal in the Bronze Age in China', *The Holocene* 24(5): 525–30. DOI: 10.1177/0959683614523155.

5. Noble, Mark 1887, 'Letter: Coal in the North', *The Monthly Chronicle of North-Country Lore and Legend,* vol 1. p. 33.

6. 'Sacoles Lane', in the spelling of the time. Brimblecombe, Peter 1976, 'Attitudes and Responses Towards Air Pollution in Medieval England', *Journal of the Air Pollution Control Association* 26:10: 941–5, DOI: 10.1080/00022470.1976.10470341.

7. Cavert, William M. 2016, ibid, p. 15.

8. Cavert, William M. 2016, ibid, p. 24, citing Harding, Vanessa 1990, 'The Population of London, 1550–1700: A Review of the Published Evidence', *London Journal* 15: 111–28.

9. Russell, J.M. 1887, 'Coal in the North', *Monthly Chronicle Of North-Country Lore And Legend*; Newcastle-upon-Tyne, March 1887, 1(1):33–5.

10. Goodman, Ruth 2020, *The Domestic Revolution: How the Introduction of Coal into Our Homes Changed Everything*, Michael O'Mara Books, London.

11. Cavert, William M. 2016, ibid, p. 16.

12. Brome, Alexander 1664, '*Epistle to C.C.* Esquire' from *Songs And Other Poems*, quoted in Cavert, William M. 2016, ibid. p.185.

13. Miller, Gifford H., et al. 2012, 'Abrupt onset of the Little Ice Age triggered by volcanism and sustained by sea-ice/ocean feedbacks', *Geophysical Research Letters* 39: L02708, DOI: 10.1029/2011GL050168.

14. Brimblecombe 1976, ibid, p. 943 – who is quoting Dugdale, William 1658, *History of St. Paul's Cathedral from its Foundations*.

15. Evelyn, John 1661, *Fumifugium: or, The inconveniencie of the aer and smoak of London dissipated: Together with some remedies humbly proposed by J.E., Esq., to His Sacred Majestie and to the Parliament now assembled*, printed by W. Godbid for Gabriel Bedel and Thomas Collins. Available online at the Internet Archive and widely elsewhere.

16. Tormey, Warren 2011, 'Milton's Satan and Early English Industry and Commerce: The Rhetoric of Self-Justification', *Interdisciplinary Literary Studies* 13(1/2): 127–159.

17. Brimblecombe, Peter 1976, ibid, p. 943.

18. Boyle, Robert 1661, *The Sceptical Chymist: or Chymico-Physical Doubts & Paradoxes*, London, available on Project Gutenberg.

19. Cavert, William M. 2016, ibid, p. 37, drawing on statistics from Brimblecombe & Grossi 2008.

20. Brimblecombe, Peter & Grossi, Carlotta 2008, 'Millennium-long damage to building materials in London', *Science of the Total Environment* 407:1354–61.

21. Cavert, William M. 2016, ibid, p. 35.

22. Centre for Cities 2020, *Cities Outlook 2020*, published 27 January 2020. See 'Section 01 Holding our breath – How poor air quality blights cities'.

23. Graunt, John 1939 (1662), *Natural and Political Observations Made upon the Bills of Mortality*, John Hopkins Press: Baltimore, available on the Internet Archive (archive.org).

24. Lai T., Chiang C., Wu C., et al. 2016, 'Ambient air pollution and risk of tuberculosis: a cohort study', *Occupational and Environmental Medicine* 73:56–61. DOI: 10.1136/oemed-2015-102995

25. Cavert, William M. 2016, ibid, p. 175.

26. Raupach, Michael R. and Canadell, Canadell, Josep G., 2010, 'Carbon and the Anthropocene', *Current Opinion in Environmental Sustainability* 2: 210–218. DOI: 10.1016/j.cosust.2010.04.003.

27. Cavert, William M. 2016, ibid, p. xviii.

28. Burke III, Edmund, 2008, 'The Big Story: Human History, Energy Regimes, and the Environment' in: E. Burke III and K. Pomeranz, eds., *The Environment and World History 1500 BCE to 2000 CE*, University of California Press.

29. Wrigley, E.A. 2010, *Energy and the Industrial Revolution, Cambridge University Press.* DOI: 10.1017/CBO9780511779619.
30. My own calculation – a rough estimate.
31. 'Ghost acres' first coined by Georg Borgström – see 1965 (1953) 'Ghost acreage', Chapter 5 in *The Hungry Planet: The Modern World at the Edge of Famine,* Macmillan – then developed by Karl Pomeranz in *The Great Divergence: China, Europe and the Making of the Modern World Economy* (2000), Princeton University Press.
32. Oak Taylor, Jesse 2016 *The Sky of Our Manufacture: The London Fog in British Fiction from Dickens to Woolf,* University of Virginia Press, Chapter 2.
33. Burke III, Edmund 2008, ibid.
34. Malm, Andreas 2013, 'The Origins of Fossil Capital: From Water to Steam in the British Cotton Industry', *Historical Materialism* 21(1):15–68.
35. Moore, Jason W. 2017, 'The Capitalocene, Part I: on the nature and origins of our ecological crisis', *The Journal of Peasant Studies* 3: 594–640. DOI: 10.1080/03066150.2016.1235036.
36. Malm, Andreas 2013, ibid.
37. Pomeranz, Karl 2000, ibid. p. 62, citing Hartwell, Robert 1967, 'A Cycle of Economic Change in Imperial China: Coal and Iron in Northeast China, 750–1350', *Journal of the Economic and Social History of the Orient* 10(1): 102–159. DOI: 10.2307/3596361.
38. Gaskell, Elizabeth 1855, *North and South,* online at Project Gutenberg.
39. For London in 1890, see Brimblecombe, Peter and Grossi, Carlotta 2009, 'Millennium-long damage to building materials in London', *Science of the Total Environment* 407, table 1. UK Air Quality Standards Regulations 2010 require that concentrations of PM10 must not exceed an annual average of 40 µg/m^3. In the last 25 years, Beijing annual PM10 peaked c. 160 µg/m^3 in 2006; Delhi PM10s averaged 221 µg/m^3 in 2021.
40. Ruskin, John 1884, *The Storm-Cloud of the Nineteenth Century: Two Lectures delivered at the London Institution February 4th and 11th, 1884,* online at Project Gutenberg.
41. Gov.uk (undated), 'History of 10 Downing Street', online at https://www.gov.uk/government/history/10-downing-street, section 'Restoration and Modernisation' (accessed 25 May 2023).
42. Brahic, Catherine 2008, 'Cleaner air to turn iconic buildings green', *New Scientist,* 4 December 2008.
43. Searle, Adrian 2016, 'The Ethics of Dust: a latex requiem for a dying Westminster', *The Guardian,* 29 June 2016.

Chapter 2

1. McWilliams, Carey 1946a, 'The Los Angeles Archipelago', *Science & Society* 10(1): 43–51. https://www.jstor.org/stable/40399736.
2. McWilliams, Carey 1946b, *Southern California Country: An Island on the Land,* Duell, Sloan & Pearce: New York, p. 200.

3. McClung, William Alexander 2000, *Landscapes of Desire: Anglo Mythologies of Los Angeles*, University of California Press, p.7.
4. Austin, Mary 1903, *The Land of Little Rain*, online at Project Gutenberg.
5. Kathy Jefferson Bancroft was interviewed by Charlotte Cotton. Metabolic Studio, undated, 'The Sacred Owens Valley: An Interview with Kathy Jefferson Bancroft', online at https://metabolicstudio.org/89 (accessed 25 May 2023).
6. Madley, Benjamin, 2016, *An American Genocide: The United States and the California Indian Catastrophe, 1846–1873*, Yale University Press.
7. Quoted in Kahrl, William M. 1982, *Water And Power: The Conflict over Los Angeles' Water Supply in the Owens Valley*, University of California Press, p. 27.
8. Reisner, Marc 1993 (1986), *Cadillac Desert: The American West and its Disappearing Water*, Penguin, pp. 58–62.
9. Kahrl, William M. 1982, ibid, p. 47.
10. Kahrl, William M. 1982, ibid, p. 48.
11. Margaret Leslie Davis speaking on Los Angeles radio. KPCC 2013, 'Take Two' programme, series 'LA Aqueduct 100 Years Later'. Interview excerpted online as 'William Mulholland's rise from ditch-digger to controversial LA power player', kpcc.org.
12. Kahrl, William M. 1982, ibid, p. 49.
13. Kahrl, William M. 1982, ibid, p. 64.
14. Quotes and figures in this section taken from Kahrl, William M. 1982, ibid, pp. 61–78.
15. Widely quoted, e.g. Piper, Karen 2006, *Left in the Dust: How Race and Politics Created a Human and Environmental Tragedy in L.A*, St Martin's Press, p. 30.
16. Mulholland's line recalled years later by surveyor Frederick C. Cross in 'My Days on the Jawbone', *Westways*, May 1968, pp.6–7. The 'Jawbone' was a particularly hard stretch north of Mojave in Kern County.
17. Reisner, Marc 1993, ibid, p. 84.
18. Nadeau, Remi A. 1950, *The Water Seekers*, Doubleday, p. 62 – available on the Internet Archive (archive.org)
19. McWilliams, Carey 1946b, *Southern California Country: An Island on the Land*, Duell, Sloan & Pearce: New York.
20. Reisner, Marc 1993, ibid, pp. 86–7.
21. Piper, Karen 2006, ibid. p. 68 – citing Mike Davis 1999, *Ecology of Fear: Los Angeles and the Imagination of Disaster*, Vintage, p. 162.
22. Nadeau, Remi A. 1950, *The Water Seekers*, Doubleday, p. 75 – available on the Internet Archive (archive.org).
23. Unrau, Harlan D. 1996, 'The Evacuation and Relocation of Persons of Japanese Ancestry During World War II: A Historical Study of the Manzanar War Relocation Center' National Park Service, available online at https://www.nps.gov/parkhistory/online_books/manz/hrs.htm (accessed 25 May 2023) – which cites Frederick Faulkner 1927, 'Owens Valley, Where the Trail of the Wrecker Runs', *Sacramento Union*, 28 March–2 April 1927.
24. Nadeau, Remi A. 1950, ibid, p. 81.
25. Walton, John 1993, *Western Times and Water Wars: State, Culture, and Rebellion in California*, University of California Press, p. 170.

26. Nadeau, Remi A. 1950, ibid, p. 76.
27. This paragraph sourced from Denslow, William R. 1957, *10,000 Famous Freemasons*, Macoy Publishing & Masonic Supply Co. Inc, Richmond, VA, available on the Internet Archive (archive.org), and Nadeau, Remi A. 1950, ibid.
28. Walton, John 1993, ibid, p. 160.
29. Nadeau, Remi A. 1950, ibid, pp. 84–9.
30. The movie in question – *Riders of the Purple Sage* (1925) – is, as the observant may realise from the date, a silent one. Nonetheless, both Ralph Nadeau and Marc Reiser claim that the movie had an orchestra and that orchestra was duly dispatched. At first I thought it a tall tale – but it turns out silent movies did sometimes have orchestras on set to add atmosphere and emotional cues.
31. Nadeau, Remi A. 1950, ibid, p. 89.
32. Ginwala quoted in Malm, Andreas 2021, *How to Blow Up a Pipeline: Learning to Fight in a World on Fire*, Verso p. 71.
33. Stelloh, Tim et al. 2016, 'Dakota Pipeline: Protesters Soaked With Water in Freezing Temperatures', *NBC News*, 21 November 2016.
34. Nadeau, Remi A. 1950, ibid, p. 101.
35. Reisner, Marc 1993, ibid, p. 97.
36. Sahagun, Louis 2019, 'Owens Valley gushes over Department of Water and Power land plan', *Los Angeles Times*, 23 January 2019.
37. Nadeau, Remi A. 1950, ibid. p.. 129–131.
38. Los Angeles County Department of Coroner, 1928, 'Transcript of Testimony and Verdict of the Coroner's Jury In the Inquest Over Victims of St. Francis Dam Disaster', p. 378. Available online at https://scvhistory.com/scvhistory/sfdcoronersinquest.htm (accessed 27 May 2023).
39. Nadeau, Remi A. 1950, ibid.
40. Unrau, Harlan D. 1996, 'The Evacuation and Relocation of Persons of Japanese Ancestry During World War II: A Historical Study of the Manzanar War Relocation Center', *National Park Service*, available online at https://www.nps.gov/parkhistory/online_books/manz/hrs.htm (accessed 25 May 2023).
41. Kahrl, William M. 1982, ibid, p. 371 – citing Dorothy Swaine Thomas & Richard S. Nashimoto 1946, *The Spoilage: Japanese American Evacuation and Resettlement*, University of California Press p. 368, and James D. Houston & Jeanne Wakatsuki Houston 1973, *Farewell to Manzanar: A True Story of Japanese American Experience During and After the World War II Internment*.
42. Hinkley, Todd K. undated, *Mineral Dusts in the Southwestern U.S.*, material prepared for the Southwest Regional Climate Change Symposium and Workshop in Tucson, Arizona, on September 3–5, 1997. Available online at https://geochange.er.usgs.gov/sw/impacts/geology/dust/ (accessed 27 May 2023).
43. Numbers given for the amount of dust vary, but this can be explained. The '900,000 – 8 million-ton' figure is from TE Gill & DA Gillette 1991, 'Owens Lake: A Natural Laboratory for Aridification, Playa Desiccation and Desert Dust', in *Geological Society of America* 23(5), and refers to dust in

general. The 300,000-ton figure is for PM10s specifically, i.e. excluding larger, sand-grain sized particles.

44. Cahill, Thomas A. et al. 1996, 'Saltating Particles, Playa Crusts And Dust Aerosols At Owens (Dry) Lake, California', *Earth Surface Processes and Landforms* 21: 621–39.

45. From Piper, Karen 2006, ibid. p. 8, who cites 'Meeting to Adopt the PM-10 Demonstration of Attainment SIP (PM-10 Amendment Plan)', at Inyo County Courthouse, 2 July 1997.

46. Piper, Karen 2006, ibid, pp. 3, 13.

47. Crystal science from Buck, Brenda et al. 2011, 'Effects of Salt Mineralogy on Dust Emissions, Salton Sea, California', *Soil Science Society of America Journal* 75(5), DOI: 10.2136/sssaj2011.0049. Salts in the Owens Lake bed from Reheis, Marith C. 1997, Dust deposition downwind of Owens (dry) Lake, 1991–1994: Preliminary findings', *Journal of Geophysical Research Atmospheres* 1022(D22):25999–26008, DOI:10.1029/97JD01967.

48. U.S. Environmental Protection Agency (EPA) figure; also '300–400,000' in – TE Gill & DA Gillette, 1991. 'Owens Lake: A Natural Laboratory for Aridification, Playa Desiccation and Desert Dust', *Geological Society of America* 23(5), DOI: 10.1029/97JD01967.

49. U.S. Environmental Protection Agency (EPA) 2010, *Region 9 Air Programs: Particulate Matter Pollution in California's Owens Valley*, retrieved from https://19january2017snapshot.epa.gov/www3/region9/air/owens/qa.html (accessed 27 May 2023).

50. Piper, Karen 2006, ibid, p. 4.

51. Emergency physician Dr Bruce Parker to journalist Marla Cone, 'Owens Valley Plan Seeks L.A. Water to Curb Pollution', *Los Angeles Times*, 17 December 1996.

52. Saint Amand et al. 1986. Mentioned in M.C. Reheis 1997, 'Dust Deposition Downwind of Owens (Dry) Lake, 1991–1994: Preliminary Findings', in *Journal of Geophysical Research* 102: 25–6.

53. Knudson, Tom 2014, 'Outrage in Owens Valley a century after L.A. began taking its water', *The Sacramento Bee*, 1 May 2014.

Chapter 3

1. Oklahoma Historical Society Audio Archives, 1984, oral history interview with Ada Kearns, conducted 13 September 1984 by Joe L. Todd. Online at https://www.youtube.com/watch?v=e1FLpjdpHRk.

2. Oklahoma Historical Society Audio Archives, 1984, oral history interview with Logan Gregg, conducted 13 September 1984. Online at https://www.youtube.com/watch?v=V5skbWLrYJA.

3. Oklahoma Historical Society Audio Archives, 1984, oral history interview with Nellie Goodner Malone & John Goodner, conducted 13 September 1984. Online at https://www.youtube.com/watch?v=V5skbWLrYJA.

4. Svobida, Laurence 1986 (1940), *Farming The Dust Bowl: A First-Hand Account from Kansas*, University of Kansas Press, p. 41.

5. History Nebraska, undated, 'Timeline Tuesday: Drought and Depression in 1890s Nebraska', history.nebraska.gov (accessed 20 May 2023).

6. Svobida, Laurence 1986, ibid, p. 234. 'Three million' figure aggregate from Donald Worster, p.49: 'Almost a million' left in the first half of the decade, and 2.5 million after 1935. See Donald Worster 2004, *Dust Bowl: The Southern Plains in the 1930s*, Oxford University Press.

7. Svobida, Laurence 1986, ibid, p. 137.

8. Egan, Timothy 2006, *The Worst Hard Time: The Untold Story of Those Who Survived the Great American Dust Bowl*, Mariner Books, pp. 2–3. My supposition that it's the Hartwell diary Egan is talking about – based on Stephen C. Behrendt's interview with Don Hartwell, which talks of a diary being rescued by 'Charlotte Lambrecht, daughter of Henry and Edith Lambrecht of Inavale and friend of Verna'. Stephen C. Behrendt 2015, 'One Man's Dust Bowl: Recounting 1936 with Don Hartwell of Inavale, Nebraska', *Great Plains Quarterly*, 35(3), 229–47. http://www.jstor.org/stable/24465605.

9. Works Progress Administration Federal Writers' Project, 1936–39, 'Range Lore and Negro Cowboy Reminiscences before and after 1875', p.17, interviewer Florence Angermiller. https://www.loc.gov/item/wpalh002179/ (accessed 20 May 2023).

10. Powell, John Wesley, 1878, *Report on the Lands of the Arid Region of the United States*, DOI: 10.3133/70039240.

11. Brown, Walter Lee 1997, *A Life of Albert Pike*, University of Arkansas Press, p.15.

12. Wilber, Charles Dana 1881, *The Great Valleys and Prairies of Nebraska and the Northwest*, Daily Republican Print, Omaha (widely digitised online). It's also Wilber who coined the 'rain follows the plow' phrase, building on the work of Samuel Aughey Jr, professor of natural sciences at the University of Nebraska.

13. Thomas (III), William G. 2011, 'Railroads, the Making of Modern America, and the Shaping of the Great Plains', Paul Olsen Seminar 13 April 2011 at the Center for Great Plains Studies, University of Nebraska-Lincoln. Available online at https://railroads.unl.edu/blog/?p=433 (accessed 20 May 2023).

14. A phrase 'buried halfway through the third paragraph of a long essay in the July–August issue of *The United States Magazine, and Democratic Review*,' the Britannica encyclopaedia notes in its entry on Manifest Destiny.

15. Turner, Frederick J. 1893, 'The Significance of the Frontier in American History', *Annual Report of the American Historical Association*, pp. 197–227.

16. Hudson, John C. 1986, 'Who Was 'Forest Man?' Sources of Migration to the Plains', in *Great Plains Quarterly* 967, p. 80.

17. Freedman, John F. 2008, *High Plains Horticulture: A History*, University Press of Colorado.

18. Bennet, Hugh H. 1970 (1936), *Soil Conservation*, Ayer Press, Manchester, pp. 894–5.

19. Henderson, Caroline 2003, *Letters from the Dust Bowl*, ed. Alvin O. Turner, University of Oklahoma Press, p. 140.

20. Svobida, Laurence 1986, ibid, p. 68.

21. Holleman, Hannah 2018, *Dust Bowls of Empire: Imperialism, Environmental Politics, and the Injustice of "Green" Capitalism*, Yale Agrarian Studies series, Yale University Press, p. 71.
22. Worster, Donald 2004, ibid, pp. 90–92.
23. Low, Ann Marie 1984, *Dust Bowl Diary*, University of Nebraska Press, p. 33.
24. Oklahoma Historical Society Audio Archives, 1984, oral history interview with Mr and Mrs Elmer Shackleford, conducted 13 September 1984 by Joe L. Todd. https://www.youtube.com/watch?v=JVUT8mBjoZw. Mrs Shackleford's first name isn't given by OHS or in the dialogue.
25. Oklahoma Historical Society Audio Archives, 1984, oral history interview with Esther Reiswig conducted 13 September 1984 by Joe L. Todd. https://www.youtube.com/watch?v=e1FLpjdpHRk.
26. Worster, Donald 2004, ibid, p. 15.
27. Worster, Donald 2004, ibid, pp. 16–17.
28. Oklahoma Historical Society Audio Archives, 1984, oral history interview with Esther Reiswig, ibid.
29. Henderson, Caroline 2003, ibid. pp. 140–1.
30. O'Hanlon, Larry 2004, 'Dust storms are truly electric', *ABC Science*, 18 August 2006 (abc.net.au).
31. Ford County Dust Bowl Oral History Project, 1998, interview with Juanita Wells, interviewer Brandon Case, 27 February 1998, Ford County Historical Society, Dodge City (kansashistory.us).
32. Ford County Dust Bowl Oral History Project, 1998, interview with Arthur W. Leonard, interviewer Brandon Case, 23 June 1998, Ford County Historical Society, Dodge City (kansashistory.us).
33. Ford County Dust Bowl Oral History Project, 1998, interview with Juanita Wells, ibid.
34. Oklahoma Historical Society Audio Archives, 1984, oral history interview with Mr and Mrs Elmer Shackleford, ibid.
35. Low, Ann Marie 1984, ibid.
36. Henderson, Caroline 2003, ibid, p. 142.
37. Worster, Donald 2004, ibid, p. 12.
38. Worster, Donald 2004, ibid, p. 107.
39. Ford County Dust Bowl Oral History Project, 1998, interview with Juanita Wells, ibid.
40. Svobida, Laurence 1986, ibid, pp. 195, 199.
41. Svobida, Laurence 1986, ibid, pp. 199–201.
42. Worster, Donald 2004, ibid, p. 21, who cites Earl Brown, Selma Gottlieb and Ross Laybourne 1938, 'Dust Storms and their Possible Effects on Health', *Public Health Reports* 50, 4 October 1938, pp. 1369–83.
43. Worster, Donald 2004, ibid, p. 21, who cites the *Kansas City Star* 1935, 'Effect of Dust Storms: Replies of County Health Officers', 27, 30 April & 1, 2 May 1935, National Archives Record Group 114.
44. Riney-Kehrberg, Pamela 1994, *Rooted in Dust: Surviving Drought and Depression in Southwestern Kansas*, University Press of Kansas, p. 33 – who cites the Kansas State Board of Health, 'Biennial Report of the State Board of Health of the State of Kansas', vols. 16–20.

45. Worster, Donald 2004, ibid, p. 21, citing Earl Brown, Selma Gottlieb and Ross Laybourne 1938, ibid.
46. Riney-Kehrberg, Pamela 1994 ibid, p. 32.
47. Svobida, Laurence 1986, ibid, p. 139.
48. Bennet, Hugh H. 1970 (1936), *Soil Conservation*, Ayer Press, Manchester, p. 120.
49. *Chicago Tribune* editorial 2012, 'During 1934 drought, dust storm swarmed Chicago', *Chicago Tribune*, 15 July 2012.
50. 80 per cent of 480M tonnes total was airborne tephra and fine ash, i.e. circa 384M tonnes. Gudmundsson, M et al. 2012, 'Ash generation and distribution from the April–May 2010 eruption of Eyjafjallajökull, Iceland', *Nature Scientific Reports* 2(572), DOI: 10.1038/srep00572.
51. Worster, Donald 2004, ibid, p. 28.
52. Svobida, Laurence 1986, ibid.
53. Hansen, Zeynep K., and Gary D. Libecap 2004, 'Small Farms, Externalities, and the Dust Bowl of the 1930s', *Journal of Political Economy* 112(3): 665–94. DOI: 10.1086/383102.
54. Svobida, Laurence 1986, ibid, p. 255.
55. Duncan, Dayton and Burns, Ken 2012, *The Dust Bowl: An Illustrated History,* Chronicle Books, p. 54.
56. Worster, Donald 2004, ibid, p. 31 – citing Albert Law, *Dalhart Texan*, 17 June 1933.
57. Worster, Donald 2004, ibid, pp. 39–40.
58. Lambert, C. Roger 1971, 'The Drought Cattle Purchase, 1934–1935: Problems and Complaints', *Agricultural History*, 45(2), 85–93. http://www.jstor.org/stable/3742072.
59. Worster, Donald 2004, ibid, pp. 12, 35.
60. Henderson, Caroline 2003 ibid, pp. 143–4.
61. Svobida, Laurence 1986 ibid, pp. 108–10.
62. Whitney, Milton 1909, *Soils of the United States: Based Upon the Work of the Bureau of Soils to January 1, 1908*, Bulletin 55, United States Bureau of Soils.
63. Bennett quoted in Egan 2006, ibid, p. 126.
64. Bennett, Hugh H. 1970 (1936), *Soil Conservation,* Ayer Press, Manchester, pp. 894–5.
65. Oklahoma Historical Society Audio Archives, 1984, oral history interview with Robert Howard, conducted 26 September 1984 by Berenice Jackson. Online at https://www.youtube.com/watch?v=33n2R-JhSag.
66. Svobida, Laurence 1986, ibid, p. 98, p. 15.
67. Svobida, Laurence 1986, ibid, p.. 116–17.
68. Don Hartwell quoted in Egan, Timothy 2006, ibid., p. 297.
69. Duncan, Dayton and Burns, Ken 2012, *The Dust Bowl: An Illustrated History*, Chronicle Books, p. 163.
70. PBS, undated, 'Timeline: The Dust Bowl', supporting material for the 1998 PBS documentary *Surviving the Dust Bowl*, online at https://www.pbs.org/wgbh/americanexperience/features/dust-bowl-surviving-dust-bowl/ (accessed 27 May 2023).
71. Svobida, Laurence 1986, ibid, pp. 232–3.

72. Svobida, Laurence 1986, ibid, p. 234. 'Almost a million' left in the first half of the decade, and 2.5 million after 1935 – Worster, Donald 2004, ibid, p. 49. The Great Migration of the mid-twentieth century saw 6 million Black Americans move to northern states, but over a period nearly ten times as long (1916–70); as such, I think it's fair to say that the dust bowl is the largest 'concentrated' movement.

73. Worster, Donald 2004, ibid, p. 49.

74. Ford County Dust Bowl Oral History Project, 1998, interview with Juanita Wells, ibid.

75. Inch per year of recharge: Hornbeck, Richard & Keskin, Pinar 2011, 'Farming the Ogallala Aquifer: Short-run and Long-run Impacts of Groundwater Access', working paper, Department of Economics, Harvard University.

76. Sanderson & Frey 2014, 'From desert to breadbasket...to desert again? A metabolic rift in the High Plains Aquifer', *Journal of Political Ecology* 21(1):516–32. DOI:10.2458/v21i1.21149.

77. Marx, Karl 1967 (1867), *Capital: A Critique of Political Economy, vol. I*, New York: International Publishers Co. Inc., pp. 505–6. NB Hannah Holleman makes this argument in '*Dust Bowls of Empire*' (2018), which I recommend to readers interested in a substantial consideration of this idea.

78. Wise, Lindsay 2015, 'A drying shame: With the Ogallala Aquifer in peril, the days of irrigation for western Kansas seem numbered', *Kansas City Star*, 26 July 2015.

79. Worster, Donald 2004, ibid, p. 253.

80. Lambert, Andrew et al. 2020, *Dust Impacts of Rapid Agricultural Expansion on the Great Plains. Geophysical Research Letters* 47(20), DOI: 10.1029/2020GL090347.

81. Re. cardiovascular & respiratory illnesses, see: Achakulwisut, P., Mickley, L. J., & Anenberg, S. C. 2018, 'Drought sensitivity of fine dust in the US Southwest: Implications for air quality and public health under future climate change', *Environmental Research Letters*, 13, 054025. DOI: 10.1088/1748 9326/aabf20.

82. Williams, A. Park et al. 2020, 'Large contribution from anthropogenic warming to an emerging North American megadrought', *Science* 368(1488): 314–318, 17 April 2020. DOI: 10.1126/science.aaz9600.

83. Heslin, Alison et al. 2020, 'Simulating the Cascading Effects of an Extreme Agricultural Production Shock: Global Implications of a Contemporary US Dust Bowl Event', Frontiers in Sustainable Food Systems, vol. 4. DOI : 10.3389/fsufs.2020.00026.

84. Glotter, Michael & Elliott, Joshua 2017, 'Simulating US agriculture in a modern Dust Bowl drought', *Nature Plants* 3, 16193. DOI: 10.1038/nplants.2016.193.

85. Momaday, N. Scott 1976, 'Native American Attitudes toward the Environment', in *Seeing with a Native Eye: Essays on Native American Religion*, ed. Walter Holden Capps, Harper & Row, New York.

86. Worster, Donald 2004, ibid, p. 77, who cites Levy, J. E. 1961, 'Ecology of the Southern Plains: The Ecohistory of the Iowa, Comanche, Cheyenne,

and Arapaho, 1830–1870', *Symposium: Patterns of Land Utilization and Other Papers*, ed. Viola E. Garfield, Proceedings of the 1961 Spring Meeting of the American Ethnological Society, University of Washington Press pp. 18–25.

87. Jacks, Graham Vernon & Whyte, Robert Orr 1938, *Vanishing Lands: A World Survey of Soil Erosion*, Faber & Faber.

88. Marx, Karl 1981 (1867), *Capital*, Vol. III. New York: Vintage Books, p. 959. Widely available open access online.

89. Popper, Deborah E. and Popper, Frank J. 1987, 'The Great Plains: From Dust to Dust', *Planning* 53(12): 12–18, and Great Plains Restoration Council (undated), 'Buffalo Commons', https://gprc.org/research/buffalo-commons/ (accessed 20 May 2023).

90. Popper, Deborah E. & Popper, Frank J. 1999, 'The Buffalo Commons: Metaphor as Method', *Geographical Review*, 89(4), 491–510. DOI: 10.2307/216099.

91. Death threats and Kansas newspapers points, from Young, Gordon 2010, 'Interview with Frank Popper about Shrinking Cities, Buffalo Commons, and the Future of Flint', on *Flint Expatriates* blog (flintexpats.com) (accessed 20 May 2023).

92. 1.1 million acres for the former Dust Bowl region and Yellowstone combined; no figures for Dust Bowl region alone. Farm Services Agency press release, 2021 'USDA Accepts More than 2.5 Million Acres in Grassland CRP Signup, Double Last Year's Signup' (fsa.usda.gov).

Chapter 4

1. 'Sawles Warde' ('The Guardianship of the Soul') in *The Katherine Group MS Bodley 34* (2006), eds. Emily Rebekah Huber and Elizabeth Robertson, Medieval Institute Publications, Kalamazoo MI. DOI: 10.2307/j.ctvndv757.

2. Scanlan, John 2005, *On Garbage*, Reaktion Books. See also Jørgensen, Dolly 2014, 'Modernity and Medieval Muck', *Nature and Culture* 9(3): 225–37.

3. Hoy, Suellen 1997, *Chasing Dirt: The American Pursuit of Cleanliness*, Oxford University Press.

4. Hoy, Suellen 1997, ibid, pp. 21–2; Catherine Beecher 1841, *A Treatise on Domestic Economy for the Use of Young Ladies at Home and at School*, available on Project Gutenberg.

5. Faraday, Michael 1855, letter to the editor of *The Times* newspaper, London, 7 July 1855. Subsequently republished by the Royal Institution.

6. Dickens, Charles 1903 (1839), *Sketches by Boz: Illustrative of Every-Day Life and Every-Day People*, Chapman & Hall, London; available on Project Gutenberg.

7. Quoted in Davies, Stephen 2004, 'The Great Horse-Manure Crisis of 1894', Foundation for Economic Education, 1 September 2004 (fee.org).

8. Quoted widely, e.g. Corbett, Christopher 1988, 'The 'Charm City' Of H. L. Mencken', *The New York Times*, 4 September 1988.

9. Varro, Marcus Terentius 30 BCE, *De Re Rustica*, I.12.2.
10. Past Tense 2019, 'Today in London's infra(re)structural history, 1845: New Oxford Street opens – built to socially cleanse the St Giles Rookery', 10 June 2019 (pasttense.co.uk).
11. Quote via Horsfield, Margaret 1997, *Biting The Dust: The Joys of Housework*, Fourth Estate, p. 64.
12. Nightingale, Florence 1860, *Notes on Nursing*.
13. Obligatory Walter Benjamin citation: Benjamin, Walter 2002 (1927–40), *The Arcades Project*, Harvard University Press, p. 103.
14. Horsfield, Margaret 1997, ibid, p. 60.
15. Haweis, Mary Eliza 1889, *The Art of Housekeeping: A Bridal Garland*, Sampson Low, Marston, Searle & Rivington: London. Available via the Internet Archive (archive.org).
16. Philip, Robert Kemp (ed.), *Enquire Within upon Everything*, London: Houlston and Wright, p. 241. Available via the Wellcome Collection and Internet Archive.
17. Centers for Disease Control and Prevention (CDC) 1999, 'Achievements in Public Health, 1900–1999: Healthier Mothers and Babies', *Morbidity and Mortality Weekly Report (MMWR)* 48(38);849–58.
18. Data from Google Ngram, a tool for analysing word usage in a large corpus of books published in English from 1500 to 2019. Both British and American English show very similar patterns.
19. Tomes, Nancy 1998, *The Gospel of Germs: Men, Women and the Microbe in American Life*, Harvard University Press.
20. Institute of Health Visiting, undated, 'History of Health Visiting', online at https://ihv.org.uk/about-us/history-of-health-visiting (accessed 27 May 2023). They quote Elizabeth Hardie 1893, *The Ladies' Health Society of Manchester and Salford*.
21. Quoted in Horsfield, Margaret 1997, ibid, p. 95.
22. Prudden, Theophilus Mitchell 1890, *Dust and its Dangers*, GP Putnam: London. Available via the Wellcome Collection, https://wellcomecollection.org/works/e2yy6adk/items (accessed 27 May 2023).
23. This misapprehension of fomite transmission over aerosol gained new relevance as I wrote this during covid times. Quotes from Tomes, Nancy 1998, ibid, pp. 8–9, 144.
24. Frederick, Christine 1915, *Household Engineering: Scientific Management in the Home*, Home Economics Association: Chicago. Available via the Internet Archive (archive.org).
25. Frederick, Christine 1913, *The New Housekeeping: Efficiency Studies in Home Management*, Doubleday. Available via the Internet Archive (archive.org).
26. Frederick, Christine 1915, ibid, pp. 513–4.
27. Harvey, David 1989, *The Condition of Postmodernity*, Wiley-Blackwell: Oxford. He is citing Cecilia Tichi's *Shifting Gears* (1987), who is in turn quoting *Good Housekeeping* from 1910.
28. Wharton, Edith 1934, *A Backward Glance*, D. Appleton-Century: London.
29. JB Priestley quoted in Delap, Lucy 2011, *Knowing Their Place: Domestic Service in Twentieth-Century Britain*, Oxford University Press, p. 2.
30. Tomes, Nancy 1998, ibid.

31. Advertisement included as backmatter in the *Journal of Geology* 18(3), April–May 1910. DOI: 10.1086/621739.
32. Quoted in Horsfield, Margaret 1997, ibid, p. 146.
33. Christine Frederick quoted in Horsfield, Margaret 1997, ibid.
34. Hoover advertisement quoted in Horsfield, Margaret 1997, ibid, p. 75 – from *Woman*, 5 June 1937.
35. Betty Friedan quotes this in *The Feminine Mystique* (2010 / 1963), Penguin.
36. Lupton, Ellen 1993, *Mechanical Brides: Women and Machines from Home to Office*, Princeton Architectural Press, p. 11.
37. Frederick, Christine 1915, ibid, p. 164.
38. Quoted in Horsfield, Margaret 1997, ibid, p. 136 – originally in the *Ladies Home Journal*, 1930.
39. Ehrenreich, Barbara 1993, 'Housework Is Obsolescent', *TIME*, 25 October 1993.
40. Tyler May, Elaine 1988, *Homeward Bound: American Families in the Cold War Era*, Basic Books: New York, pp. ix–xix.
41. Smith, Elinor Goulding 1957, *The Complete Book of Absolutely Perfect Housekeeping*, Frederick Muller: London, p.44.
42. Tyler May, Elaine 1988, ibid, p. 18.
43. Friedan, Betty 2010 (1963), *The Feminine Mystique*, Penguin, p. 16.
44. *Mad Men* series 1, episode 9, first broadcast 13 September 2007, AMC. See Rine, Abigail 2013, 'The Postfeminist Mystique: Or, What Can We Learn From Betty Draper?' *PopMatters*, 14 April 2013.
45. Le Corbusier 2007 (1923), *Toward An Architecture*, trans. John Goodman, Getty Publications: Los Angeles, p. 124.
46. Le Corbusier 1925, *Plan Voisin* – accessed via Fondation Le Corbusier (fondationlecorbusier.fr).
47. Le Corbusier 1967 (1933), *The radiant city : elements of a doctrine of urbanism to be used as the basis of our machine-age civilization*, Faber, p. 42.
48. Schachtman, Tom 1999, *Absolute Zero and the Conquest of Cold*, Houghton Mifflin: Boston.
49. Iddon, Chris 2015, 'Florence Nightingale: nurse and building engineer', *CIBSEJournal* (Chartered Institute of Building Engineers), May 2015.
50. Le Corbusier 1987 (1925), 'The Law of Ripolin' in *The Decorative Arts of Today*, trans. James I. Dunnett, Architectural Press: London. Italics his.
51. Personal communication, Twitter, 30 March 2021. https://twitter.com/entschwindet/status/1376989997312700420.
52. Le Corbusier 1987 (1925), ibid.
53. Sully, Nicole 2009, 'Modern Architecture and Complaints about the Weather, or, "Dear Monsieur Le Corbusier, It is still raining in our garage..."', *Media/Culture Journal* 12(4). DOI: 10.5204/mcj.172.
54. Murphy, Douglas 2012, *The Architecture of Failure*, Zero Books.
55. *TIME* magazine (unbylined), 1962, 'Science: Mr. Clean', 13 April 1962.
56. Different measurement regimes make direct comparisons a bit tricky. Outdoor air pollution measures PM2.5s (>= 2.5μm) and PM10s (>=10μm), whereas cleanroom air standard ISO9 (room air) gives figures for particles over 0.5μm, 1μm and 5μm – so it's not directly comparable. ISO9 is defined

as 35.2 million particles >=0.5μm/m3, but this appears to be a metric conversion of the more memorable 'one million particles per cubic foot'. The key thing to remember is, the smaller the scale you look at, the number of particles increases exponentially.

57. Morton, Corn 1961, 'The Adhesion of Solid Particles to Solid Surfaces, a Review', *Journal of the Air Pollution Control Association*, 11:11, 523–8, DOI: 10.1080/00022470.1961.10468032.

58. Sandia Labs 2012, 'Modern-day cleanroom invented by Sandia physicist still used 50 years later' news release, https://newsreleases.sandia.gov/cleanroom_50th/ (accessed 27 May 2023). How 1960s to be smoking in a science lab...

59. International Standards Organisation (ISO) 2015, 'Part 1: Classification of air by particle concentration' in *ISO14644-1 Cleanrooms and associated controlled environments* 2nd edition, 15 December 2015.

60. JPL Planetary Protection Center of Excellence, undated, 'Mission Implementation' webpage https://planetaryprotection.jpl.nasa.gov/mission-implementation (accessed 27 May 2023).

61. NASA, undated, Genesis Mission website: https://solarsystem.nasa.gov/genesismission/index.html and Genesis Solar Wind Samples webpage: https://curator.jsc.nasa.gov/genesis/index.cfm (both accessed 27 May 2023).

62. One football (soccer) field: *7,140* sq m; TSMC's cleanroom is 'over 160,000 sq m'. Cleanroom Technology (unbylined) 2018, 'TSMC starts construction of Fab 18 in Taiwan', CleanroomTechnology.com.

63. NASA Jet Propulsion Lab 2002, 'JPL High Bays Give a Whole New Meaning to 'Clean Your Room'' webpage, https://www.jpl.nasa.gov/news/jpl-high-bays-give-a-whole-new-meaning-to-clean-your-room (accessed 27 May 2023).

64. Douglas, Mary 1984 (1966), *Purity and Danger: An Analysis of Concepts of Pollution and Taboo*, ARK edition, Routledge, pp. 36, 3.

65. Forty, Adrian 1986, *Objects of Desire: Design and Society from Wedgwood to IBM*, Pantheon Books, p. 159.

66. Beeton, Isabella (Mrs) 1861, *The Book of Household Management*, S.O. Beeton: London, Chapter 41 Servants, paragraph 2153. Available online via Project Gutenberg.

67. Leaver, Eric 1997, 'Precious stones: Another tradition fades away', *Lancashire Telegraph*, 7 April 1997.

68. Nottstalgia forum, 2012, 'Polishing the step' discussion thread, online at https://nottstalgia.com/forums/topic/9788-polishing-the-step/ (accessed 27 May 2023).

69. Boy Scouts of America, 1911, *Boy Scouts Handbook* first edition, available online via Project Gutenberg.

70. Douglas, Mary 1984 (1966), ibid, p. 5.

71. Bonnett, Alistair 1999, 'Dust (1)' in *City A–Z: Urban Fragments*, eds. Nigel Thrift & Steve Pile, Routledge, p. 63.

72. Douglas, Mary 1984 (1966), ibid, pp. 160–5.

73. Here Douglas is quoting Mircae Eliade 1958, *Patterns in Comparative Religion*, p. 194.

Chapter 5

1. Elpiner, Leonid I. 1999, 'Public Health in the Aral Sea coastal region and the dynamics of changes in the ecological situation', Chapter 7 in *Creeping Environmental Problems and Sustainable Development in the Aral Sea Basin*, ed. Michael H. Glantz, Cambridge University Press.
2. One fully grown saxaul tree can fix up to 10 tons of soil around its roots,' Orazbay Allanazarov, a forestation specialist, told the BBC. Qobil, Rustam & Harris, Paul 2018, 'Restoring life to the Aral Sea's dead zone', *BBC Uzbek*, 1 June 2018.
3. Mirovalev, Mansur 2015, 'Uzbekistan: A dying sea, mafia rule, and toxic fish', *Al Jazeera*, 11 June 2015.
4. UNFAO 2013, *Irrigation in Central Asia in Figures – AQUASTAT Survey 2012*, ed. Karen Frenken, FAO Water Reports 39, Food and Agriculture Organization of the United Nations.
5. Peak figures of 62.8 km³ in 1985 and 62.5 km³ in 1990 – from UNFAO 2013, ibid. For comparison, this is sufficient to submerge an area the size of Greater London (1,572 km²) 40m underwater, or New York City (778 km²) 80m.
6. Quoted by Zonn, Igor S. 1999, 'The impact of political ideology on creeping environmental changes in the Aral Sea Basin', Chapter 8 in *Creeping Environmental Problems and Sustainable Development in the Aral Sea Basin*, ed. Michael H. Glantz, Cambridge University Press.
7. Dates, quotes and decrees in this section are taken from Zonn, Igor S. 1999, ibid, unless otherwise specified.
8. Brain, Stephen 2010, 'The Great Stalin Plan for the Transformation of Nature', *Environmental History* 15(4): 670–700.
9. Kafikov, Asomitdin A., 1999, 'Desertification in the Aral Sea region', Chapter 4 in *Creeping Environmental Problems and Sustainable Development in the Aral Sea Basin*, ed. Michael H. Glantz, Cambridge University Press.
10. 43 million tons: Micklin, Philip P. 1998, 'Desiccation of the Aral Sea: A Water Management Disaster in the Soviet Union', *Science*, 241: 1170–6. 140 million tons: Tursunov, A.A. 1989, 'The Aral Sea and the ecological situation in Central Asia and Kazakhstan', *Hydrotechnical Construction* 23 pp. 319–25.
11. Figures an average for 1960–79. Source: OE Semenov 2011, 'Dust Storms and Sandstorms and Aerosol Long-Distance Transport', Chapter 5 in *Aralkum – a Man-Made Desert*, Ecological Studies Volume 218 eds. Siegmar-W. Breckle et al., Springer: Berlin. DOI: 10.1007/978-3-642-21117-1_16.
12. Micklin, Philip P. 1998, ibid.
13. Elpiner, Leonid I. 1999, ibid.
14. Medicins Sans Frontiers 2003, 'The Aral Sea disappears while tuberculosis climbs', 19 March 2003 https://www.msf.org/aral-sea-disappears-while-tuberculosis-climbs (accessed 27 May 2023).
15. Premiyak, Lisa 2019, 'Selfie sticks at the Aral Sea: another example of disaster tourism's self-indulgence?', *Calvert Journal*, 21 June 2019.

16. Lorenz, Taylor 2019, 'There's Nothing Wrong With Posing for Photos at Chernobyl', *The Atlantic*, 12 June 2019.
17. Sophocles c. 441 BCE, 'Antigone' in *The Three Theban Plays*, trans. Robert Fagles, Penguin 1984.
18. Novitskiy, Z.B. 2012a, 'Phytomelioration in the Southern Aralkum', pp. 13–35 in *Aralkum – a Man-Made Desert*, Ecological Studies Volume 218 eds. Siegmar-W. Breckle et al., Springer: Berlin. DOI: 10.1007/978-3-642-21117-1_16.
19. Sheraliyev, Normukhamad 2015, 'Restoration of degraded land through afforestation of the dried out Aral Sea bed', presentation by International Fund for Saving the Aral Sea at 37th Session Joint FAO/UNECE Working Party on Forest Statistics, Economics and Management, 18–20 March 2015.
20. Sheraliyev, Normukhamad 2015, ibid.
21. Novitsky, Z.B. 2012a, 'Afforestation of the Aral Sea dried floor', presentation at International Workshop, Konya, Turkey, 28–30 May 2012. Online at http://www.fao.org/3/a-bl687e.pdf (accessed 27 May 2023).
22. Lipton, Gabrielle 2019, 'In Uzbekistan, a lowly tree afforests a lost sea', *Landscape News*, Global Landscapes Forum, 23 April 2019.
23. Human Rights Watch report into violence from Rickleton, Chris 2022, 'Uzbekistan's Deadly Crackdown In Karakalpakstan 'Unjustifiable,' Says Rights Group', Radio Free Europe / Radio Liberty (rferl.org), 7 November 2022. Numbers of missing: ODF 2022, 'The Shooting of Peaceful Protesters in Karakalpakstan', Open Dialogue Foundation, 3 August 2022. Shamshetov's death: RFE/RL 2022, 'Karakalpak Activist Dies In Custody Four Days After Being Sentenced Over Protests', Radio Free Europe / Radio Liberty (rferl.org), 6 February 2023.
24. Statements via diplomatic news websites: Abuturov, Victor 2022, 'Uzbekistan: In the center of special attention', *Diplomatist.com*, 2 June 2022; *Diplomat Magazine* editorial 2022, 'On the Policy of the Uzbek leadership to support the development of the Republic of Karakalpakstan', *DiplomatMagazine.eu*, 12 July 2022.
25. Lillis, Joanna 2023, 'Storied rave to skip Karakalpakstan after violence', *Eurasianet*, 7 February 2023.

Chapter 6

1. The description in this section is written from three main sources: (1) Rhodes, Richard 2012 (1986), *The Making of the Atomic Bomb* 25th anniversary edition, Simon & Schuster; (2) Bruce Cameron Reed 2015, 'Atomic Bomb: The Story of the Manhattan Project', Morgan & Claypool, DOI: 10.1088/978-1-6270-5991-6; and (3) Bethe, Hans A. 1964, 'Theory of the Fireball', Los Alamos Scientific Laboratory.
2. Bainbridge, Kenneth T. 1976, 'Technical Report: *Trinity*', Los Alamos Scientific Laboratory, p. 60. DOI: 10.2172/5306263
3. Rhodes, Richard 2012, pp. 668–9 – who is quoting from Teller, Edward 1962, *The Legacy of Hiroshima*, Doubleday, p. 17.

4. Rhodes, Richard 2012, p. 672 – who is quoting from Rabi, Isidor 1970, *Science: The Centre of Culture*, pub. World, p. 138.

5. Frisch, Otto R. 1980, *What Little I Remember*, Cambridge University Press.

6. Groves, Leslie Richard, 1945, 'Report on the Trinity Test', War Department, Washington DC, 18 July 1945.

7. Oppenheimer's hat is often written up as a fedora, but this is incorrect: it's too wide-brimmed. The Fedora Lounge online forum has scrutinised it and report that it is most likely a western hat with a 'rebashed' telescope crease. He wore it pretty hard – but, the hat guys reckon, with a certain élan. https://www.thefedoralounge.com/threads/j-robert-oppenheimer-fedora.47932/ (accessed 21 May 2023).

8. Rhodes, Richard 1986, p. 675 – who is quoting from Rabi, Isidor 1970, *Science: The Centre of Culture*, pub. World p. 138.

9. Rhodes, Richard 1986, ibid, p. 675 – who is quoting from Szasz, Ferenc Morton 1984, *The Day the Sun Rose Twice*, University of New Mexico Press, p. 91.

10. Rhodes, Richard 1986, ibid, p. 676 – who is quoting from Giovanitti, Len and Freed, Fred 1965, *The Decision to Drop the Bomb*, Coward-McCann, p. 197.

11. Rhodes, Richard 1986, ibid – who is quoting from Oppenheimer, J. Robert 1946, 'The atom bomb and college education', *The General Magazine and Historical Chronicle*, University of Pennsylvania General Alumni Society, p. 265.

12. Rhodes, Richard 1986, ibid, p. 675 – who is quoting from Wilson, Jane (ed.) 1975, *All in our Time: The Reminiscences of Twelve Nuclear Pioneers*, Bulletin of the Atomic Scientists, p. 230.

13. See discussion in Wellerstein, Alex 2015, 'The First Light of Trinity', *New Yorker*, 16 July 2015.

14. Interviewed by Robert J. Lifton in his 1962 study of *hibakusha* (bomb survivors) and quoted widely in his subsequent writing, e.g. Lifton, Robert J. 2011, *Witness to an Extreme Century: A Memoir*, Free Press, New York, p. 11.

15. Groves, Leslie R. 1963, *Now It Can Be Told: The Story of the Manhattan Project*, André Deutsch, London, p. 286.

16. Wellerstein, Alex 2020. 'Counting the dead at Hiroshima and Nagasaki', *Bulletin of the Atomic Scientists*, 4 August 2020 – citing the Radiation Effects Research Foundation's Life Span Study.

17. Perrow, Charles 2013. 'Nuclear denial: From Hiroshima to Fukushima', *Bulletin of the Atomic Scientists*, 69(5), 56–67. DOI: 10.1177/0096340213501369.

18. IPPNW (International Physicians for the Prevention of Nuclear War and the Institute for Energy and Environmental Research) 1991, *Radioactive Heaven and Earth The health and environmental effects of nuclear weapons testing in, on, and above the earth*, Apex Press, New York.

19. Hales, Peter B. 1991, 'The Atomic Sublime', *American Studies* 32(1): 5–31. http://www.jstor.org/stable/40642424.

20. Congressional Record 1995, 'THE NUCLEAR AGE'S BLINDING DAWN', vol. 141 issue 115 pages S10082-S10085, 17 July 1995, US Government Publishing Office (govinfo.gov).

21. Vincent, Bill 1984, 'Mine Strafed, Bombed By Air Force', attributed to *Western Sportsman* newspaper August 1984 (I believe actually *Western Outdoors: The Magazine for the Western Sportsman*) – accessed via p. 134 of United States Senate Committee on Energy and Natural Resources, 1987, 'Land Withdrawals from the Public Domain for Military Purposes', Hearing before the Subcommittee on Public Lands, Reserved Water and Resource Conservation, 99th Congress Second Session on S. 2412 a Bill to Withdraw and Reserve Certain Public Lands, 17 July 1986 (available online).

22. I am paraphrasing Joseph Masco's insight here: 'As Blanchot (1995) suggests, this effort to think through the disaster colonized everyday life as well as the future, while fundamentally missing the actual disaster.' From '"Survival Is Your Business": Engineering Ruins and Affect in Nuclear America', *Cultural Anthropology*, May 2008, 23(2): pp. 361–39. https://www.jstor.org/stable/pdf/20484507.pdf.

23. Stafford Warren to Major General Groves, quoted in 'Report on Test II at Trinity, 16 July 1945', 21 July 1945, Department of Energy Open-Net, Nuclear Testing Archive. Available online via the National Security Archive, Document 9 in '77th Anniversary of Hiroshima and Nagasaki Bombings: Revisiting the Record' collection, online at https://nsarchive.gwu.edu/document/28687-document-9-stafford-warren-major-general-groves-report-test-ii-trinity-16-july-1945 (accessed 11 June 2023).

24. 'Downwinder' is a name adopted by nuclear affected people exposed or presumed to be exposed to radiation from the explosion of nuclear devices at the Nevada Test Site, and subsequently others exposed to ionising radiation from nuclear mining and waste disposal as well. Quote from Downwinders of Utah Archive 2017, 'Mary Dickson Downwinders interview' 11 January 2017, interviewers Tony Sams and Justin Sorensen, J. Willard Marriott Library, University of Utah. Online at https://collections.lib.utah.edu/details?id=1246433 (accessed 11 June 2023).

25. Fuller, John G. 1984, *The Day We Bombed Utah: America's Most Lethal Secret*, New American Library: New York & Ontario, p. 6.

26. Letter from Daniel & Martha R. Sheahan to Attorney General William Rogers, 2 March 1959, reproduced in full in Rogoway, Tyler 2015, 'The Unlikely Struggle Of The Family Whose Neighbor Is Area 51', *Jalopnik* magazine, 9 November 2015.

27. Carr Childers, Leisl 2015, *The Size of the Risk: Histories of Multiple Use in the Great Basin*, University of Oklahoma Press, p. 66.

28. Lauriston Taylor quoted in Caulfield, Catherine 1990, *Multiple Exposures: Chronicles of the Radiation Age*, University of Chicago Press, p. 67.

29. Fuller, John G. 1984, ibid, p. 6.

30. Carr Childers, Leisl 2015, ibid, p. 79.

31. National Cancer Institute 1997, *Exposure of the American People to Iodine-131 from Nevada Nuclear-Bomb Tests: Review of the National Cancer Institute Report and Public Health Implications*. National Academies Press, Washington DC. Available from: https://www.ncbi.nlm.nih.gov/books/NBK100833/.

32. Carr Childers, Leisl 2015, ibid, pp. 79–80.

33. Fuller, John G. 1984, ibid, p. 3.

34. Dialogue from PRX 2004, 'Downwinder Diaries Part 2: Son of Bitchin Bomb Went Off'. Produced by Claes Andreasson, 28 May 2004, online at https://beta.prx.org/stories/1451 (accessed 11 June 2023).

35. Fuller, John G. 1984, ibid, pp.15–16.

36. Downwinders of Utah Archive 2017, 'Mary Dickson Downwinders interview' 11 January 2017, ibid.

37. Masco, Joseph 2008, '"SURVIVAL IS YOUR BUSINESS": Engineering Ruins and Affect in Nuclear America', *Cultural Anthropology*, 23: 361–398. DOI: 10.1111/j.1548-1360.2008.00012.x.

38. United States Department of Energy 1955, *Atomic Tests In Nevada: The Story of AEC's Continental Proving Ground,* film, 25 minutes, colour. Available on the Internet Archive https://archive.org/details/AtomicTestsInNevada.

39. Masco, Joseph 2008, pp. 370, 376.

40. Fjelsted's recollections of distance conflict with military accounts. The military factsheet for Operation Plumbob says that troops in trenches were 'no closer than 2,600 yards from Ground Zero', or 2,377m (7,800 ft). Defense Nuclear Agency undated, 'Fact Sheet: Operation Plumbob', See https://www.nrc.gov/docs/ML1233/ML12334A808.pdf (accessed 11 June 2023).

41. Downwinders of Utah Archive 2017, 'Russell Fjelsted Downwinders interview' 13 July 2017, interviewers Anthony Sams, J. Willard Marriott Library, University of Utah. Online at https://collections.lib.utah.edu/ark:/87278/s6vm8nbv (accessed 11 June 2023).

42. U.S. Federal Civil Defense Administration (FCDA) 1995, 'Operation Cue', film, 15 minutes, colour. Quote at 03:11. Widely available online, e.g. https://youtu.be/-w78tdFxog8.

43. Downwinders of Utah Archive 2017, 'Sherre Finicum-H Downwinders interview' 14 July 2017, interviewers Anthony Sams, J. Willard Marriott Library, University of Utah. Online at https://collections.lib.utah.edu/ark:/87278/s60c94t4 (accessed 11 June 2023).

44. Hales, Peter B. 1991, ibid.

45. 'The History of the Bikini' photo essay, *Time* magazine, 3 July 2009.

46. National Nuclear Security Administration 2013, Nevada National Security Site information sheet 'Miss Atomic Bomb' August 2013. Available online at https://www.nnss.gov/docs/fact_sheets/doenv_1024.pdf (accessed 11 June 2023).

47. IPPNW (International Physicians for the Prevention of Nuclear War) 1991, *Radioactive Heaven and Earth The health and environmental effects of nuclear weapons testing in, on, and above the earth*, Apex Press, New York.

48. National Research Council 1999. *Exposure of the American people to Iodine-131 from Nevada nuclear bomb tests: review of the National Cancer Institute Report and Public Health Implications*, Washington, DC.

49. See also Keith Meyers for the US ('340,000 to 460,000 excess deaths from 1951 to 1973'), in 'Some Unintended Fallout from Defense Policy: Measuring the Effect of Atmospheric Nuclear Testing on American Mortality Patterns', 24 October 2017, available online https://cms.qz.com/wp-content/

uploads/2017/12/6043f-meyers-fallout-mortality-website.pdf (accessed 11 June 2023), and Jun Takada's analysis for China (at least 194,000 people dead from acute radiation exposure alone, and 1.2 million cancer cases) – quoted in Merali, Zeeya 2009, 'Did China's Nuclear Tests Kill Thousands and Doom Future Generations?', *Scientific American* 1 July 2009.

50. Gillies, Michael & Haylock, Richard 2022, 'Mortality and cancer incidence 1952–2017 in United Kingdom participants in the United Kingdom's atmospheric nuclear weapon tests and experimental programmes', *Journal of Radiological Protection* 42(2). DOI: 10.1088/1361-6498/ac52b4.

51. Discussion at the LABRATS seminar (Manchester, 19 February 2022) and past media coverage, e.g. Townsend, Mark 2008 'Dying crew of atomic test ship battle MoD for compensation', *The Guardian*, 6 January 2008.

52. See media coverage: *Warrington Guardian* 2012, 'Radiation claim denied for war veteran', 22 March 2012, and Boniface, Susie 2021, 'UK's nuclear test veterans "were victims of a crime" with one suffering 100 tumours', *Daily Mirror*, 15 December 2021.

53. Gillies, Michael & Haylock, Richard 2022, ibid.

54. Alexis-Martin, Becky 2019, *Disarming Doomsday: The Human Impact of Nuclear Weapons since Hiroshima*, Pluto Press, p. 67.

55. Downwinders of Utah Archive 2017, 'Alva Matheson Downwinders interview' 13 July 2017, interviewers Anthony Sams, J. Willard Marriott Library, University of Utah. Online at ttps://collections.lib.utah.edu/ark:/87278/s6zd2bzs (accessed 11 June 2023).

56. Yong, Ed 2022, 'How Did This Many Deaths Become Normal?' *The Atlantic*, 8 March 2022.

57. Sodaro, Amy 2018, *Exhibiting Atrocity: Memorial Museums and the Politics of Past Violence*, Rutgers University Press, p. 116. DOI: 10.36019/9780813592176.

58. Between 1948 and 1976, New Mexico mines produced 40% of America's national production of U_3O_8 – from Rautman, Christopher 1977, 'The Uranium Industry in New Mexico', New Mexico Bureau of Mines and Mineral Resources, March 1977, p. 11.

59. Figures from USEPA, undated, 'Superfund site: Jackpile-Paguate Uranium Mine, Laguna Pueblo, NM', Environmental Protection Agency, online at https://cumulis.epa.gov/supercpad/cursites/csitinfo.cfm?id=0607033 (accessed 27 May 2023).

60. Purley, Dorothy Ann 1995, 'The Jackpile Mine: Testimony of a Miner', *Race, Poverty & the Environment* 5(3/4): 16–17. https://www.jstor.org/stable/41554896.

61. Fox, Sarah Alizabeth 2014, *Downwind: A People's History of the Nuclear West*, University of Nebraska Press, p. 27.

62. Lorenzo, June 2019, 'Gendered Impacts of Jackpile Uranium Mining on Laguna Pueblo', *International Journal of Human Rights Education*, 3(1). Retrieved from https://repository.usfca.edu/ijhre/vol3/iss1/3.

63. Fox, Sarah Alizabeth 2014, ibid, p. 49.

64. Dawson, Susan E. 1992, 'Navajo uranium workers and the effects of occupational illness: a case study', *Human Organisation* 51(4):389–97, p. 392.

65. ACRO 2021, 'Nuage de sable du Sahara: une pollution radioactive qui

revient comme un boomerang', *L'Association Pour le Contrôle de la Radioactivité Dans l'Ouest*, Communiqué 24 February 2021.

66. Hawa, Maïa Tellit 2022, 'Sahara Mining: The Wounded Breath Of Tuareg Lands', *The Funambulist*, 18 October 2022.

67. Collin, Jean-Marie and Bouveret, Patrice 2020, *Radioactivity Under the Sand: The Waste From French Nuclear Tests in Algeria*, Heinrich Böll Foundation, July 2020, pp.19–20.

68. Bryant Paris, Elizabeth 2020, 'French nuclear tests in Algeria leave bitter fallout', *Deutsche Welle* 13 February 2020.

69. The documentary films 'At(h)ome' (Elisabeth Leuvrey, 2013), and 'Algeria, De Gaulle and the Bomb' and 'Sand Storm: the Sahara of the Nuclear Tests' (Larbi Benchiha, both 2010). See Collin & Bouveret 2020, ibid. p. 39.

70. Collin & Bouveret 2020, ibid, p. 14.

71. Two sites, those of the Amethyst test and the Pollen series, were decontaminated, which entailed the most radioactive ground being covered with sand and tarmacked over – though Collin & Bouveret cast doubts as to whether this was done sufficiently well (pp. 38–9).

72. *The Maghreb Times* (unbylined) 2021, 'Nuclear Tests: In Algeria Too, France Called to Act', 29 July 2021.

73. Collin & Bouveret 2020, ibid, p. 14.

74. Bannerjee, Mita 2017, 'Nuclear testing and the 'terra nullius doctrine': from life sciences to life writing', in *Property, Place, and Piracy*, ed. James Arvanitakis and Martin Fredriksson, Routledge.

75. Hawad, Maïa Tellit 2022, 'Sahara Mining: The Wounded Breath of Tuareg Lands' in *The Funambulist* issue 44 'The Desert', 18 October 2022.

76. Johnnye Lewis, Melissa Gonzales, Courtney Burnette, Malcolm Benally, Paula Seanez, Christopher Shuey, Helen Nez, Christopher Nez & Seraphina Nez 2015, 'Environmental Exposures to Metals in Native Communities and Implications for Child Development: Basis for the Navajo Birth Cohort Study', *Journal of Social Work in Disability & Rehabilitation*, 14(3–4): 245–269. DOI: 10.1080/1536710X.2015.1068261.

77. Lin, Yan, Hoover, J., Beene, D. et al. 2020, 'Environmental risk mapping of potential abandoned uranium mine contamination on the Navajo Nation, USA, using a GIS-based multi-criteria decision analysis approach', *Environmental Science Pollution Research* 27: 30542–57. DOI: 10.1007/s11356-020-09257-3.

78. Various publications from the team – one being: Hayek EE, Medina S, Guo J, Noureddine A, Zychowski KE, Hunter R, Velasco CA, Wiesse M, Maestas-Olguin A, Brinker CJ, Brearley AJ, Spilde MN, Howard T, Lauer FT, Herbert G, Ali AS, Burchiel SW, Campen M, Cerrato JM. 2021, 'Uptake and toxicity of respirable carbon-rich uranium-bearing particles: Insights into the role of particulates in uranium toxicity', *Environmental Science & Technology* 55(14):9949–57. DOI: 10.1021/acs.est.1c01205.

79. UNM METALS 'Modulation of Uranium and Arsenic Immune Dysregulation by Zinc', investigators Debra MacKenzie, Laurie Hudson, & Eszter Erdei. Details online: https://hsc.unm.edu/pharmacy/research/areas/metals/projects.html (accessed 11 June 2023).

80. See US Department of Energy, Legacy Management 2014, *Defense-Related*

Uranium Mines Cost and Feasibility Topic Report: Final Report, Doc. No. S10859, June 2014, online at https://www.energy.gov/sites/prod/files/2017/07/f35/S10859_Cost.pdf (accessed 11 June 2023).

81. See Malcom Ferdinand's *Decolonial Ecology: Thinking from the Caribbean World*, 2022, Polity Press on the 'colonial and environmental double fracture of modernity.'

82. James, Robin 2016, tweet on 4 October 2016, https://twitter.com/doctaj/status/783293051712237568. Abbreviated for Twitter character limit; written out in full here. 'This also changes what justice looks like: not access to governing, but fair distribution of care,' James added.

Chapter 7

1. Reported by British naval officer Sir John Scott's 1818 expedition to find the Northwest Passage – widely quoted, e.g. Durey, Michael 2008, 'Exploration at the Edge: Reassessing the Fate of Sir John Franklin's Last Arctic Expedition', *The Great Circle* 30(2): 3–40.

2. Story told by the Greenlander Arnajaq to explorer Knud Rasmussen (recorded in his 'The People of the Polar North: a Record', 1908, pp. 104–5) and also reported by the turn-of-the-twentieth century ethnographers Franz Boas and Singe Rink. The wider context is a story known as 'The Girl and the Dogs', which is the Greenlandic origin story for white people.

3. Reporting as iron: 'Red snow from the Arctic regions', *The Times*, 4 December 1818, p. 2. Discovery as algae: Sir John Ross, 1819, 'A Voyage of Discovery Made Under the Orders of the Admiralty, in His Majesty's Ships Isabella and Alexander, for the Purpose of Exploring Baffin's Bay, and Enquiring Into the Probability of a North-west Passage. Volume 2', Appendix V pp. 181–4 (pub. Strahan & Spottisbrooke), available open access on Google Books.

4. Nordenskiöld, A.E. 1875, 'Cryoconite found 1870, July 19th–25th, on the inland ice, east of Auleitsivik fjord, Disco Bay, Greenland', *Geology* magazine 2(2): 157–162.

5. Nansen, Fritjof 1906, *The Norwegian North Polar Expedition 1893–1896: Scientific Results*, Longmans, Green and Co., London. Cited in Cook, Joseph et al. 2016, 'Cryoconite: The dark biological secret of the cryosphere', *Progress in Physical Geography: Earth and Environment*, 40(1): 66–111. DOI: 10.1177/0309133315616574.

6. Maurette, M., Jéhanno, C., Robin, E. et al. 1987, 'Characteristics and mass distribution of extraterrestrial dust from the Greenland ice cap', *Nature* 328: 699–702. DOI: 10.1038/328699a0.

7. Tedesco, Mario 2020, *Ice: Tales from a Disappearing Continent*, Headline, Chapter 7, p. 91 onwards.

8. The original website (darksnow.org) is now down, but information about the project is available via the Internet Archive (http://web.archive.org/web/20150109120350/http://darksnow.org/about-the-august-2014-dark-

greenland-photos) and NASA: https://earthobservatory.nasa.gov/images/84607 /dark-snow-project (accessed 11 June 2023).

9. Holthaus, Eric 2016, 'Why Greenland's "Dark Snow" Should Worry You', *Slate*, 16 September 2014.

10. Tedesco, Mario 2020, ibid, p. 38.

11. Cuffey, Kurt & Paterson, WSB 2010, *The Physics of Glaciers* 4th edition, Elsevier, table 5.2 – cited in Box, Jason E. et al. 2012, 'Greenland ice sheet albedo feedback: thermodynamics and atmospheric drivers', *The Cryosphere* 6: 821–39, DOI: 10.5194/tc-6-821-2012

12. Bullard, Joanne E et al. 2016, 'High-latitude dust in the Earth system', *Reviews of Geophysics* 54: 447– 85, DOI: 10.1002/2016RG000518.

13. Cuffey, Kurt & Paterson, WSB 2010, ibid.

14. Marie Dumont to Christa Marshall (Climatewire) 2014, 'Wind-Blown Dust Darkens Greenland, Speeding Meltdown', *Scientific American*, 10 June 2014.

15. Jim McQuaid to Harry Baker, 'Mystery of Greenland's expanding 'dark zone' finally solved', *LiveScience*, 27 January 2021.

16. Dumont, Marie et al. 2014, 'Contribution of light-absorbing impurities in snow to Greenland's darkening since 2009', *Nature Geoscience* 7: 509–12. DOI: 10.1038/ngeo2180.

17. McCutcheon, Jenine et al. 2021, 'Mineral phosphorus drives glacier algal blooms on the Greenland Ice Sheet' *Nature Communications* 12, article no. 570. DOI: 10.1038/s41467-020-20627-w.

18. National Snow & Ice Data Center (NSIDC), 2013, 'An intense Greenland melt season: 2012 in review', 5 February 2013, University of Colorado Boulder. Online at http://nsidc.org/greenland-today/2013/02/greenland-melting-2012-in-review/ (accessed 11 June 2023).

19. Sato, Yaosuke et al. 2016, 'Unrealistically pristine air in the Arctic produced by current global scale models', *Nature Scientific Reports* 6, article no. 26561. DOI: 10.1038/srep26561 and public communications, 'Current atmospheric models underestimate the dirtiness of Arctic air', *ScienceDaily*, 25 May 2016.

20. Keegan, Kaitlin M. et al. 2012, 'Climate change and forest fires synergistically drive widespread melt events of the Greenland Ice Sheet', *PNAS* 111(22): 7964–7. DOI: 10.1073/pnas.140539711.

21. Sea level rise: IPCC, 2019, Chapter 4 'Sea Level Rise and Implications for Low-Lying Islands, Coasts and Communities', *IPCC Special Report on the Ocean and Cryosphere in a Changing Climate*. Ice sheet contribution: Thomas Slater et al., 2020, 'Ice-sheet losses track high-end sea-level rise projections', *Nature Climate Change*, vol 10, pp. 879–81. DOI: 10.1038/ s41558-020-0893-y.

22. Rahmstorf, S., Box, J., Feulner, G. et al. 2015, 'Exceptional twentieth-century slowdown in Atlantic Ocean overturning circulation', *Nature Climate Change* 5: 475–80. DOI: 10.1038/nclimate2554 (especially the final paragraph).

23. U.S. National Oceanic and Atmospheric Administration (NOAA) 2020, *Arctic Report Card: Update for 2020*, section 'Vital Signs: Greenland Ice Sheet'. DOI: 10.25923/ms78-g612.

24. Lorenz, Edward N. 1972, 'Predictability: Does the Flap of a Butterfly's Wings in Brazil Set Off a Tornado in Texas?' Presentation to the American Association for the Advancement of Science, 29 December 1972. Lorenz had apparently first used the example of a seagull causing a storm – but made the example more poetic following a suggestion from colleagues.

25. Sermeq Kujalleq has slowed down from an annual average high of ~52 Gt yr^{-1} in 2012 to ~45 Gt yr^{-1} in 2016 and ~38 Gt yr^{-1} in 2017, likely due to ocean cooling. See Khazendar, A. et al. 2019, 'Interruption of two decades of Jakobshavn Isbrae acceleration and thinning as regional ocean cools, *Nature Geoscience* 12: 277–83, DOI: 10.1038/s41561-019-0329-3. Ice is a little less dense than water (that's why it floats) – so a gigaton of ice (a billion tons) is about 1.09 km^3 in volume. It's a staggering amount.

26. Ruskin, John 1857, *Modern Painters* vol. 1, Chapter 3 'Of the Sublime', pub. Smith, Elder, and Company, p. 40.

27. Morton, Timothy 2013, *Hyperobjects: Philosophy and Ecology after the End of the World*, University of Minnesota Press.

28. Greene, Mott T. 2015, *Alfred Wegener: Science, Exploration, and the Theory of Continental Drift*, John Hopkins University Press.

29. McCoy, Roger M. 2006, *The Ending in Ice: The Revolutionary Idea and Tragic Expedition of Alfred Wegener*, Oxford University Press.

30. Schwartzbach, Martin 1986, *Alfred Wegener, the Father of Continental Drift*, Scientific Revolutionaries series, Springer, p. 44.

31. Details in this paragraph from Langway, Chester C. 2008, 'The history of early polar ice cores', *Cold Regions Science and Technology* 52(2): 101–17, DOI: 10.1016/j.coldregions.2008.01.001 and Talalay, Pavel G. 2016, *Mechanical Ice Drilling Technology*, Springer Geophysics series, Springer (particularly pp. 59–64 and 76–77).

32. Regarding pollution: Colgan, William et al. 2016, 'The abandoned ice sheet base at Camp Century, Greenland, in a warming climate', *Geophysical Research Letters* 43: 8091–6, DOI: 10.1002/2016GL069688. Re. the ice core: Dansgaard, Willi 2005, *Frozen Annals – Greenland Ice Sheet Research*, Niels Bohr Institute, Copenhagen, p. 54 (online: https://icedrill.org/sites/default/files/FrozenAnnals.pdf, accessed 11 June 2023).

33. Ice core initially thought to cover 740,000 years, then subsequent work suggested 800,000 years. EPICA community members 2004, 'Eight glacial cycles from an Antarctic ice core', *Nature* 429, 623–8, DOI: 10.1038/nature02599 and Lambert, F. et al 2008, 'Dust-climate couplings over the past 800,000 years from the EPICA Dome C ice core', *Nature* 452: 616–19, DOI: 10.1038/nature06763.

34. Castellani, Benjamin B. et al. 2015, 'The annual cycle of snowfall at Summit, Greenland', *Journal of Geophysical Research: Atmospheres* 120: 6654–68. DOI: 10.1002/2015JD023072.

35. Voosen, Paul 2017, 'Record-shattering 2.7-million-year-old ice core reveals start of the ice ages', *Science*, 15 August 2017.

36. Alley, Richard B. 2000b, *The Two–Mile Time Machine – Ice Cores, Abrupt Climate Change & Our Future: Ice Cores, Abrupt Climate Change, and Our Future*, Princeton University Press, p. 7.

37. From conversation with Pete Neff of University of Minnesota, Twitter (https://twitter.com/icy_pete/status/1486444367158161418), 26 January 2022. At proof date in June 2023, EastGRIP had made it to 100,000-year-old ice but was still '50m from bedrock', team member Nico Stoll reported on Twitter (https://twitter.com/stoll_nico/status/1666883921970188288).

38. Interview with Eric Wolff, 7 April 2020. See also Neff, Pete 2014, 'A review of the brittle ice zone in polar ice cores', *Annals of Glaciology* 55(68): 72–82. DOI: 10.3189/2014AoG68A023.

39. Alley, Richard B. 2000b, ibid, p. 44.

40. Alley, Richard B. 2000b, ibid.

41. Unsurprisingly this varies by location and time. For 2x to 20x more dust: Claquin et al. 2003, 'Radiative forcing of climate by ice-age atmospheric dust', Climate Dynamics 20: 193–202. DOI: 10.1007/s00382-002-0269-1. The EPICA core in Antarctica has reported 25x more dust: Lambert, F. et al. 2008, ibid.

42. McConnell, Joseph R. et al. 2018, 'Lead pollution recorded in Greenland ice indicates European emissions tracked plagues, wars, and imperial expansion during antiquity', *PNAS* 115 (22): 5726–31, DOI: 10.1073/pnas.172181811.

43. Niels Bohr Institute Centre for Ice and Climate, undated, 'Continuous Flow Analysis (CFA) impurity measurements' webpage, University of Copenhagen, Denmark. Now offline, accessed via the Internet Archive (URL https://www.iceandclimate.nbi.ku.dk/research/drill_analysing/cutting_and_analysing_ice_cores/cfa/, page date 7 May 2022).

44. Peter Neff, email correspondence, 5 February 2022.

45. Dansgaard, Willi et al. 1969, 'One Thousand Centuries of Climatic Record from Camp Century on the Greenland Ice Sheet', *Science* 166(3903): 377–81, DOI: 10.1126/science.166.3903.377.

46. Hays. J. D. et al., 1976 'Variations in the Earth's Orbit: Pacemaker of the Ice Ages', *Science* 194(4270): 1121–32, DOI: 10.1126/science.194.4270.1. A somewhat overenthusiastic write-up of this research in *Newsweek* in 1975 – alongside findings on the impact of black carbon soot on shielding sunlight – produced a panic about 'Global Cooling' and an incipient new ice age that has dogged climate science as misinformation for almost half a century. The original authors (JD Hays et al.) had been explicit that 'extensive Northern Hemisphere glaciation' was only a trend 'ignoring anthropogenic effects'. See Doug Struck 2014, 'How the "Global Cooling" Story Came to Be', *Scientific American*, 10 January 2014.

47. Buckley, Brendan M. et al. 2010, 'Climate as a contributing factor in the demise of Angkor, Cambodia', *PNAS* 107(15) 6748–52, DOI: 10.1073/pnas.0910827107.

48. Alley, Richard B. 2000a, 'Ice-core evidence of abrupt climate changes', *PNAS* 97(4): 1331–34. DOI: 10.1073/pnas.97.4.1331.

49. Information in this and the next paragraph taken from Richard B. Alley 2000b ibid. and KC Taylor et al, 1997, 'The Holocene-Younger Dryas Transition Recorded at Summit, Greenland', *Science* 278(5339): 825–7, DOI: 10.1126/science.278.5339.825. NB Before Present dating conventions set the 'present' as 1950 CE.

50. Dansgaard, Willi et al. 1993, 'Evidence for general instability of past climate from a 250-kyr ice-core record', *Nature* 364 (6434): 218–20. DOI: 10.1038/364218a0.

51. Christ, Andrew J. et al. 2021, 'A multimillion-year-old record of Greenland vegetation and glacial history preserved in sediment beneath 1.4 km of ice at Camp Century', *PNAS* 118 (13) DOI: 10.1073/pnas.2021442118. See also Andrew Christ & Paul Bierman 2021, 'Ancient leaves preserved under a mile of Greenland's ice – and lost in a freezer for years – hold lessons about climate change', *The Conversation*, 15 March 2021.

52. IPCC 2001, *Climate Change 2001: Impacts, Adaptation, and Vulnerability: Contribution of Working Group II to the Third Assessment Report of the Intergovernmental Panel on Climate Change*, eds. James J. McCarthy et al, Cambridge University Press.

53. Lenton, Timothy M. et al., 2019, 'Climate tipping points – too risky to bet against', *Nature* 575: 592–5, DOI: 10.1038/d41586-019-03595-0.

54. Boers, Niklas & Rypdal, Martin 2021, 'Critical slowing down suggests that the western Greenland Ice Sheet is close to a tipping point', *PNAS* 118(21), DOI: 10.1073/pnas.2024192118.

55. Hooijer, A., Vernimmen, R. 2012, 'Global LiDAR land elevation data reveal greatest sea-level rise vulnerability in the tropics', Nature Communications 12, 3592. DOI: 10.1038/s41467-021-23810-9.

56. For the Torres Strait Islanders case, see 'Daniel Billy and others v Australia (Torres Strait Islanders Petition' decided 22 September 2022. See also 'Ioane Teitiota v. The Chief Executive of the Ministry of Business, Innovation and Employment' regarding a man from the Pacific nation of Kiribati who applied for refugee status in New Zealand in 2013, denied in 2015, which he then contested. In 2020, the UN Human Rights Committee stated that countries may not deport individuals who face climate change-induced conditions that violate the right to life.

Chapter 8

1. 'Then, touch'd with grief, the weeping Heavens distill'd / A shower of blood o'er all the fatal field' in Homer's *Iliad*, book 16, as translated by Alexander Pope. Pliny, *Historia Naturalis*, Book II, LVII. Plutarch, *The Parallel Lives: The Life of Romulus*.

2. Smith, Oli 2018, "BLOOD RAIN" pours down in Russian freak weather event prompting "biblical plague" panic", *Daily Express*, 5 July 2018.

3. Cicero, *De divinatione* ('Concerning Divination') Book II, 27. He's marvellously skeptical: '"Reports," you say, "were made to the Senate that there was a shower of blood, that the river Atratus actually flowed with blood and that the statues of the gods dripped with sweat." You do not think for a moment that Thales, Anaxagoras, or any other natural philosopher would have believed such reports? Sweat and blood you may be sure do not come except from animate bodies. An effect strikingly like blood is produced by the admixture of water with certain kinds of soil.'

4. McCafferty, P. 2008, 'Bloody rain again! Red rain and meteors in history and myth' in the *International Journal of Astrobiology*, 7(1), 9–15. DOI: 10.1017/S1473550407003904.

5. '24 million tons': NASA's Hongbin Yu to Warren Cornwall, 2020, "Godzilla' dust storm traced to shaky northern jet stream', *Science*, 7 December 2020.

6. Chadwick, Jonathan 2020, 'Amazing satellite imagery shows giant dust plume known as 'Godzilla' sweeping across the Atlantic from the Sahara to the Caribbean', *Daily Mail*, 14 July 2020, and Irfan, Umair 2020, 'The "Godzilla" Saharan dust cloud over the US, explained', *Vox*, 1 July 2020.

7. von Humboldt, Alexander and Bonpland, Aimé 1807, '*Essay on the Geography of Plants*, ed. Stephen T. Jackson, trans. Sylvie Romanowski, University of Chicago Press.

8. Steffen, Will et al. 2020 'The Emergence and Evolution of Earth System Science' in *Nature Reviews Earth & Environment* 1, pp. 54–63. DOI: 10.1038/s43017-019-0005-6.

9. Inoue, Cristina Yumie Aoki & Moreira, Paula Franco 2016. 'Indigenous knowledge systems and the Earth System governance project's epistemological dimension', presented at Nairobi Conference on Earth System Governance, December 2016.

10. Cited as '5 x10^3 Tg yr^{-1}' of particles up to 20 micrograms (PM20), in Kok, Jasper F. et al., 2021, 'Improved representation of the global dust cycle using observational constraints on dust properties and abundance', *Atmospheric Chemistry & Physics* 21: 8127–67. DOI: 10.5194/acp-21-8127-2021.

11. Ginoux, Paul et al. 2012, 'Global-scale attribution of anthropogenic and natural dust sources and their emission rates based on MODIS Deep Blue aerosol products', Reviews of Geophysics 50. DOI: 10.1029/2012RG000388.

12. Darwin, Charles 1846, 'An account of the fine dust which often falls on vessels in the Atlantic Ocean', *Quarterly Journal of the Geological Society* 2: 26–30. DOI: 10.1144/GSL.JGS.1846.002.01-02.09.

13. Barth cited in deMenocal, Peter B. & Tierney, Jessica E. 2012, 'Green Sahara: African Humid Periods Paced by Earth's Orbital Changes', *Nature Education Knowledge* 3(10):12.

14. Alverson, Andrew 2022, 'The Air You're Breathing? A Diatom Made That', *LiveScience*, 14 October 2022.

15. Africa produces '58 per cent of emissions and 62 per cent of total dust in the atmosphere' in Tanaka, Taichu Y. & Chiba, Masaru 2006, 'A numerical study of the contributions of dust source regions to the global dust budget', Global and Planetary Change 52(1–4): 88–104, DOI: 10.1016/j.gloplacha.2006.02.002 – or '55 per cent' of emissions in Ginoux et al. 2012, ibid.

16. NOAA, undated, 'How much oxygen comes from the ocean?' National Ocean Service website (accessed 14 May 2023).

17. Lenes et al. 2001, 'Iron fertilization and the Trichodesmium response on the West Florida shelf', *Limnology and Oceanography*, 46(6): 1261–77. DOI: 10.4319/lo.2001.46.6.1261

18. 'Diversity' from Yamaguchi, Nobuyasu et al. 2012, 'Global dispersion of bacterial cells on Asian dust', *Scientific Reports* 2(525), DOI: 10.1038/

srep00525. 'Adaptation': Isabel Reche et al. 2018, 'Deposition rates of viruses and bacteria above the atmospheric boundary layer. *ISME Journal* 12: 1154–62, DOI: 10.1038/s41396-017-0042-4.

19. Shinn, Eugene A. et al, 2000, 'African Dust and the Demise of Caribbean Coral Reefs', *Geophysical Research Letters* 27(19): 3029–3032, DOI: 10.1029/2000GL011599.

20. NASA Earth Observatory 2001, 'When the dust settles', 18 May 2001, https://earthobservatory.nasa.gov/features/Dust (accessed 14 May 2023).

21. Prospero, Joseph M. et al. 2021, 'The Discovery of African Dust Transport to the Western Hemisphere and the Saharan Air Layer: A History', *Bulletin of the American Meteorological Society* 102: E1239–E1260, DOI: 10.1175/BAMS-D-19-0309.1.

22. IPCC 2007, *Climate Change 2007: Synthesis Report. Contribution of Working Groups I, II and III to the Fourth Assessment Report of the Intergovernmental Panel on Climate Change*, eds. / core writing team Pachauri, R.K and Reisinger, A., IPCC, Geneva, Switzerland, p. 168. Specifically, 'The 90 per cent confidence level is estimated to be ±0.2 W m^{-2}, reflecting the uncertainty in total dust emissions and burdens and the range of possible anthropogenic dust fractions. At the limits of this uncertainty range, anthropogenic dust RF is as negative as -0.3 W m^{-2} and as positive as $+0.1$ W m^{-2}.'

23. Froyd, Karl D. et al. 2022, 'Dominant role of mineral dust in cirrus cloud formation revealed by global-scale measurements', *Nature Geoscience* 15: 177–83, DOI: 10.1038/s41561-022-00901-w.

24. NASA, 2011, 'Glory Promises New View of Perplexing Particles' webpage, 16 February 2011, https://www.nasa.gov/mission_pages/Glory/news/Perplexing_particles.html, and NASA 2010, 'Aerosols and Clouds (Indirect Effects)' webpage, 2 November 2010, https://earthobservatory.nasa.gov/features/Aerosols/page4.php (both accessed 11 June 2023).

25. Kami, Aseel 2012, 'Iraq battles dust from marshes drained by Saddam', *Reuters*, 21 June 2012.

26. Bou Karam Francis, D. et al. 2017, 'Dust emission and transport over Iraq associated with the summer Shamal winds', *Aeolian Research* 24:15–31, DOI: 10.1016/j.aeolia.2016.11.001.

27. '8 million tonnes' as 8 Tg in Hoesley, Rachel M. et al. 2018, 'Historical (1750–2014) anthropogenic emissions of reactive gases and aerosols from the Community Emissions Data System (CEDS)', Geoscientific Model Development 11: 369–408, DOI: 10.5194/gmd-11-369-2018. For 'second most dangerous', see Bond, TC et al. 2013, 'Bounding the role of black carbon in the climate system: A scientific assessment', *Journal of Geophysical Research*: Atmospheres 118: 5380– 5552, DOI:10.1002/jgrd.50171.

28. Regarding clouds: Le Treut, H. et al. 2007, 'Historical Overview of Climate Change Science' in Climate Change 200y6: The Physical Science Basis, eds. Solomon, Qin, Manning et al., p. 114. Regarding soil and vegetation: Xue, Yongkang 1996, 'Impact of vegetation properties on U.S. summer weather prediction', *Journal of Geophysical Research: Atmospheres 101*(D3), DOI: 10.1029/95JD02169.

29. David Thomas to David Shukman 2015, 'Clearing up dust's effect on climate', *BBC News*, 9 December 2015.

30. IPCC, 2007: Summary for Policymakers. In *Climate Change 2007: The Physical Science Basis. Contribution of Working Group I to the Fourth Assessment Report of the Intergovernmental Panel on Climate Change*, eds S. Solomon et al., Cambridge University Press.

31. 'The only tools we have' from project website, 'DO4 Models: Dust Observation', University of Oxford School of Geography and the Environment, online at https://www.geog.ox.ac.uk/research/climate/projects/do4models/ (accessed 14 May 2023).

32. Ryder, Claire L. et al. 2019, 'Coarse and giant particles are ubiquitous in Saharan dust export regions and are radiatively significant over the Sahara', *Atmospheric Chemistry & Physics* 19: 15353–76. DOI: 10.5194/acp-19-15353-2019.

33. Dimitropoulos, S. 2020, 'Have We Got Dust All Wrong?', *Eos*, 25 September 2020.

34. Carnegie Institution for Science, 2018, 'Particulate pollution's impact varies greatly depending on where it originated', *ScienceDaily*, 17 August 2018.

35. Chaibou, Abdoul Aziz Saidou; Ma, Xiaoyan and Sha, Tong 2020, 'Dust radiative forcing and its impact on surface energy budget over West Africa', *Scientific Reports* 10: 12236. DOI: 10.1038/s41598-020-69223-4.

36. Zhao, Alcide; Ryder, Claire L. and Wilcox, Laura J. 2022, 'How well do the CMIP6 models simulate dust aerosols?' *Atmospheric Chemistry & Physics*, 22: 2095–2119. DOI: 10.5194/acp-22-2095-2022.

37. 40 megatons is from Koren 2006 – there are lower figures in other studies, hence 'up to'. Koren, Ilan et al. 2006, 'The Bodélé depression: a single spot in the Sahara that provides most of the mineral dust to the Amazon forest', *Environmental Research Letters* 1: 014005, DOI: 10.1088/1748-9326/1/1/014005.

38. Swap, R. et al. 1992, 'Saharan dust in the Amazon Basin', *Tellus B: Chemical and Physical Meteorology*, 44:2, 133–149, DOI: 10.3402/tellusb.v44i2.15434

39. Koren, Ilan et al. 2006, ibid.

40. Hongbin Yu quoted in Andrew Buncombe 2015, 'Amazon Rainforest: Nasa satellite measures remarkable 1,600 mile journey of dust from Sahara desert to jungle basin', *The Independent*, 26 February 2015.

41. Yu, Yan; Kalashnikova, Olga et al. 2020, 'Disproving the Bodélé depression as the primary source of dust fertilizing the Amazon Rainforest', *Geophysical Research Letters* 47, DOI: 10.1029/2020GL088020.

42. Nogueira, Juliana et al. 2020, 'Dust arriving in the Amazon basin over the past 7,500 years came from diverse sources'. *Communications Earth & Environment* 2(5), DOI: 10.1038/s43247-020-00071-w.

43. Barkley, Anne E. et al. 2019, 'African biomass burning is a substantial source of phosphorus deposition to the Amazon, Tropical Atlantic Ocean, and Southern Ocean', *PNAS* 116 (33) 16216–21, DOI: 10.1073/pnas.1906091116, and Prospero, Joseph M. et al. 2020, 'Characterizing and quantifying African dust transport and deposition to South America: Implications for the phosphorus budget in the Amazon Basin', *Global Biogeochemical Cycles*, 34, DOI: 10.1029/2020GB006536.

44. Biomass burning produces 35% of northern hemisphere black carbon, and

50% of southern hemisphere – from Qi, Ling and Wang, Shuxiao 2019, 'Fossil fuel combustion and biomass burning sources of global black carbon from GEOS-Chem simulation and carbon isotope measurements', *Atmospheric Chemistry and Physics*, 19: 11545–57, DOI: 10.5194/acp-19-11545-2019.

45. Thompson, Erica & Smith, Leonard A. 2019, 'Escape From Model Land', *Economics E-Journal* 13(1), DOI: 10.5018/economics-ejournal.ja.2019-40, and Thompson's subsequent book, '*Escape from Model Land: How Mathematical Models Can Lead Us Astray and What We Can Do about It*', Basic Books, 2022.

46. Law, John 2004, *After Method: Mess in Social Science Research*, Routledge: Oxford.

Chapter 9

1. Eastern Sierra Audubon Society, undated, 'Owens Lake Important Bird Area' webpage (accessed 14 May 2023). Text from National Audubon Society IBA Database Site Profile.

2. Lave, Lester B. & Seskin, Eugene P. 1973, 'An Analysis of the Association between U.S. Mortality and Air Pollution', *Journal of the American Statistical Association*, 68:342, 284–90, DOI: 10.1080/01621459.1973.10482421.

3. Dockery, Douglas W. et al. 1993, 'An Association between Air Pollution and Mortality in Six U.S. Cities', *New England Journal of Medicine* 329:1753–1759, DOI: 10.1056/NEJM199312093292401.

4. Saint-Armand, Pierre et al. 1986, 'Dust Storms From Owens and Mono Valleys, California: Summary report 1975–86'. Naval Weapons Center China Lake CA, September 1986.

5. National Academies of Sciences, Engineering, and Medicine (NASEM) 2020, *Effectiveness and Impacts of Dust Control Measures for Owens Lake*, The National Academies Press, Washington, DC, DOI: 10.17226/25658 – see table on p. 60.

6. NASEM 2020, '*Effectiveness and Impacts of Dust Control Measures for Owens Lake*', ibid, p. 74.

7. Figures from contractor websites: KDG Construction Consulting, 2014 'Owens Lake Dust Mitigation Project, Phase 7A', https://kdgcc.com/projects/owens-lake-dust-mitigation-project-phase-7a/ (now offline; accessed via Internet Archive) and 711 Materials Inc, 2020, 'Dust Pollution: The Transformation Of Owens Lake', https://www.711materials.com/post/dust-pollution-the-transformation-of-owens-lake (accessed 14 May 2023).

8. 'Two months' figure from according to LA Mayor Eric Garcetti in 2014, widely quoted in news, e.g. Sierra Wave news staff 2014, 'LADWP/GBAPCD settle lawsuits over dust control and look forward to more use of waterless control methods without harming wildlife', sierrawave.net, 14 November 2014.

9. 'An average of 31 per cent since 2007' from p.1 of NASEM 2020, '*Effectiveness and Impacts of Dust Control Measures for Owens Lake*', ibid. 'Thirty billion' from Knudson, Tom 2014, 'Outrage in Owens Valley a century after L.A. began taking its water', *The Sacramento Bee*, 1 May 2014.

10. NASEM 2020, '*Effectiveness and Impacts of Dust Control Measures for Owens Lake*', ibid.

11. Owens Valley Committee, 'The Rainshadow' bulletin, vol 4 no. 2, Winter 2008/09.

12. p. 60 of NASEM 2020, '*Effectiveness and Impacts of Dust Control Measures for Owens Lake*', ibid, and Louis Sagahun 2018, 'Owens Lake: Former toxic dust bowl transformed into environmental success', *Los Angeles Times*, 28 April 2018.

13. Maisel, David 2001–2 and 2015, *The Lake Project* (davidmaisel.com).

14. Cronon, William 1995, 'The Trouble With Wilderness', *New York Times*, 13 August 1995.

15. Figures as of April 2019, from p.14, NASEM 2020, '*Effectiveness and Impacts of Dust Control Measures for Owens Lake*', ibid.

16. Non-woven FFP2 masks, the type I had (similar to the US N95 standard), should filter out 94 per cent of solid particles down to around 0.1–0.3 micrometres in size – making them potentially well-suited against this fine PM10 and smaller dust.

17. Davidson, Captain J. W. 1859, 'Report of the Results of an Expedition to Owens Lake, and River' (accessed via p.10 of the *Cultural Landscape Report: Manzanar Historic Site*, 2006, National Park Service (NPS) Pacific West Region, San Francisco).

18. Sagahun, Louis 2013, 'DWP archaeologists uncover grim chapter in Owens Valley history', *Los Angeles Times*, 2 June 2013.

19. NASEM 2020, '*Effectiveness and Impacts of Dust Control Measures for Owens Lake*', ibid, pp. 30 and 54–5, citing previous work by Kathy Bancroft.

20. Land stats from Inyo County Planning Department 2015, 'Managed Land in Inyo County' dataset (https://databasin.org/datasets/edeb-5c36a529484b854ff95ce5aeea5c, accessed 14 May 2023). Historic acreage statistics from Teri Red Owl 2015, 'Payahuunadü Water Story' in *Wading Through the Past: Infrastructure, Indigeneity & the Western Water Archives*, ed. Char Miller, 2021 Western Water Symposium, The Claremont Colleges Library.

21. Office of the Attorney General, 1995, *Memorandum on Indian Sovereignty*, Department of Justice, Washington, DC.

22. Wei, Clarissa 2017, 'After Long Delay, LADWP Fixes Broken Pipeline on Big Pine Paiute Reservation', *KCET*, 28 June 2017.

23. Kahrl, William M. 2013, 'The long shadow of William Mulholland', *Los Angeles Times*, 3 November 2013.

24. Kahrl, William M. 1982, *Water And Power: The Conflict over Los Angeles' Water Supply in the Owens Valley*, University of California Press, p. 439.

25. Mika, Katie et al. 2018, *LA Sustainable Water Project: Los Angeles City-Wide Overview*. UCLA: Sustainable LA Grand Challenge. Retrieved from https://escholarship.org/uc/item/4tp3x8g4.

26. Source: then Los Angeles Mayor Eric Garcetti to Chris Hayes, 2015, 'Mayor Garcetti on adapting to historic drought', *MSNBC*, 15 July 2015.
27. Garcetti, Eric 2019, *L.A.'s Green New Deal Sustainable City pLAn 2019*, City of Los Angeles (plan.lamayor.org), p. 44.
28. Quoted in Simon, Matt 2010 'LA Is Doing Water Better Than Your City. Yes, That LA', *Wired* (US), 12 June 2010.
29. Simon, Matt 2010, ibid.
30. Smith, Hayley 2022, 'L.A. is taking a different path on severe watering restrictions. Here's how it will work', *Los Angeles Times*, 11 May 2022.
31. Melbourne, see Wright, Ian A. 2019, 'Why Sydney residents use 30% more water per day than Melburnians', *The Conversation*, 23 May 2019. UK, 145 litres per capita, DEFRA data for the period 2018/19 to 2020/21, indicator E8 (https://oifdata.defra.gov.uk/5-8-1/). Germany, 127 litres per capita in 2021, source BDEW, 'Profile of the German Water Sector 2020' (in English).
32. 'Currently, outdoor irrigation makes up about 35% of L.A.'s total water use.' Stone, Erin 2022, 'Water Restrictions Have Started In Southern California. Here's What You Need To Know', *LAist.com*, 3 June 2022.
33. Owens Valley Committee, undated. 'Solutions' (https://owensvalley.org/solutions/, accessed 14 May 2023).
34. Owens Valley Indian Water Commission, undated. 'A History of Water Rights and Land Struggles', online at http://oviwc.org/water-crusade (accessed 14 May 2023).
35. Owens Valley Committee, undated. 'Solutions', online at https://owensvalley.org/solutions/ (accessed 14 May 2023).
36. Pfeiffer, Jeanine 2021, 'Honoring A Water Warrior: How Harry Williams Fought for Paiute Water Rights in Owens Valley', *KCET*, 8 July 2021.

Coda

1. Rosa, Hartmut 2020, *The Uncontrollability of the World,* trans. James C. Wagner. Polity Press, Cambridge.
2. Weber, Max 1946 (1922), *'Science as a Vocation'*. Collected in *Max Weber: Essays in Sociology*, translated and edited by H.H. Gerth and C. Wright Mills, Oxford University Press.
3. Harvey, David 1989, *The Condition of Postmodernity*, Wiley-Blackwell: Oxford.
4. Southern Environmental Law Centre 2019, 'Tip of the Ashberg', *Broken Ground* podcast, season 1 episode 1, April 2019.
5. Fortun, Kim 2014, 'From Latour to late industrialism', *HAU* 4(1). DOI: 10.14318/hau4.1.017.
6. American Coal Ash Association (AACA), *2021 Coal Combustion Products Survey*, 6 December 2022. 77.3 million tons of coal combustion products produced in 2021 – though 60% is recycled, e.g. in concrete and wallboard production.
7. Calhoun quoted in Southern Alliance for Clean Energy 2013, 'Five years

since Kingston: Perry County, Alabama's Toxic Tragedy', 22 December 2013 (cleanenergy.org).

8. Miéville, China 2018, 'A Strategy for Ruination: An interview with China Miéville', *Boston Review*.

9. Benjamin W. Abbot et al. 2023, 'Emergency measures needed to rescue Great Salt Lake from ongoing collapse', Brigham Young University, 4 January 2023. DOI: 10.13140/RG.2.2.22103.96166.

10. Saltzman, Hannah 2023, 'The health of the lake is the health of our kids', *Salt Lake Tribune*, 26 January 2023.

11. American Lung Association 2023, 'State of the Air Report' for Salt Lake City, UT. Data available online at https://www.lung.org/research/sota/city-rankings/msas/salt-lake-city-provo-orem-ut (accessed 11 June 2023).

12. Errigo I. et al. (2020), Human health and the economic costs of air pollution in Utah', *Atmosphere* 2020, 11(11): 1238. DOI: 10.3390/atmos11111238.

13. American Lung Association 2023, ibid.

14. Utah Department of Natural Resources, undated. 'Great Salt Lake' webpage (https://water.utah.gov/great-salt-lake/).

15. Williams, Terry Tempest 2023, 'I Am Haunted by What I Have Seen at Great Salt Lake', *New York Times*, 25 March 2023.

16. 39 per cent decline: Null, S.E., Wurtsbaugh, W.A. 2020, 'Water Development, Consumptive Water Uses, and Great Salt Lake', in *Great Salt Lake Biology*, eds B. Baxter & J. Butler, Springer. DOI: 10.1007/978-3-030-40352-2_1.

17. *Salt Lake Tribune* editorial board, 2022, 'Why it's time for Utah to buy out alfalfa farmers and let the water flow', *Salt Lake Tribune*, 12 December 2022.

18. Quoted by Christopher Flavelle 2022, 'As the Great Salt Lake Dries Up, Utah Faces an "Environmental Nuclear Bomb"', *New York Times*, 7 June 2022.

19. *Salt Lake Tribune* editorial board, 2022, ibid.

20. Abbot, Benjamin W. et al. 2023, ibid.

21. Evans, Monica 2021, 'Everything you need to know about drylands', *Global Landscapes Forum*, 25 March 2021.

22. Paragraph A.1.5 in IPCC, 2019, 'Summary for Policymakers', in *Climate Change and Land: an IPCC special report on climate change, desertification, land degradation, sustainable land management, food security, and greenhouse gas fluxes in terrestrial ecosystems*.

23. Evans, Monica 2021, ibid.

24. IPBES 2018, 'Media Release: Worsening Worldwide Land Degradation Now 'Critical', Undermining Well-Being of 3.2 Billion People', IPBES (Intergovernmental Science-Policy Platform on Biodiversity and Ecosystem Services), 23 March 2018.

25. Schauenberg, Tim 2022, 'How to stop deserts swallowing up life on Earth', *Deutsche Welle* (DW.com), 16 February 2022.

26. Watts, Jonathan 2011, 'China makes gain in battle against desertification but has long fight ahead', *The Guardian*, 4 January 2011.

27. While livestock numbers have risen significantly, many ecologists would argue that the blame does not rest on pastoralists' 'backwardness' or lack

of care for the land, but rather the state policies that have pushed them into debt, reduced available grazing land, and forced short-termist stocking behaviours. See Kolås, Åshild 2014, 'Degradation Discourse and Green Governmentality in the Xilinguole Grasslands of Inner Mongolia', *Development and Change*, 45: 308–328. DOI: 10.1111/dech.12077 and Ruxin Zhang, Emily T. Yeh, Shuhao Tan, 2021. 'Marketization induced overgrazing: The political ecology of neoliberal pastoral policies in Inner Mongolia', *Journal of Rural Studies*, 86: 309–17, DOI: 10.1016/j.jrurstud.2021.06.008.

28. Beiser, Vince 2018, *The World in a Grain: The Story of Sand and How It Transformed Civilization*, Riverhead, New York City.

29. NB this does also include trees dying of 'old age' (trees grown from cuttings are apparently short-lived) as well as pests, drought and so on. Cao, Shixiong 2008. 'Why large-scale afforestation efforts in China have failed to solve the desertification problem', *Environmental Science & Technology* 42(6):1826–31. DOI: 10.1021/es0870597.

30. Elkin, Rosetta S. 2022, *Plant Life: The Entangled Politics of Afforestation*, University of Minnesota Press.

31. Wang, X.M. Zhang, C.X. Hasi, E. & Dong, Z.B. 2010. 'Has the Three Norths Forest Shelterbelt Program solved the desertification and dust storm problems in arid and semiarid China?', *Journal of Arid Environments* 74(1): 13–22. DOI: 10.1016/j.jaridenv.2009.08.001. Alongside the pandemic barring all access, and the genocide in key dust source region Xinjiang, this is the third major reason I haven't written more about China in this book: the data is difficult to evaluate.

32. Wu, C., Lin, Z., Shao, Y. et al. 2022, Drivers of recent decline in dust activity over East Asia. *Nature Communications* 13(7105). DOI: 10.1038/s41467-022-34823-3.

33. Voiland, Adam 2023, 'A Dusty Day in Northeastern China', NASA Earth Observatory, 22 March 2023.

34. Example memes: Jinfeng Zhou (@zhou_jinfeng) on Twitter https://twitter.com/Zhou_jinfeng/status/1371283896013189125 and Sony Movies on Weibo https://weibo.com/2526171271/K6irscCc7.

35. 1-hour PM2.5 levels from Zhicong Yin, Yu Wan, Yijia Zhang, Huijun Wang, 2022, 'Why super sandstorm 2021 in North China?', *National Science Review* 9(3), March 2022, DOI: 10.1093/nsr/nwab165.

36. Via Myers, Stephen Lee 2021, 'The Worst Dust Storm in a Decade Shrouds Beijing and Northern China', *New York Times*, 15 March 2021.

37. UNCCD, 2020, 'The Great Green Wall Implementation Status And Way Ahead To 2030' executive summary', prepared by Climatekos gGmbH, Berlin.

38. FAO, undated webpage, 'Action Against Desertification: Great Green Wall', Food and Agriculture Organisation of the United Nations (fao.org).

39. Morrison, Jim 2016, 'The "Great Green Wall" Didn't Stop Desertification, but it Evolved Into Something That Might', *Smithsonian* magazine, 23 August 2016.

40. Reij, C., Tappan, G., Smale, M. 2009, 'Agroenvironmental Transformation in the Sahel: Another Kind of "Green Revolution"'. IFPRI Discussion

Paper 00914. International Food Policy Research Institute, Washington, DC.

41. Watson, Cathy 2018, 'Farmer-managed natural regeneration: the fastest way to restore trees to degraded landscapes?', *Mongabay News*, 29 June 2018.

42. Reuters staff, 2021, 'Farmer coaxes forest from the desert in Burkina Faso', *Reuters*, 26 March 2021.

43. Quoted in Morrison, Jim 2016, ibid.

44. Anon, 'The Ruin' trans. Michael Alexander 1991, in *The Earliest English Poems*, Penguin Classics.

45. WNYC, PRX and WGBH 2022, 'How Indigenous Water Protectors Paved Way for Future Activism', *The Takeaway* radio programme, 22 April 2022.

46. Ackoff, Russell L. 1974, *Redesigning the future: a systems approach to societal problems*. Wiley, New York.

47. Prather, Michael 2023, post in the 'Owens Valley History' Facebook Group, 15 April 2023. https://www.facebook.com/groups/392735971601693/permalink/1284929745715640/ (accessed 11 June 2023).

Index

Index

Index

Index

Index